THE PICTURE OF THE TAOIST GENII PRINTED ON THE COVER of this book is part of a painted temple scroll, recent but traditional, given to Mr Brian Harland in Szechuan province (1946). Concerning these four divinities, of respectable rank in the Taoist bureaucracy, the following particulars have been handed down. The title of the first of the four signifies 'Heavenly Prince', that of the other three 'Mysterious Commander'.

At the top, on the left, is Liu *Thien Chün*, Comptroller-General of Crops and Weather. Before his deification (so it was said) he was a rain-making magician and weather forecaster named Liu Chün, born in the Chin dynasty about +340. Among his attributes may be seen the sun and moon, and a measuring-rod or carpenter's square. The two great luminaries imply the making of the calendar, so important for a primarily agricultural society, the efforts, ever renewed, to reconcile celestial periodicities. The carpenter's square is no ordinary tool, but the gnomon for measuring the lengths of the sun's solstitial shadows. The Comptroller-General also carries a bell because in ancient and mediaeval times there was thought to be a close connection between calendrical calculations and the arithmetical acoustics of bells and pitch-pipes.

At the top, on the right, is Wen *Yüan Shuai*, Intendant of the Spiritual Offcials of the Sacred Mountain, Thai Shan. He was taken to be an incarnation of one of the Hour-Presidents (*Chia Shen*), i.e., tutelary deities of the twelve cyclical characters (see Vol. 4, pt 2, p. 440). During his earthly pilgrimage his name was Huan Tzu-Yü and he was a scholar and astronomer in the Later Han (b. +142). He is seen holding an armillary ring.

Below, on the left, is Kou *Yüan Shuai*, Assistant Secretary of State in the Ministry of Thunder. He is therefore a late emanation of a very ancient god, Lei Kung. Before he became deified he was Hsin Hsing, a poor woodcutter, but no doubt an incarnation of the spirit of the constellation Kou-Chhen (the Angular Arranger), part of the group of stars which we know as Ursa Minor. He is equipped with hammer and chisel.

Below, on the right, is Pi *Yüan Shuai*, Commander of the Lightning, with his flashing sword, a deity with distinct alchemical and cosmological interests. According to tradition, in his early life he was a countryman whose name was Thien Hua. Together with the colleague on his right, he controlled the Spirits of the Five Directions.

Such is the legendary folklore of common men canonised by popular acclamation. An interesting scroll, of no great artistic merit, destined to decorate a temple wall, to be looked upon by humble people, it symbolises something which this book has to say. Chinese art and literature have been so profuse, Chinese mythological imagery so fertile, that the West has often missed other aspects, perhaps more important, of Chinese civilisation. Here the graduated scale of Liu Chün, at first sight unexpected in this setting, reminds us of the ever-present theme of quantitative measurement in Chinese culture; there were rain-gauges already in the Sung (+12th century) and sliding calipers in the Han (+1st). The armillary ring of Huan Tzu-Yü bears witness that Naburiannu and Hipparchus, al-Naqqash and Tycho, had worthy counterparts in China. The tools of Hsin Hsing symbolise that great empirical tradition which informed the work of Chinese artisans and technicians all through the ages.

SCIENCE AND CIVILISATION IN CHINA

Joseph Needham
(1900–1995)

'Certain it is that no people or group of peoples has had a monopoly in contributing to the development of Science. Their achievements should be mutually recognised and freely celebrated with the joined hands of universal brotherhood.'

Science and Civilisation in China VOLUME I, PREFACE.

*

Joseph Needham directly supervised the publication of seventeen books in the *Science and Civilisation in China* series, from the first volume, which appeared in 1954, through to Volume 6.3, which was in press at the time of his death in March 1995.

The planning and preparation of further volumes will continue. Responsibility for the commissioning and approval of work for publication in the series is now taken by the Publications Board of the Needham Research Institute in Cambridge, under the chairmanship of Dr Christopher Cullen, who acts as general editor of the series.

SCIENCE AND CIVILISATION IN CHINA

The Duke of She asked Tzu Lu what he thought about Confucius, but Tzu Lu returned him no answer. 'Why did you not say', said the Master, 'he is simply a man so eager for improvement . . . that he forgets his sorrows and does not observe that old age is at hand?'

Confucius, *Analects*, VII.xviii, tr. Soothill

李約瑟著

中國科學技術史

冀朝鼎

JOSEPH NEEDHAM

SCIENCE AND CIVILISATION IN CHINA

VOLUME 7

PART II: GENERAL CONCLUSIONS AND REFLECTIONS

BY

JOSEPH NEEDHAM F.R.S., F.B.A.

WITH THE COLLABORATION OF

KENNETH GIRDWOOD ROBINSON AND

RAY HUANG (HUANG JEN-YÜ)

AND INCLUDING CONTRIBUTIONS BY THEM

WITH AN INTRODUCTION BY

MARK ELVIN

EDITED BY

KENNETH GIRDWOOD ROBINSON

PUBLISHED BY THE PRESS SYNDICATE OF THE UNIVERSITY OF CAMBRIDGE
The Pitt Building, Trumpington Street, Cambridge, United Kingdom

CAMBRIDGE UNIVERSITY PRESS
The Edinburgh Building, Cambridge, CB2 2RU, UK
40 West 20th Street, New York, NY 10011–4211, USA
477 Williamstown Road, Port Melbourne, VIC 3207, Australia
Ruiz de Alarcón 13, 28014 Madrid, Spain
Dock House, The Waterfront, Cape Town 8001, South Africa

http://www.cambridge.org

First published 2004

Printed in the United Kingdom at the University Press, Cambridge

Typeface Baskerville MT 11.25/13 pt. *System* LaTeX 2ε [TB]

A catalogue record for this book is available from the British Library

ISBN 0 521 08732 5 hardback

This book is dedicated to the many scholars who have worked as collaborators on the Science and Civilisation in China Project over the last half century.

CONTENTS

ILLUSTRATIONS

TABLES

ABBREVIATIONS

A/AIHS	*Archives Internationales d'Histoire des Sciences*
AHOR	*Antiquarian Horology*
AJS	*American Journal of Sociology*
AM	*Asia Major*
ARB	*Annual Review of Biochemistry*
ARO	*Archiv Orientalni* (Prague)
ASTJ	*Astronomical Journal*
AX	*Ambix*
BIHM	*Bulletin of the (Johns Hopkins) Institute of the History of Medicine*
BLSOAS	*Bulletin of the School of Oriental and African Studies*
BMFEA	*Bulletin of the Museum of Far Eastern Antiquities* (Stockholm)
CEN	*Centaurus*
CHS	*Chhien Han Shu*
CNOW	*China Now*
COMP	*Comprendre* (Soc. Eu. de Culture, Venice)
CR/MSU	*Centennial Review of Arts and Science* (Michigan State University)
CRR	*Chinese Recorder*
CS	*Current Science*
CSCI	*Chinese Science*
CSMJ	*Chinese Medical Journal*
DAE	*Daedalus*
EASTM	*East Asian Science, Technology and Medicine*
EW	*East and West* (Quaert. Rev. Pub. Instituto Ital. per il Medio e Estremo Oriente, Rome)
HEJ	*Health Education Journal*
HJAS	*Harvard Journal of Asiatic Studies*
HOR	*History of Religions*
HOSC	*History of Science*
IJE	*International Journal of Ethics*
IMPAC	*Impact of Science on Society (UNESCO)*
IQB	*Iqbal* (Lahore), later *Iqbal Review* (Journal of the Iqbal Academy or Bazm-i Iqbal)
JAN	*Janus*
JAS	*Journal of Asian Studies*
JBASA	*Journal of the British Astronomical Association*
JCE	*Journal of Chemical Education*
JCP	*Journal of Comparative Philosophy*
JEH	*Journal of Economic History*
JESHO	*Journal of the Economic and Social History of the Orient*

JHI	*Journal of the History of Ideas*
JOSHK	*Journal of Oriental Studies* (Hong Kong University)
JPOS	*Journal of the Peking Oriental Society*
JRAI	*Journal of the Royal Anthropological Institute*
JRSA	*Journal of the Royal Society of Arts*
JUB	*Journal of the University of Bombay*
JWCI	*Journal of the Warburg and Courtauld Institutes*
JWH	*Journal of World History*
LH	*L'Homme (Revue Française d'Anthropologie)*
MBLB	*May & Baker Laboratory Bulletin*
MH	*Medical History*
MJA	*Medical Journal Australia*
MXTD	*Marxism Today*
N	*Nature*
NS	*New Scientist*
OSIS	*Osiris*
PAAAS	*Proceedings of the American Academy of Arts and Sciences*
PHR	*Philosophical Review*
PHY	*Physis* (Florence)
PRMS	*Proceedings of the Royal Microscopical Society*
PRSB	*Proceedings of the Royal Society (Ser. B)*
PRSM	*Proceedings of the Royal Society of Medicine*
PTKM	*Pen Tshao Kang Mu*
QRB	*Quarterly Review Bio.*
RBS	*Revue Bibliographique de Sinologie*
RHSID	*Revue d'Histoire de la Siderurgie* (Nancy)
RKP	*Routledge and Kegan Paul*
SAM	*Scientific American*
SBE	*Sacred Books of the East series*
SCC	*Science and Civilisation in China*
SCI	*Scientia*
SOAS	*School of Oriental and African Studies*
SOB	*Sobornost*
SPCK	*Society for the Promotion of Christian Knowledge*
SPPY	*Ssu Pu Pei Yao*
SS	*Science and Society* (New York)
STIC	*Science and Technology in China*
TCKM	*Thung Chien Kang Mu*
TCULT	*Technology and Culture*
TT	*Tills and Tillage*
UNASIA	*United Asia* (India)
VA	*Vistas in Astronomy*
W	*Weather*
ZVRW	*Zeitschrift f. d. vergleichende Rechtswissenschaft*

SERIES EDITOR'S PREFACE

This is the concluding part of the final volume in the formal sequence of *Science and Civilisation in China*, although it is by no means the last book in the series to be published. But since it will stand in the last place on the shelves of many libraries, something by way of a provisional concluding word is called for.

By now readers of this series will be well aware of the distinction between what Joseph Needham used to call the 'seven heavenly volumes' amongst which the topic areas of the series are distributed, and the much larger number of physical 'earthly volumes' published as separate parts. Behind that lies the original conception of a work to be published in a single volume, which was the proposal first agreed between its author and Cambridge University Press over half a century ago. When that proved inadequate, a seven-volume plan was adopted, and it was on that basis that the moderately sized Volume 1 appeared in 1954. But when the immense bulk of Volume 3 had passed through the press, it became clear that further subdivision was a practical necessity, and so Volume 4 appeared in three parts, each one being physically and intellectually weighty enough to stand as a life's work for a less ambitious and energetic scholar.

Volumes 5 and 6 continued the process of subdivision. To some extent this was the result of Needham's own creativity, increasingly seconded by a team of talented collaborators. But as time went by, the role of these collaborators tended to change from that of co-workers gratefully acknowledged on a title-page to that of independent authors in their own right. The first example of these to have a whole book to herself was Francesca Bray, author of Volume 6, part 2, on Agriculture. In that case the plan from the outset was that a whole book needed to be written on such a major topic. But this was not always the case: there were instances when a topic originally planned as the equivalent of a major chapter grew in the hands of an enthusiastic collaborator until another physical part had to be added to the corresponding volume. Anyone who looks at the sequence of varied and fascinating scholarly publications that resulted from this process will surely agree that on the whole this process of growth was an overwhelmingly positive phenomenon.

Volume 7 was subject to this process in its turn, but with less obvious effects. As the editor of this volume, Kenneth Robinson, explains in his preface, once the actual publication of Volume 7 began to be discussed the initial eagerness of collaborators exercised a continual pressure towards expansion of the author's original modest plan for a concluding survey of social factors in the development of science, medicine and technology in China. Given Needham's increasing age, the end result of this would have been a number of 'earthly volumes' by other hands, with which Needham himself would have had no very close authorial connection. But as Kenneth Robinson recounts, the end result of what was at times a convoluted process was much simpler than at one time seemed likely. Volume 7 appears in two parts, the first of which was published some years ago with Christof Harbsmeier's discussion of language and logic, and the second of which, largely by Joseph Needham himself, is now before us.

There is however something missing in that short summary, and indeed in the account given by Kenneth Robinson himself. Anyone who knew Joseph Needham will remember him as a fountain of physical and intellectual energy, a man who continued regular work at his desk until the last day of his life. But that life began in 1900, and did not end until 1995. Although much of the material for this volume had been in draft for several decades, a great deal of work on it still remained to be done in the closing years of Needham's life, and it was clear that without the help of a devoted collaborator much would have been left in a state unready for the press. Without Kenneth Robinson to fill that role this book could not have appeared. He was in fact the first of Needham's collaborators to be credited as author of a separate part of the text of *Science and Civilisation in China*, in this case the section on Acoustics in Volume 4, part 1, which was begun in 1949, and his association with the project continued for much of the second half of the 20th century. It was therefore very fitting that he should be the last of Needham's collaborators to work directly with him in producing a text and in readying it for press, as he has done so devotedly in recent years. There is more to be learned from such a close working association with a great mind than can be expressed in conventional academic writing. We are fortunate, therefore, that in the Soliloquy that concludes the main text, Kenneth Robinson has given us some of the lasting impressions made by many conversations with Joseph Needham in the closing years of his life. This is not the place for a biographical sketch that might help to explain the circumstances that so obviously produced the right collaborator at the right time. For the moment it is sufficient to say that Kenneth Robinson's help with this volume was yet another of the strokes of good fortune that combined with Joseph Needham's personal dedication to make *Science and Civilisation in China* possible in its present form. Few others could have filled this role so effectively, so tactfully, and so tirelessly, and readers of this volume owe him a considerable debt of gratitude. As Series Editor, I will simply say that the job could never have been done without him.

When the author of such a vast work as this says his last word to his readers, some historical reflection on the significance of his achievement is surely appropriate. Those of us who have worked to bring this volume to press in the years since Joseph Needham's death are perhaps too close to the whole enterprise to provide this. What was needed was to find a scholar who was intimately concerned with and deeply learned in many if not all of the areas covered by Needham's writing, but was distant enough from the work of the *Science and Civilisation in China* project to view it objectively, while acknowledging it as one of the great intellectual phenomena of the 20th century. I am sure that the ideal candidate has been found in the person of Professor Mark Elvin. The reader who wants to know, in the end, what difference it all made can turn to 'Vale atque ave' with confidence.

Christopher Cullen

VOLUME EDITOR'S PREFACE

This is formally the last volume in the *Science and Civilisation in China* series, by Joseph Needham, although volumes that precede it in numerical sequence are still in press or in preparation at the time of writing. When he first thought seriously about the *Science and Civilisation in China* project in the intervals of his work in war-time China, Needham originally aimed to write a single book. By 1948 this plan had expanded to seven volumes, but later he realised that it would probably require some thirty volumes to cover all the topics envisaged, not so much because of the number of topics, but because of the wealth of information his wide-ranging researches revealed.

He aimed to write eighteen of these himself and to delegate twelve to specialists. But of these eighteen volumes there were six which he was not able to finish, or in some instances even to start. These were three of the four volumes initially planned on the institutes of medicine, and two on the social background of China in relation to science, the second of which would have ended with his General Conclusions.

The volumes which he actually wrote were: Volume 1: Introductory Orientations; Volume 2: History of Scientific Thought; Volume 3: Mathematics and the Sciences of the Heavens and the Earth; Volume 4, part 1: Physics; Volume 4, part 2: Mechanical Engineering which concluded with some thirty pages concerning Chinese inventions and discoveries relevant to aeronautics; Volume 4, part 3: Civil Engineering and Nautics; Volume 5, part 2: the first of four volumes on alchemy and the beginning of chemistry, this one being mainly concerned with 'the magisteries of gold and immortality'. Part 3 of Volume 5 traced the development of synthetic insulin from cinnabar elixirs. Part 4 was concerned with the development of chemical apparatus and theories from their alchemical beginnings, and the last of the four, part 5 of Volume 5, was concerned with physiological alchemy and its contribution to modern psychology. This made ten volumes. Two more were Volume 5, part 7, on 'The Gunpowder Epic' which was, however, in large measure the work of Ho Peng-Yoke (Ho Ping-Yü), one of his most productive collaborators, and Volume 6, part 1, the first of two volumes on botany. Volume 6, part 6, represents what he completed on the subject of medicine.

The average length of time spent in bringing a volume to publication was 3.3 years, though this takes no account of the time spent on preliminary work and research before the final writing began. Very often portions of the work were first published as scientific papers. Needham was, for example, working on the text of Volume 4, part 1, in 1948, fourteen years before it was published, and six years before even Volume 1 was published by Cambridge University Press. The eighteen volumes which Needham intended to write himself were therefore likely to require approximately sixty years of work, plus the immense labour of preliminary research. The extent of this labour can be judged by the fact that when I asked him in the mid-1970s how he was getting on with the volume on botany, he replied, 'At present it's a green blur'. He once wrote to a friend, Mr Hu Tao-Ching 胡道靜[1]: 'It has been my experience all along that whenever I embark on a new chapter, we find that the subject is full of misunderstandings, mistranslation, erroneous ideas, and what are just plain errors. Gradually the true picture emerges as one works on.' By 1983, however, when we saw his botany text being made ready for printing, it was no longer a green blur. On the contrary, Needham had sorted out the different Chinese systems of plant classification, comparing them with European systems to such good effect that this volume on botany made an important contribution to the writing of the present volume in the article: 'Literary Chinese as a Language for Science'.

As Needham was forty-seven when he started writing the final text for *Science and Civilisation in China*, Volume 1, he could not have hoped to complete the eighteen volumes he eventually envisaged before he had reached the age of 107. What he was not to know until much later was that the scientific treasures to be discovered in the mountain of Chinese literature were so great that they could not be brought completely to light in seven volumes, or even in eighteen, but that eventually twenty-nine volumes would be needed to contain them, despite the fact that some intended subjects were abandoned as impracticable.

Needham was granted ninety-four years for his amazingly productive life. He continued working on his General Conclusions till only two days before he died. But *Science and Civilisation in China* was only part of his total output. He had some 250 works to his credit in addition to the volumes of *Science and Civilisation in China*. Gregory Blue lists some 385 'Publications' in his very useful *Joseph Needham – A Publication History* (*Chinese Science*, 14, 1997, pp. 90–132). These cover new books of his own, translations, contributions with other authors, books he had edited, and well over 150 scientific papers.

We may now consider his dilemma. As Needham delved ever deeper into his subject, more and more sources for the precious ore which he was seeking kept opening up. Was he to leave it, as an archaeologist finding King Priam's golden treasure might bury it again for fear that he would not have time to assess it properly, or was he to bring it into the light of day and trust that future archaeologists would finish his work where he had left off? All Needham's instincts were for bringing what was hidden into the light of day, and trusting posterity to make a true evaluation of it. Therefore each volume in turn became more bulky than the last. Volume 1 was a reasonably slender book of 248 pages of text. But Volume 2 had more than doubled with 503 pages, and Volume 3, with its 683 pages and very extensive supplementary pages, had become so heavy that, as Needham said, 'it was too big to read comfortably in one's bath'. So it was decided that in future each volume should be a reasonable size. This good resolution was followed in Volume 4, part 1, which had only 334 pages of text, but by Volume 4, part 2, this good intention was abandoned, for it now had 602 pages, and Volume 4, part 3, broke all records with 699 pages of text.

[1] See Li Guohao *et al.* eds. (1982), frontispiece.

Then it came to Volume 5 whose subject was chemistry and its applications. The introduction to chemistry through alchemy in four volumes made it quite certain that Needham would not be able to finish the work himself, for the alchemical volumes would require thirteen years for their publication, by which time Needham would be eighty-four. Gunpowder, botany, four medical volumes and two concluding volumes on the social background still barred the way to a quiet retirement. Delegation became essential.

He had first delegated to me the sub-section in Volume 4, part 1, on acoustics in 1949 with the words: 'acoustics is my blind spot. I'd like you to fill it with perhaps twenty pages.' In the end it came to just over 100, and that was before the subject had been opened up by archaeologists with the wonderful discovery of complete sets of bells establishing the frequencies of the notes of the traditional scale.

This expansion persisted throughout. Once something previously unknown in the West was discovered Needham was reluctant to let it go again. Consequently several collaborators who were asked to contribute portions of a volume ended up by writing complete volumes themselves. Eventually sixteen volumes would be completed by fifteen collaborators.

Fresh problems began to arise. That whole books were being produced by authors who had been invited to contribute just one section of a book was not in itself necessarily contrary to Needham's wishes, but a different situation arose when the work submitted by that author did not fit into the general plan or where it contained ideas which Needham found unacceptable.

Needham had originally listed a number of factors that he felt formed part of the social background enabling or disabling the rise of science. These included the geography of China, fiscal and economic circumstances, language, logic, concepts of time, the role of religion, the consequences of class attitudes, competition, nature and man, and many others. He hoped to retain overall control of the final volumes in which these factors were considered, but to be helped in the task by specialists who would contribute sections that would fit into the general plan. This was not possible. By the time he was eighty and ready to give his mind to the writing of the last volume, while also writing the Institutes of Medicine, a wide generation gap had opened up between himself and the best of the young sinologists. In the year 1980, I was invited to come from Hamburg to help Needham and his collaborators finish the remaining volumes. Needham at that time still intended to be in strict control of Volume 7. There were difficulties, however. It was found impossible, for example, to persuade a modern geographer to write about his subject in such a way that the geography of China could be seen to be influencing the way scientific thinking developed. A volume was produced, but while excellent in itself it was a historical geography of China, not part of the *Science and Civilisation in China* series as Needham saw it.

A more difficult situation arose, however, when Professor Derk Bodde, one of a group of six scholars who, Needham hoped, would help him in the writing of Volume 7, unexpectedly presented him with a substantial manuscript which followed very completely and precisely the outline which Needham had drawn up for Section 49 of Volume 7, 'Intellectual and Social Factors'. In it Bodde put forward many views which Needham found quite unacceptable. Yet he hated saying 'No', and, as he explained later, when the leading American sinologist offers you as a gift a work on which he has spent more than a year, you can hardly refuse to accept it. Needham felt that Bodde's work would be far better published as an independent contribution to scholarship, but sought a compromise, and said he would be pleased to have it published as part of Volume 7, provided that certain changes could be made in the writing. There was the rub. I was called in to sort out the differences of view between the two giants of the sinological world. This situation in which I found myself is neatly summed up in the Burmese proverb: 'Where elephants fight the grass is trampled'.

The two main differences in points of view were (1) the role of the classical language of China in relation to science, and (2) whether the Chinese could be said to have any real science at all. These were areas in which compromise was not possible. When it became obvious that compromise would not be possible, Needham had a new idea. H. G. Wells' *Outline of History*, which had been written in 1920, was met with a great deal of criticism. In 1931 Wells revised it, but following an original method. When he came to a passage with which a critic disagreed, Wells included the criticism in his text as a quotation, and then added his own riposte. Needham felt that this method would admirably suit Derk Bodde's text. Where he disagreed with Bodde, he would insert the reasons for his disagreement, and if so desired would include Bodde's further argument. A beginning was made on these lines, and it would certainly have made an unusual and interesting book, but the crisis came when Bodde asked Needham to withdraw a passage of close argument as it would upset the friends he had been quoting. Needham refused and Bodde then withdrew his entire contribution. In this way the Gordian Knot was cut, and the work of parcelling out portions of Volume 7 to specialised writers began.

At first it looked as if it would be possible to do this following Needham's original plan. The role of logic was an important part of it, and the great Polish logician Janusz Chmielewski had not only agreed to contribute a section, but had already submitted the first forty-two pages of his text. During the year 1983 it became clear that Professor Chmielewski would not be able to complete his undertaking due to the pressures he was under in Warsaw. It was suggested that he might have an assistant to work with him. Perhaps he would suggest and recommend a suitable younger logician. He recommended Dr Christoph Harbsmeier, who visited Chmielewski, discussed the work and agreed to take entire responsibility for it. He started work in September 1983. His contribution, like so many others to *Science and Civilisation in China*, rapidly increased from a section of a volume to a complete book that in 1998 became Volume 7, part 1.

An article on Language and Science in traditional China had been begun as soon as disagreement on this topic with Derk Bodde had become clear. It began as a 'riposte' in the H. G. Wells tradition, was first conceived as forming part of a volume together with the work on logic, then became separated, and was to form a volume on its own. But after Needham's death it became important to get the whole work published as soon as possible. The text on language has therefore been included in the present volume in a shorter form than originally envisaged.

The redesigning of Volume 7 had by the mid-1990s resulted in the following plan, which represents the concept of the volume at its greatest length:

Part 1 'The Nature of Chinese Society in Comparative Perspective.' This would fall into six sections by different authors.

Part 2 'Language and the Foundations of Scientific Civilisation in Ancient China.' This eventually resolved itself into 'Language and Logic in Traditional China', written by Christoph Harbsmeier.

Part 3 'Language and Science in Traditional China', to contain 'a reconsideration of some characteristics of Chinese Science' by Christopher Cullen, and 'Literary Chinese as a Language for Science' by Kenneth Robinson working closely with Needham.

Part 4 'The World View of the Literati', to cover concepts of Time, History, Religion, Education and Political Organisation.

One by one items in this ambitious scheme of things had to be abandoned, either because Needham found the work that was submitted was unacceptable, or because the author was unable to complete it, or decided to withdraw it. In some cases work that had been completed and accepted had to be abandoned when the failure of other contributions withdrew vital elements from the structure. One decisive factor was that the history and sociology of science were in a state of turmoil from which they have not yet emerged, and if a work was not published soon after it was written it was likely to be out of date. The old secure concept of 'world science' which had been part of Needham's background had vanished.

It was about this time that HRH The Duke of Edinburgh, Chancellor of the University of Cambridge, visited The Needham Research Institute, and interested himself in the progress of the project. 'And how long will it take to finish it?' he enquired. On being given a rather conservative answer, 'At least ten years', he exclaimed, 'Good God, man, Joseph will be dead before you've finished', a very true appreciation of the situation. If it had been possible for Needham effectively to write Volume 7 himself, using others to reinforce or supplement his own writing as he had done earlier in his life, Volume 7 might have been presented with the consistency of a scientist writing in the mid-20th century. But by the time he was eighty it was too late. Much had to be abandoned. Western views on China itself were changing, as Needham stresses on page 203 of the present volume, and these views have continued to change not only in the West but in China itself. There is, therefore, no possibility of presenting any definitive judgement on the role of China in world history, but only the views of a great scientist and polymath in the mid-20th century.

The staff of the Needham Research Institute now found themselves in the position of the young man in Pope's *Essay on Criticism* for whom – Hills peep o'er Hills, and Alps on Alps arise.

The policy agreed on was not to attempt some of the major hills, on which Needham had written only briefly, and which would require much work from him if they were to be updated and brought to the level of his published work. Time, Religion and Education fell into this category. But some of his existing writings could be brought together into Volume 7, part 2, preserving as much as possible of what he had originally intended Volume 7 to be, and some could be updated if he had the help of an academically trained secretary; for it must be remembered that in his closing years, though his mind remained lucid and his memory astonishing, Needham had great difficulty even in moving from one chair to another, and even more difficulty in speaking and in making himself understood, due to the effect of the medicines he took to control Parkinsonism. But a secretary, working closely with him day by day, could often understand what he had said, and could read what he had written, when others were baffled.

When Needham died in March 1995 none of the material that he had intended for Volume 7 was in a condition ready for the press. In some cases there were in existence a number of inconsistent drafts without a clear indication of which was to be the final text, and in other cases a section of text required the searching out of a great mass of references that were no doubt in Needham's capacious memory, but which had only been entered into his typescript in an abbreviated and sometimes cryptic form. As Series Editor and Volume Editor respectively, Christopher Cullen and I sat down together to decide what was to be done. Of decisive importance for us in deciding what should be included in Volume 7 was a document which Needham had composed in 1981 and revised in 1987 in the form of a 'testament' in which he had set out his wishes as regards those of his writings which should be included in the event that he died without being able to see Volume 7 through the press. It is substantially this material that now forms Volume 7, part 2. To this has been added the interesting chronological list of Chinese discoveries and inventions that Needham extracted from earlier volumes of *Science and Civilisation in China* and elsewhere during his work for this volume in the last years of his life. Needham himself acknowledged the provisional nature of any such list, and no doubt some items in it are open to discussion. But many readers will find this compilation stimulating and perhaps useful in beginning investigation of other volumes of *Science and Civilisation in China*. I should like to point out that in editing Needham's text we have not made any attempt to eliminate occasional evidence that this material was not written yesterday. China has changed rapidly over the last decade, but Needham's views expressed in these writings largely relate to a China before these changes occurred. Likewise, respect for the integrity of the author's text has made us reluctant to make cuts when some parts of Needham's argument are common to various sections of this book.

As editor of this volume, I was fortunate in having been frequently engaged in close discussion with Needham in the closing years of his life on all the topics with which Volume 7 deals, so that it was possible, though not always easy, to decide which of the variant versions of his text was the one he intended as his last word. It was also easier than it might have been (though not very easy) to locate the material needed for some of the lacunae in his references. The resulting task has been time-consuming, but I hope the reader will consider it worthwhile: I am confident that apart from the section of this volume in which I and others are specifically named as co-authors what the reader has here is emphatically and authentically Joseph Needham, speaking here in the closing stages of one of the greatest scholarly projects ever undertaken.

Kenneth Robinson

VALE ATQUE AVE

Coming to the end of Joseph Needham's *Science and Civilisation in China* can only be compared, as an experience, to reaching the final page of Gibbon's *Decline and Fall of the Roman Empire*.[1] One looks back at the journey taken, almost disbelieving its immensity. Parts of the landscape stand out clearly in the sunlight of memory. Others are already covered over by the mists of uncertain recollection. One knows, however, that one could return there easily enough if the need arose. What is harder to recall is the different state of mind with which one started out long ago.

One's conception of the world has been transformed. In the case of Gibbon, by a vision of the slow passage from antiquity to the Middle Ages, from our ancient Mediterranean *imperium* to the doubly divided inheritance of today: eastern and western Europe, and Christendom – or ex-Christendom – and Islam. In the case of Needham, by the revelation of a Chinese cultural universe whose triumphs in mathematics, the sciences, and technology were often superior, and only rarely inferior, to those of western Europe until about 1600. That contributed astonishing riches of practical invention to the origins of the modern age, the proper understanding of whose nature subverts the analytical logic of the standard model of modern Western history for those interested enough, and historically modest enough, to see why.

The present volume, edited by Kenneth Robinson, and partly co-authored by him and Huang Jen-Yü, offers us a last chance to look a little deeper at how Needham thought about his work on the history of Chinese science in the wider context both of Chinese society and culture, including the surprising capacity of the old Chinese literary language for technical precision when it needed it, and of comparisons between China and Europe. Some of this material is new, notably that on language. Some of it, for the *cognoscenti* at least, is old though not easily available, and revised and updated here in varying degrees.

It has to be said that, as scholarship has advanced, not everything that Needham wrote, forty or more years ago, on the social and economic history of China now seems as solidly based as the greater part of his reconstructions of Chinese technical practice and scientific theory. The reader needs to exercise a certain caution here, searching at times less for information than for inspiration. But the challenge of arguing with him is invariably well worth taking up, in these areas as elsewhere. He saw a number of pivotal problems before anybody else did, and struggled to discover answers with such psychological energy that even those we now tend to judge as partial failures are illuminating. In the scientific domain, of course, he grasped that the question of why something like 'modern' science did not arise in imperial China was a serious one, even if it has since provoked efforts either to bypass it or to recast it

[1] Which I read in 1954–5, mostly while travelling on the London Underground. It took over a year to finish. Reading *SCC* has taken much longer as the volumes have appeared *seriatim* since 1954, and, even so, I certainly would not claim to have looked at every page.

in a fashion more amenable to testing.[2] But even his famous theory that Chinese social structure in *all* of imperial times could be described as 'bureaucratic feudalism' – on which I already differed with him in the early 1960s on the grounds that the changes over 2,000 years were too great to allow any single such label, if rigorously defined, to be applied equally to all periods – has a heuristic value. To put a complex point a little too simply, the phrase embodies a long-lasting characteristic tension in the cultured classes of Chinese society: that between a deep belief, universalistic at least within the Chinese oecumene, in general ideals applied generally, and a comparably intense attachment to certain specific particularistic and personalistic loyalties.

Perhaps, to use metaphorical language, one can say that some of the mountains that grew from his work have been more eroded with the passing of time than others. But the orogeny was extraordinary.

The Chinese Perspective on the West

What is hard to come to terms with, almost half a century after the appearance of the first volume in 1954, is the limited assimilation of Needham's work into the bloodstream of the history of science in general; that is, outside the half-occluded universe of East Asian specialists and a handful of experts sensitive to the decisive contributions of comparisons.[3] For these to be useful, there has to be enough in common between two domains to make comparisons and contrasts relevant, and enough different to make such juxtapositions reveal critically distinctive aspects of one or the other. Though one should acknowledge the claims of Islam and India, it is China, outstandingly, that has this quality *vis-à-vis* Europe, and vice versa. It is ungracious to start what should be a celebration with negative comments, but this continuing neglect needs insisting upon. Opening an issue of *Nature* not so long ago,[4] I found an essay by Adrian Johns that argues that 'the social structures of [modern European] science were invented to cope with an explosion of printed information'. No one aware that printing was invented in China in or around the +9th century[5] could possibly countenance such an argument, at least in such a simple form, yet it seems that no one on the editorial staff of *Nature* knew or remembered enough to raise the question of the Chinese counter-case with the author.[6] One might say, with wry respect to Morris Low, that the only fault in his otherwise excellent recent

[2] As by Sivin (1982) and Elvin (1993–4), respectively.

[3] Pre-eminently, at the time of writing, Geoffrey Lloyd. See Lloyd (1996). Floris Cohen's survey of historiography of the Scientific Revolution is also exceptional in devoting a long section to Needham's work. While his analysis depends on an unqualified distinction between 'science' and 'technology' that I, at least, find open to question, his appreciation of the key role of the Chinese case in helping to disentangle probable patterns of causation is as exemplary as it is rare. See Cohen (1994).

[4] Johns (2001). Johns' *The Nature of the Book* (Johns (1998)) is subtler. He rightly insists that it was not the technology of printing as such that was crucial, but the interactions of the potential of this technology with a variety of changing social and cultural circumstances (pp. 19–20). But the absence of any understanding of, or even awareness of, the partial parallels and probable differences between Europe and the other, and older, great print culture, that of China, makes even this rich work seem one-dimensional.

[5] Barrett (1998).

[6] The underlying idea is derived from Eisenstein (1980).

special issue of *Osiris* is its title, *Beyond Joseph Needham*.[7] The majority of historians of science are still, usually wilfully, on the wrong side or at best, as in the case of my old friend Alistair Crombie, daunted by the immensity of the task of coming to serious terms with Needham's work even though aware of its importance.[8]

Addendum: The Practical Knowledge of Probabilities

Liu Tun 劉鈍 and Wang Yang-Tsung 王揚宗 have recently published an anthology of pieces by both Chinese and non-Chinese scholars on the issue that concerned Needham.[9] Liu Tun cites and includes recent analyses that discuss the two main pre-Needham debates on the absence in imperial China of a home-grown modern science. The first was that among the Jesuits and interested French scholars in the 18th century; the second was that started by Chinese intellectuals during the first third of the 20th century in the context of concern with China's modernisation. This latter played a role in shaping Needham's own early ideas on the subject, and he knew most of the participants. Liu also gives references to, and some items from, the extensive discussions on the 'Needham Problem' that have re-emerged in China since about 1980. The collection further features many translations of recent Western scholarship, including pieces by distinguished historians of Chinese science such as Sivin, Blue and Hart, some sociology, and pages from contemporary mainstream Western historians of Western science such as John Schuster and Floris Cohen, plus Alexandre Koyré from an earlier generation. This book, which is a most valuable resource, appeared too late to be properly incorporated into the argument of this introductory survey. Overall, though, it tends if anything to reinforce one's impression that points of contact through which a live intellectual current is flowing between members of this mainstream and those inspired by the discoveries of Needham and his colleagues have remained, to date, disappointingly few.

Crucially, intermittent inadequacy, or even the collapse, of the arguments of historians, the high quality of whose contributions one is generally happy to acknowledge, can often be linked with a disregard of the Chinese case. A simple example occurs in David Lindberg's *The Beginnings of Western Science*.[10] He sees the 'burst in creativity in lyric poetry and philosophy' in ancient Ionia, the fountainhead of later European sensibilities, as being first and foremost the result of a 'critical factor', namely 'the availability of fully alphabetic writing and its wide dissemination among the Greek population'. Yet ancient China, beginning at much the same time, had superb lyric poetry and philosophers some of whose work is still alive today, especially, perhaps, that of Chuang-tzu. And China used a *non*-alphabetic script.

[7] *Osiris*, **13** (1998).

[8] A comment based on years of participation in his seminar at Trinity College, Oxford, and several talks at his home in Boar's Hill shortly before his death.

[9] *Chung-Kuo Kho Hsüeh Yü Kho Hsüeh Ko Ming. Li Yüeh-Se Nan Thi Chi Hsiang Kuan Wen Thi Yen Chiu Lun Chu Hsüan.* 中國科學與科學革命。李約瑟難題及相關問題研究論著選 [Science and scientific revolutions in China: Selected research on the Needham Problem and related topics] from the Liaoning Education Press in Shenyang (2002).

[10] Lindberg (1992), p. 27.

While this argument as stated is thus dead, there is still an important issue to be examined. Science, more than any other human pursuit, is situated at the interface between symbols and reality. The power or inadequacy of a symbolic system is not a trivial question. We all know how a new notation can transform our ability to grasp a difficult concept.[11] What the comparison with China can help us do is to look for subtler characterisations than the crude and unhelpful 'alphabetic' or 'non-alphabetic' when exploring this interface. Numerals such as '1', '2', '3', etc., are not alphabetic. No one would, I think, essay the argument that this makes them inherently less useful than 'one', 'two' and 'three', etc. It has recently been maintained, however, that the Chinese '一', '二', '三', etc., were inferior to '1', '2' and '3', etc., which, given that they were similarly used in a decimal place-value system, seems more than a little unconvincing.[12] The European algebraic notation developed by Viète, Recorde (inventor of the '=' sign), Stevin and others is a different matter.[13]

Subtler, and more interesting, is the oversimplification committed by Edward Grant in *The Foundations of Modern Science in the Middle Ages*.[14] Grant uses the 'university' as a magic variable that set medieval western Europe apart from the rest of the world as regards the origins of modern science. This is not a trivial argument, but thus stated it is wrong. There were analogues to universities in China, though not many. Perhaps the best-known example, though not the earliest, is the *Thai-hsüeh* 大學, or 'Great School', run by the government during the Sung dynasty. It had both mathematics and medicine in its curriculum, and examinations.[15] Both in the Southern Sung and later dynasties the 'academies' (*shu-yüan* 書院) also offered a mixture of instruction, debate and training for the imperial examinations that changed over time.[16] What Grant needs to do, if his argument is to carry conviction, is to sharpen up the focus and ask what it was about the *Western* university that was significantly different from these Chinese institutions as regards fostering the growth of scientific styles of thinking. My personal view is that this would not necessarily prove impossible to do. My point is that Grant does not see that it needs to be done. So we still cannot see clearly what it was, *precisely*, that might have made the difference.

I have chosen the two foregoing authors because I have found their work in general interesting and valuable. My criticism is motivated by a friendly dismay, and not by a sense of hostility.

Familiarity with the patterns of Chinese premodern science also helps disentangle the confusions that surround our conceptualisation of the coalescence of scientific 'modernity'. Most of us have enjoyed the opening sentence of the introduction to Steven Shapin's recent little paperback: 'There was no such thing as the Scientific Revolution, and this is a book about it.'[17] But such a delightful, and insightful, witticism is no substitute for analysis. If we use the analytical framework of Crombie's

[11] Nowhere more spectacularly in medieval times than in the creation of a musical notation that permitted singers to sing directly from a score music that they had not previously heard. See Langellier (2000), on Guido d'Arezzo.

[12] In Huff (1993), p. 288. [13] Crombie (1994), p. 519. [14] Grant (1996).

[15] Wang Chien-Chhiu (1965). [16] Grimm (1977). [17] Shapin (1996), p. 1.

'styles of scientific thinking',[18] it is possible to compare China and Europe in a controlled way that avoids the sillinesses at times associated with arguments as to whether some idea was 'prescientific', or 'protoscientific', or perhaps 'modern', or, then again, not. Saying this does not imply any unconditional adherence to Crombie's ideas.[19] It is drawing attention to a method that offers a fruitful way of *disaggregating* a question into more manageable subquestions, namely how far premodern Chinese thinkers had developed the various styles of thinking that have in the long run, as they have combined with each other, proved crucial to the growth of a distinctively 'modern' science. This combination might even be used as the basis of a secure definition of the development of increasing degrees of such 'modernity'.

As I have shown,[20] as of about 1600, China possessed in varying degrees *all* of the styles of thought identified by Crombie as the eventual key components of science – the 'postulational' (like Euclidean geometry), the 'experimental', 'hypothetical modelling', 'taxonomy', the 'probabilistic', and 'historical derivation' (the prototype of which, in Europe, was the study of the genesis and diversification of the Indo-European languages) – with the apparent exception of the probabilistic, which hardly yet existed at this time even in Europe. By this date these styles were mostly less strongly formulated in China than in Europe, but they were there. The revolution in Europe after 1600, in so far as there was one, lay mainly in the *acceleration* with which these styles of thought both developed, and interconnected, rather than in any fundamental qualitative innovation – probability excepted. By 1750 China was far behind in almost every one of these domains – the 'historical' being perhaps the only doubtful case if we recall the sustained scholarly programme (in a more or less Lakatosian sense) to reconstruct the filiation of ancient Chinese pronunciations of characters as these pronunciations changed over time.

Another way of looking at the issue of continuity versus radical change in 17th-century European scientific thinking is to re-read Newton's *Principia* adopting the imagined mind-set of a scholar (of any nationality) sufficiently steeped in the Chinese tradition to be able to look at Europe, even if only momentarily, with the eyes of a cultural outsider. I tried this seemingly, but only seemingly, disingenuous thought-experiment recently with Bernard Cohen and Anne Whitman's lucid new translation.[21] It is a totally unrigorous procedure, but the result makes a sort of sense. Such a person's first reaction would almost certainly be that this is a book conceptually deeply rooted in *European* antiquity, as with its diagrams that seem so Euclidean at first sight and its pervasive use of ratios, combined perhaps with some later medieval notions, like that of acceleration, invented some time before 1235 in Merton College.[22] His second reaction would be one of awe at the wizardry with which the lines of these diagrams take on imaginary motions in the mind, approaching their own extinction, and conjuring up limits and the concepts of the calculus. Seen, in the imagination, from the other end of Eurasia it is both traditionally and recognisably

[18] Crombie (1994). [19] There is a lucid and careful critique in Iliffe (1998).
[20] Elvin (1993–4). [21] Newton (trans. 1999). [22] Crombie (1994), vol. 1, pp. 412–14.

European and breathtakingly new. An old and needless conflict thus to some extent dissolves. The capacity to translate ourselves at will to a different conceptual perspective on our own local history by means of an imaginative absorption into that of China is one of Needham's most fruitful, and underappreciated, gifts to the field.

These are impressionistic remarks, as befits an introduction and perhaps only permissible in such a context. What I hope they suggest is that there are, still, argument-sharpening comparisons and perspectives accessible to Western historians of Western science if they will familiarise themselves in a serious and sophisticated way with the world of premodern Chinese science opened up by Needham and his collaborators.

The adjectives 'serious' and 'sophisticated' matter. Credit must be given to scholars who have recently attempted a comparative analysis, such as David Goodman and Colin Russell in their Open University textbook, *The Rise of Scientific Europe: 1500–1800*, which is in many ways admirably multiple-angled in its approach, and Toby Huff, *The Rise of Early Modern Science. Islam, China, and the West*, which raises the majority of the key issues, even if it does not always probe deeply enough.[23] Apart from a number of avoidable factual errors,[24] they suffer, however, from two systemic defects. The first of these is an inability to use the prism of chronology to split the light from the Chinese past into periods with distinct characteristics. Thus Huff states that,

> the Chinese state in theory owned all the land and mineral wealth of the country, so that even mining operations . . . were operated as government monopolies. . . . Likewise all banking innovations such as letters of credit . . . , long-distance facilitation of exchange, and so forth, were taken over as state monopolies. There was no scope for entrepreneurial innovation, and thus disinterested learning . . . was discouraged *because* these avenues of advancement were closed without state sponsorship.[25]

Leaving aside the issue of theoretical eminent State domain, which was important up to and during the +1st millennium but rarely later, the points in the first two sentences applied to *some* sectors at *some* periods, and in *some* places, but virtually never universally. There was plenty of private mining from Sung through to Chhing times;[26] and private instruments of credit were extensively used during the Sung as they were under the Chhing, which also saw the rise of private financial institutions like the 'money shops'.[27] Under the Chhing the long-distance transfer of funds was handled above all by the Shansi banks, which were technically private though in a sort of symbiotic relationship with the government. The last sentence in the quoted passage is, however, untrue. The Chinese economy during most of the last

[23] Goodman & Russell (1991), and Huff (1993).

[24] One example will serve. Goodman & Russell (1991), p. 7, state that the Chinese empire had 'reached its furthest extension in the sixth century AD. . . . It was . . . at this time that the Muslim Empire, expanding northeast from Persia, encountered the Chinese and defeated them.' It was of course not the +6th century but the late +7th and the +8th centuries. The key battle, at the Talas River, was in +751. See Blunden & Elvin (1983), pp. 92–3. In the +6th century Islam did not even exist, and a 'Chinese empire' recrystallised only after the +580s, with the Sui dynasty.

[25] Huff (1993), pp. 312–13. My italics. It is worth repeating that, like the other works criticised above, this book has merits that justify the provision of the sinological qualification and correction that follow.

[26] Golas (1999). [27] *Chhien-Chuang* 錢莊.

millennium fizzed with entrepreneurial activity. This is apparent from, for example, Shiba Yoshinobu's *Commerce and Society in Sung China*, and the materials cited in my own books, *The Pattern of the Chinese Past* and *Another History* (1996),[28] but there is plenty of other evidence, both primary and secondary. Given this, which is a fact that cannot reasonably be disputed by anyone who is informed, the logic of Huff's 'because' becomes the crux of the issue. If it is a valid statement of a historical causal relationship, then there must have been abundant 'disinterested learning' in China during the middle and later empires. If there was not, then it is false. His and the reader's choice.

It should be noted here that the pioneering nature of Needham's work meant that he himself, half a century ago, could not develop a fully integrated chronological approach to the development (and the setbacks) of Chinese science and technology in their full historical social contexts. Practical reasons obliged him, and his collaborators, to begin by treating the subject topic by topic and so create an *Encyclopédie* in the best sense of that word, often profound and provocative, but compartmentalised. It also seems likely that he was personally uncomfortable with the prospect of exploring a logic of Chinese historical development that might prove too different from the immobile and Eurocentric formulae of the Soviet and Chinese Marxism of his time, though Marx himself, one suspects, with his independence of mind, might well have relished the task, given adequate information to work with.[29]

As I have already noted, Needham acknowledged but at the same time evaded the issue by comprehensively defining the imperial Chinese system over two millennia as 'bureaucratic feudalism', a new term in this context, and one which is discussed at several points in the present volume. The problem with this label, as I suggested to him in private correspondence and discussion in the early 1960s, is not only that any tightly constrained and testable definition that one constructs for this concept is unlikely to work equally well for all periods, but that its adoption tends to negate *a priori* the possibility of recognising significant changes.[30] The once justifiable excuse of the constraint of insufficient information has, moreover, ceased to apply for some time now; and one of the long-term objectives of present and future historians of Chinese science has to be precisely to *restructure* the mass of information made available by Needham, and by his collaborators, and his successors, and critics – by no means disjoint sets of persons – into a synthesis more sensitive to the long-term changes in Chinese society and culture.

The second defect in the two comparative studies just mentioned is a disinclination both to analyse the meanings of key terms and to show how they might fit into an explicit argument focused on the development, or non-development, of scientific thinking in various socio-intellectual contexts. Here we can take as an example the

[28] Elvin (1973) and (1996).

[29] Joan Robinson (1942), in *An Essay on Marxian Economics* long ago pointed out how extensively Marx changed his ideas from those in volume I of *Das Kapital* to those embodied in the fragments published as volume III, ending up by seeing 'underconsumption' as the most likely cause of crises in capitalist economies.

[30] Needham papers in the Needham Research Institute.

idea of 'proof'. Huff sees as foremost among 'the main defects of Chinese mathematical and scientific thought' that it 'lacked the logic of proof as well as the concept of mathematical proof'.[31] Goodman, in Goodman & Russell (1991), concurs: 'A . . . shortcoming of Chinese science is the absence of the idea of proof, so important in Western science since the time of the ancient Greeks'.[32] The issue here is not the absence, or presence, in premodern Chinese mathematics of the ideal of proof in the sense of a consistent system of primitive terms, a set of axioms, and more or less fully formalised rules for generating well-formed propositions and 'true' theorems as exemplified in a Western tradition running from before Euclid down to Hilbert. Apart from the flickers of almost Boolean logic in the Mohist canon 2,000 years before Boole,[33] it is generally agreed that this is a fair statement of the case. The issues are, rather, the following:

(1) In four out of the six of Crombie's styles of scientific thinking, 'proof' in this sharply defined sense is not relevant to scientific advance at what we might loosely call an 'early modern' level. This comment applies to modelling, experimentation, taxonomy and historical derivation, though not of course to the postulational style nor, with some reservations, to probabilistic thinking. With respect to particular fields, it clearly plays little or no part in chemistry, the earth sciences or the life sciences at this level. The case can also be made that it is of little importance even for the early history of some parts of physics, such as magnetism. If it was so crucial an element, one would have expected to see marked differences in China between the domains where the absence of 'proof' mattered and where it did not. So far as I can tell, if we examine the period 1550 to 1750, this was not the case, or only weakly so.

(2) If we take as a rough working rule, to simplify a subtle discussion, the notion that 'theorems are discovered but proofs are invented', in what specific crucial aspects as regards discovery did what we may term 'Euclidean-Hilbertian proof' in Western mathematics differ from what I would describe as the 'sequences of co-ordinated demonstrations' that we find in the best traditional Chinese mathematics? Works like the *Chiu Chang Suan Shu* 九章算術 (Nine chapters on mathematical procedures), with its +3rd-century commentaries by Liu Hui 劉徽, at least come close at times to what Karine Chemla has called 'algebraic proof within an algorithmic context'.[34] They are not just numerical cookery books. Even if we do not, yet, easily go all the way with Chemla's challenging formulation, there is a need for an examination of the particularities in this domain, rather than unexamined generalities. Moreover, Sung and Yüan mathematics was still capable of important discoveries, such as aspects of modular algebra and the theory of determinants.[35] The late imperial discovery-barrier problem remains as elusive as it is important.

[31] Huff (1993), p. 288; see also p. 278. [32] Goodman & Russell (1991), p. 12.

[33] For a summary of the debate between the present author and Professor Makeham on this see Elvin (1996), pp. 276–9. Also Elvin (1990).

[34] Chemla (1997–8).

[35] See the revised edition of Blunden & Elvin (1998), 'Principles of Mathematics', pp. 194–7, for a sketch of these and other topics. (The original English-language edition of 1983 contains misprints in this section.) See also *SCC*, vol. 3.

(3) Though it was certainly not non-existent, it seems likely that the drive to what might be called the 'mathematisation of observations' was significantly weaker in pre-modern China than in Greek antiquity or in Europe after the later Middle Ages. The Chinese artisans who built rectangular-trough pallet-chain pumps for drainage and irrigation in the lower Yangtze region in late imperial times altered the proportions of troughs and pallets by (one assumes) trial-and-error to optimise the use of energy for different angles of inclination of the trough, presumably knowing but keeping to themselves the appropriate empirical proportions. The 18th-century French hydraulic engineer De Bélidor, using simple Euclidean geometry plus simple mechanics, *calculated* and *published* the specific optimal ratios for the same type of pump – which was clearly, by a nice twist, originally borrowed from China.[36] Although 'proof' hovers in the background, because of the geometry, the crucial difference here was the early modern European desire not only to express the particular proportions (observed in both cultures to work better or worse according to the slope) in *numbers*, but, above all, to find a *general* way of *calculating* the *best* for any given circumstances. The mathematisation of observations is not the same as the use of 'proofs'. Though the distinction can be subtle, it needs to be made.

Exceptions to Non-comparability

There has been an intermittent tendency among historians and philosophers of science during the century just past to make two partial truths into absolutes. The first is that all systems of thought inevitably rest on assumptions that are formally arbitrary – which is correct[37] – and that *therefore* any one such system is just as objectively correct as any other – which is misleading in its unstated implications. It is not the case that 'anything goes', to quote Feyerabend's celebrated if not entirely intelligible phrase.[38] From a historian's point of view, at least, different systems of thought and practice have different behaviours and different capacities for particular types of action, and *distinguishing between them* is the essence of his work, just as a biologist distinguishes between species and their behaviours. The second partial truth is that cultural systems of belief, value, conception and perception are independent worlds of mutually interdetermining parts whose separate components cannot meaningfully be compared with their apparent counterparts in other systems. Anyone who has lived a substantial part of his life in different cultures is aware that this is not without its measure of truth, and that people in different societies do tend to live in what I have elsewhere called complexes of different 'stories'.[39] But there are *subsets* of effectively identical phenomena that can be *common* to two or more such societies,

[36] Discussed and documented in Elvin (1975), reprinted in Elvin (1996), pp. 98–9. The French name was 'moulin à chapelet' or 'pompe à chapelet', in other words a 'rosary-beads pump'.

[37] Thus in one formulation of mathematical logic, the system is based on 'true' and 'false' applied to atomic propositions, a universal propositional calculus connective such as 'neither . . . nor . . .', quantification (the idea of 'all'), and some suitably circumscribed notion of the 'membership of an element' in a 'set'. See Quine (1951) and also (1963) for a demonstration that such foundations are not unique.

[38] Lakatos & Feyerabend (1999), pp. 14, 116 and 324. [39] Elvin (1997).

and where 'right' or 'wrong' answers have also to be effectively *identical*. These subsets have the useful property of permitting precise comparisons, even if it is necessary to be cautious about delimiting them. There are normally cultural penumbrae around the subsets that are indeed significantly different.

Where China and Europe are concerned, during the period, broadly speaking, from +500 to +1700, the most useful such subset is probabilistic thinking relating to gambling, particularly with dice. The crucial analytical importance of probabilistic thinking for many general issues in the early history of science has been on the whole poorly appreciated by historians of science. It is also one of the domains where the divide between so-called 'science' (in the sense of, loosely, explicit abstract theoretical conceptualisation) and so-called 'technology' (in the sense of, loosely, concrete practice) – a commonplace that becomes questionable when looked at in terms of the tendency towards the close interrelationship in the styles of thinking and action that prevail in a given society at a given time – is blurred to the all but invisible. In the probabilistic domain the adequacy of conceptualisation is constantly being tested by repeated 'experiments', such as throwing dice or using other aleatoric devices such as the Sung dynasty practice of commercial gaming for prizes by means of darts being thrown in quasi-random fashion at a revolving circular target subdivided into different areas – the physical analogue of a subdivided probability space. In the premodern Chinese case, when it is contrasted with early modern Europe, the study of probabilistic thinking also shows that the factors commonly instanced as 'external' causes in Europe (commercialisation, monetisation, widespread numeracy and the like) are useless as criteria of differentiation. (They were all present in China.) We can accept them as probably necessary preconditions, but not as sufficient ones.

'Internal' criteria present different but equally tough problems. The Chinese tradition had more than sufficient skill with the arithmetic of fractions required for the analysis of probabilities. There was some familiarity with permutations, most generally in the case of the $2^6 = 64$ possible sequences of six unbroken or broken lines in the hexagrams of the *Book of Changes*. The 3^{361} possible arrangements of black and white counters and empty spaces on the 19×19 intersections of the *go* board, famously established in principle by Shen Kua, would not have been as well known, but show the capacity of Chinese mathematical culture to come up with the appropriate formula. In late imperial times there was likewise a basic familiarity with combinations, most evidently in the systematically tabulated sets of all possible distinct scoring patterns from throwing five six-sided dice without considering the order of occurrence of individual values.[40] Putting these three together would have yielded a basic calculus of probabilities, in all likelihood a lucrative discovery for its discoverer. It is also virtually certain by deduction from the values used that, at least by late imperial times, the exact relation between stakes, probabilities of outcomes, and payoffs *was* grasped by the professionals running certain gambling games

[40] Elvin (2000).

such as *fan-than* 番攤, a version of a very ancient amusement called by many other names, such as 'Lay Out the Coins' (*than-chhien* 攤錢), summarised below. Probably to safeguard valuable trade secrets the figures were never published, and so the codification, generalisation and progress usually associated with public availability never occurred. But there appears to have been no 'internal', that is conceptual, barrier to taking another step.

Analogously, by the +4th century or earlier, the Chinese had some notion of what we would today regard as a 'statistical distribution', usually imagining it in a bureaucratic context such as the unequal burden of labour-service obligations on different areas (such as carrying tax-grain for different distances). So far as we know, however, they devoted themselves *only* to methods for calculating how to *remove* the inequalities, as part of administrative fair practice, not to their analytical definition. Thus, while they knew how to interpolate a mean value between two extrema, they had no idea, then or later, of characterising a statistical distribution by means of its variance and other such measures.[41] To date, neither mainstream historians of Western science nor specialists in the premodern Chinese variety have realised just how great are the difficulties which the considerations outlined in this and the preceding paragraph make for their favoured explanatory frameworks.

This is one of the few scientific topics with which Needham did not, so far as I know, concern himself, though he opened the door to us by even thinking of pursuing such an enquiry in a Chinese context. This is testimony to the continuing vitality of the project, and the way it continues to impinge on the broader discipline of the history of science as a whole. The elusive history of clues to the presence of probabilistic thinking in premodern China is too complex to be pursued further here; but what can be said by way of a summary of the present position is that two main points seem to have emerged. The first is that in terms of what made its way into the public domain, the apparent stopping point is baffling: all the basic conceptual tools were there in a highly appropriate socio-economic setting, but nothing further happened. The second is that there are numerous indications that unpublished investigations, even if possibly only of an empirical or tabulatory sort, must have gone considerably further, whether one is looking at some of the details of the scoring system in the milfoil stalk oracle-casting technique used in association with the *Book of Changes*, at the operations of professional gamblers, or the gambling-selling methods of Sung dynasty shopkeepers.[42] Simply as an illustration of the type of information from which this unpublished capacity can be deduced, is the family of coin-games that culminated in late imperial times as *fan-than*, already alluded to and summarised here in a separate box.

[41] *Ibid.*

[42] Those interested in pursuing this topic should consult my unpublished paper, Elvin (2000), which is on deposit, with a short bibliography, in the Library of the Needham Research Institute at 8 Sylvester Road, Cambridge, England, and available to anyone who contacts the Librarian.

Practical Knowledge of Probabilities

The commonest type of coin-game was probably that known as 'Lay Out the Coins' (*than-chhien* 攤錢). It had other names: 'Mind Money' (*i-chhien* 意錢), 'Guess Numbers' (*she-shu* 射數) and 'Tricky Estimate' (*khuei-i* 詭億) among them. In the later 19th century it was the basis for *fan-than* 番攤, played in State-supported dens to provide finance for the navy. By this late date at least, the game was played as follows:

An arbitrary quantity of copper coins was drawn at random from a heap called 'the cash surface' (*chhien-phi* 錢皮) and put into a receptacle (*than-chhung* 攤盅). The participants then laid bets on the residue that would remain when the contents were counted off by fours. (Thus 30, which is $(4 \times 7) + 2$, would leave a remainder of 2.) There were thus four possibilities, or 'gates' (*men* 門): 1, 2, 3 and 4 or 0, taken as a cycle for ordering. There were four types of wager:

1. *Fan* 番. Betting on a *single* number. The pay-off for success was the stake back plus three times the stake.
2. *Nien* 捻, or *jen* 稔. Betting on two *adjacent* numbers, one of which was the principal and the other the 'support'. If the winning number was the principal, the pay-off was the stake back plus two times the stake. If the support was the number that came up, the stake was returned.
3. *Chiao* 角. Betting on two *adjacent* numbers. If either was the number that came up, the pay-off was the stake back plus one times the stake.
4. *Cheng* 正. Betting on three *sequential* numbers, the one in the *middle* being the principal. The pay-off for the principal was the stake back plus one times the stake. If either of the other two numbers came up, the stake was returned.

The gambling establishment derived its income from a 10 per cent levy on winnings. This apart, was it a fair game? It seems so. Writing 's' for 'stake', we have for the four options the following expectations:

1. $(1/4) \times 4s = 1s$
2. $(1/4 \times 3s) + (1/4 \times 1s) = 1s$
3. $(1/4 \times 2s) + (1/4 \times 2s) = 1s$
4. $(1/4 \times 2s) + (1/4 \times 1s) + (1/4 \times 1s) = 1s$

Computer simulation and analysis both show that the debarred betting options do not affect fairness. The number of coins drawn has a slight effect on the frequency of the four possible moduli, but simulation shows this can be virtually eliminated by requiring that the number of cash drawn falls within a reasonable range, such as 31 to 98. (In 10^7 trials for this range all four moduli appeared at a frequency of 2.50×10^6.) How far probabilistic thinking was needed to construct this game is an open question. It may have been enough to see that, given equal probability of outcomes, the better's *net* gains and losses sum to zero for each of the four betting patterns. These may be expressed as the cyclical groups formed from $(-1, -1, -1, +3)$, $(-1, -1, 0, +2)$ and $(-1, -1, +2, 0)$, $(-1, -1, +1, +1)$ and $(-1, 0, +1, 0)$ respectively.

Note: This reconstruction is a compromise based on conflicting accounts in Kuo & Hsiao (1995), and Lo & Hsüeh (1994). See Elvin (2000) for the details, and *Tzu Hsia Chi*, p. 157, for the earlier history.

Finally, aleatoric technology was to a surprising degree *shared* between the Chinese and European (and other major Old World) cultures after about +600, though not completely. Games, like backgammon and dominoes, travelled back and forth. The first, for example, went to China; and the second came from it. The *correct* solutions to problems involving probabilities had therefore to be essentially *identical* in both cultures, though full precision in matching stakes and odds was not always necessary if one side (like a gaming-house) had the institutional power to impose a margin in its favour. Those who got the sums wrong lost money – in both cultures. The well-known problem of the difficulty in transferring skills needed for the successful replication of technical results does not arise. The basic criteria for minimal comparability are thus met, as is the criterion of a similar definition of 'success'. The concept of fully separate Chinese and European cultural domains cannot be sustained in this area, which is what gives it much of its theoretical interest.

Possible Futures

So all-embracing is the intellectual tapestry woven over fifty years in the volumes of *Science and Civilisation in China*, ranging from the particularities of tradesmen's tools to the vast generalities that sustain the premodern Chinese versions of 'theories of everything' that it is hard to resist the question, 'What remains to be done, apart from supplementing, correcting, fine-tuning and reinterpreting?'[43] Thanks first and foremost to Wang Ling's eleven years as Needham's first research assistant, the project long ago identified the majority of the major texts,[44] even if much work has still to be undertaken on them. This aside, it might seem that what is left to be attempted is mostly to trawl a vaster ocean of less familiar literature and less densely distributed materials, cetacean-style, for intellectual nourishment and perhaps occasional *trouvailles*. Thus the publication of what is, formally, the final volume in the survey of the history of Chinese science undertaken by Needham and his collaborators is an appropriate moment to ask, 'Is this impression justified?' If not, where can substantive research advance next?

I say 'substantive' as re-evaluation and criticism have of course been going on since the 1950s.[45] In recent years French scholars in particular have been at the forefront

[43] An example of the need for fine-tuning arises from the divergence of views between Dieter Kuhn, the author of *SCC*, vol. 5, pt. 9 (1988), pp. 225–36, and the present author in Elvin (1972) on the most plausible reconstruction of the water-powered spinning machine described by Wang Chen early in the 14th century. My reply to Kuhn appears on pp. 60–3 of Elvin (1996) and highlights distinctive strengths and weaknesess in the two versions. Notably Kuhn adheres to the proportions given in Wang's *text*, but pays no attention to the best iconographic tradition, with which these are not consistent. My version, validated by professional spinning engineers as workable, is based on the superior *images* and matches each item depicted, but does not have Kuhn's scrupulous correspondence with the text. Both versions, however, require assumptions justified *neither* in the text *nor* the images in order to work, but make different choices. It would seem that the experimental reconstruction and testing of a range of close-to-full-scale working models is called for if the analysis is to be carried forward.

[44] See Needham's characteristically generous acknowledgement of Wang Ling's first seven years on p. 14 of *SCC*, vol. 1. It was only recently, when beginning to work through Wang Ling's papers and library in Canberra, courtesy of Mrs Ruth Wang, that I realised how much of the bibliography of secondary works in *SCC* was also likely to have been due to Ling's indefatigable truffle-hunting.

[45] For references to the older literature, see the references in Elvin, Peterson, Libbrecht & Cullen (1980).

of offering a different perspective on many parts of the field,[46] as well as important new contributions in fields such as mathematics and botany. The francophone Swiss sinologist Jean-François Billeter and his colleagues have also launched a plausible and provocative attack on the view of Needham and his collaborators that Shen Kua can be sensibly conceptualised as a 'scientist'.[47] If tenable, this key case implies a need to re-examine many of the assumptions built into the language that we have grown accustomed to using. The present introduction is not the place for a critique of these criticisms. The developing polyphony of multiple voices, whether one always agrees with them or not, is, however, an enrichment of the field to be cherished. Like the journal *Chinese Science*, now re-entitled *East Asian Science, Technology, and Medicine*, it bespeaks its continued vitality.

It is unlikely that these responses and initiatives would have existed but for Needham's volumes to provoke them. Many recent scholars who have to some degree taken issue with Needham's views have in fact gone generously out of their way to acknowledge this debt.[48] The only partially parallel sources of inspiration during the appropriate period were Japanese: the rich anthologies on aspects of the history of science and technology in China in various periods inspired and edited by the late Yabuuti Kiyosi 藪內清.[49] Though these are slightly later than Needham's earlier work, they have to a large extent overlapped with it chronologically.[50] Being of a more restricted scope, however, it seems fair to say that they did not present a comparable challenge to existing paradigms in the history of science.

I would identify three domains that seem to me to be promising for future substantive work. It is certain that many readers can think of others. The first domain is probabilistic thinking, already touched on above and illustrated in the accompanying box. The reader will have seen from this that, at least in a context such as this one, Needham's sometimes criticised custom of thinking in terms of an ultimately *universal* modern science cannot, perhaps surprisingly, be faulted.

The second is scientific knowledge in action in specific environments, drawing mainly on the Chinese sources for local history, which are particularly rich for the late imperial age. In no other area is the unsatisfactory nature of drawing a sharp, as opposed to what might be called a 'blurred', distinction between 'science' and 'technology' so evident. Conceptual understanding of how a landscape 'worked' regularly informed premodern Chinese actions to transform it, or control it, or restore it.[51] This was evident, for example, in the interplay between hydrological and hydraulic theorising on the one hand and hydraulic practice on the other. Irrigation systems, flood control, drainage and sea-wall construction had a large rule-of-thumb component, but, especially as regards how it was thought that water behaved and

[46] See the *Revue d'histoire des sciences*, **42**, 4 (1989) and **43**, 1 (1990), 'Problèmes d'histoire des sciences en Chine', (I) 'Méthodes, contacts et transmissions' and (II) 'Approches spécifiques'.

[47] Billeter *et al.* (1993).

[48] Notable examples are Cullen (1977) and Libbrecht (1973).

[49] Sometimes written 'Yabuuchi Kiyoshi'.

[50] For a full bibliography of Yabuuti's publications, see Tōgō (2002).

[51] There are examples in several of the chapters in Elvin & Liu (1998), notably those by Vermeer, Osborne, Will, and Elvin & Su.

sediment was deposited or entrained, there was – over more than a millennium – a vigorously debated component of theory. And more than just theory: theory constantly tested in practice.

There is a slightly Brahminical disdain among some of those scholars concerned with the history of the 'higher' sciences towards early attempts to make rational sense of such rough-and-ready complex phenomena as moving water, mud, earth, erosion and vegetation cover. It is true of course that the results were rarely clear-cut, and empirical art and experience could sometimes win out for a time over theoretical understanding. Effective action in this context is commonly required more than comprehension in the abstract, especially at the beginning. To give an example, Su Ninghu and I pointed out some time ago that although Phan Chi-Hsün 潘季馴 in the 16th century had grasped that the celerity of a moving current determines its capacity to transport suspended sediment,[52] this did not mean that Phan's attempt to use this – correct – principle to clear the silted-up lower reaches of the southern-course Yellow River by concentrating them in a single channel, and then narrowing this channel, was wholly successful. In particular, while doing this shifted an enormous volume of sediment, he overlooked that, since the speed of the current slowed as it approached the sea, the suspended materials would be largely dropped again, creating a barrier across the sea mouth.[53] In 1593 he was replaced by the more traditional and pragmatic Yang I-Khuei 楊一夔 who did the best he could to undo the damage, though only with limited success.

Even accepting that there is thus a gross and approximative aspect to most early Chinese environmental science, it seems difficult to deny it the label of science. To give just one example following on from Phan's case: by the first half of the 19th century we can find a clear understanding that as a current slowed it progressively dropped its suspended sediments in a sequence from the heavier and larger to the lighter and smaller particles, beginning with stones and gravels, then sands, then silts, and finally clays. In terms of mean diameters, this covers approximately four orders of magnitude, from centimetres down to microns. As an illustration we may take some passages from the mid-19th-century local gazetteer for Teng-chhuan department 鄧川州, which lay at the northern end of the Erh-hai lake 洱海 in Yünnan. They describe the works undertaken each year on the Mi-chü River 瀰苴河.[54] In late imperial times this river was seen as a lesser counterpart of the Yellow River, as it carried so much sediment that the first part of its lower course below the Phu-tho Gorge 蒲陀崆, on the department border, ran (and still runs) in dykes above the level of the rooftops of the houses alongside it, and its debouchment into the Erh-hai has recently built up an extended lacustrine delta. The passage shows not only the good general understanding of fluvial geomorphology, but even a rough notion of cross-channel helical flow at bends. After a description of the way in which the river, as its speed decreased, steadily dropped its sediment load in order of decreasing size, it concluded:

[52] Vermeer (1987). [53] Elvin & Su (1998), pp. 401–5. [54] *Teng-chhuan Chou Chih* (1854/5), pp. 79–80.

One can test this point when the water is dried up in the spring and winter. If one looks at the river bed it consists of large stones squatting in the upper reaches and fragmented stones *arranged in graded order* in the sections that follow. The accumulated sediments and piled up pebbles are *in continuous succession, subdivided ever more finely*.

Though not 'modern', in the sense of not being precisely quantified or based on systematic mensuration, the sharpness of perception in these hydrological and hydraulic observations is impressive. What is especially interesting, though, is that they were put to practical use.

In the later 18th and the early 19th centuries, on the west side of the upper Mi-chü River just above the Phu-tho Gorge, large amounts of sediments were intermittently washed off the mountain slopes by the wet-season rains, once these slopes had begun to be cleared for farming in the 18th century. They came down a large and normally dry valley called the Pai-han 白漢 into the river gorge where they choked it, causing back-flooding of farmland and houses. The solution developed was to divert the flash floods, with their heavy loads of stones and gravels, along an extended and almost level deposition surface, walled off from the direct course into the river by what was called 'a dry cross-dyke'. As the current slowed, the heavier sediments were dropped on this surface, and the next heaviest then further filtered by a low barrier planted with trees. Only after this was the water, now carrying only a suspension of fine sediment, redirected into the river. This was informed and deliberate engineering. The elevated causeway that has been built up by years of repeated spate-season depositions is today of massive size, and the deposited large-diameter sediments are quarried by the locals in the dry period for gravel. What is interesting about it is the successful application in practice of a hydraulic principle derived from careful local observation.

It is my impression that a large quantity of historical material of this sort remains as yet little explored in local gazetteers and analogous works. Apart from hydraulics, some of the topics that might repay investigation are botanical and zoological observations, and attempts to explain unusual local phenomena. An example of this last, from the 13th century, is Chu Chung-Yu's 朱中有 construction of a flume as an experimental model in which he tried to reproduce the tidal bore in his native Hang-chou Bay.[55] At best he seems to have confused a hydraulic jump, which may have been what he produced (or at worst just turbulence due to a rough surface), with the bore effect, which is a hydrological analogue to a sonic boom; but the idea and his description of it are worth noting. The style of scientific thinking that Crombie labelled 'hypothetical modelling' had a real, if tenuous, place in medieval Chinese culture.

The third is the hazier topic of the changing mindset of the élite. Hazy but important. There are *prima facie* grounds, for example, for suspecting that the late imperial

[55] See Elvin & Su (1988), pp. 369–70.

age saw a 'disenchantment' of the world,[56] that is, a trend towards seeing fewer dragons and miracles, not unlike the disenchantment that began to spread across the Europe of the Enlightenment – but without an accelerating science either as a cause or a consequence. The same phenomenon of a weaker but not non-existent Chinese parallel with Europe can be found in the domain of fashion. It used to be thought that the intensifying passion for continual change in early modern Europe, visible not only in shifting styles of personal clothing but in most of the arts, had no Chinese parallel. This, too, has been found to be too extreme a contrast. Fashion in women's clothes, known as *shih feng* 世風 'trends of the times', was a feature of Shanghai, for example, in the later 17th century.[57] The analytical issue here, too, is dissolving into a question of *different degrees*, and becoming correspondingly hard to handle satisfactorily.

The Present Volume

Needham never solved the 'Needham Problem'. Nor has anyone else to date, at least in any fashion that commands general acceptance. In part this is because it coexists in an ambiguous relationship with a counterpart, which may be called for convenience the 'Weber Problem': why did China not create its own endogenous industrial capitalism?[58] The two questions are evidently interrelated, but the conceptual frameworks with which specialists approach them are usually significantly different. Even the most polymathic and penetrating in one of these worlds of thought – and Needham was certainly among the most deservingly celebrated of this number – tend to use oversimplified or somewhat dated ideas from the other world. Perhaps Needham's most favoured magic variable for explaining the relative acceleration of science in early modern Europe was 'the rise of the bourgeoisie', but recent work on this area by economic and political historians has made this concept ever more elusive.[59] One could even argue – polemically, but to make a point – that one of the many factors that distinguished early modern western Europe from late imperial China was precisely the perduring in Europe of a relatively independent *aristocracy*. Just as one could also argue – equally polemically, but likewise to make a point – that the Roman Catholic Church has long outlasted any comparably identifiable Chinese institution, and was no less capable of mobilising talent from the lower levels of society than the mandarinate. Changeless Europe? – No, hardly, but we accept the old clichés too easily.

[56] Max Weber's *Entzäuberung*. I first noticed this when I found that the miracles associated with the fidelity of widows, whose brief biographies are recorded in immense numbers in local gazetteers, declined precipitously from the Ming to the Chhing. See Elvin (1996), chapter 10.

[57] Elvin (1999), pp. 152–3. [58] See Elvin (1984) and (1988).

[59] Summarised at various points in Goldstone (1991). A more subtle variant of Needham's position may be found in Billeter (2000), p. 18: 'Ce dont les savants n'ont pas conscience, c'est que la raison abstraite qu'ils manient avec tant de succès résulte de l'application au monde physique d'une forme d'abstraction qui a son origine dans la relation marchande et qui entretient avec elle un indissoluble lien'.

Conversely, Ken Pomeranz's challenging work *The Great Divergence. China, Europe, and the Making of the Modern World Economy*[60] bypasses science as an area deserving of attention.[61] Effecting an organic conjunction of the two discourses is difficult, and a task, still, for the future.

Most of the chapters of the present volume record one or another of Needham's approaches to the 'Needham Problem', and they were originally formulated at a time when hardly anyone else in the world was aware of the problem, let alone wrestling with attempted solutions. It is difficult to be one step ahead of the game, and dauntingly so to be two steps ahead, as he was for a number of years. They remain, without exaggeration, a heroic legacy, and are still capable of inspiring. This is why they are printed here, even if at moments – to speak honestly – they can also now seem naive or infuriating. I doubt, for example, if many serious economic historians of China would today see the economy of the Sung and later times as having been, in the majority of the centuries that followed, seriously undermonetised. Needham was supremely sensitive to the issue of the later near-standstill raised by China's achievements in technology and the sciences up to about 1600, but was much less sensitive than he perhaps should have been to the parallel economic aspect of the problem: that China was in many ways economically up with or even ahead of Europe for quite some time after this – as correctly stressed by Pomeranz, among others.

The problem is thus harder than Needham thought, and probably than most of us like to think even now. It requires not only the organic joining of the history of science and technology with the history of the economy and society referred to above, but also abandoning the ever-tempting and easily acquired habit of attributing virtually unchanging cultural characteristics to China and Europe as the key to discriminating between them over the long run. This erroneous habit of thought amounts to implying that because 'China' was 'this', and 'Europe' was 'that', in all but perpetuity, everything else followed – in other terms, relying on some version of Weber's ideal types. What is, *per contra*, needed is a history focused on changing conjunctures,[62] since the minimum required, as Needham himself came to emphasise, is the explanation of – seen from the Chinese side – a relative advance followed by a relative retardation; and this, so easy to specify in the abstract, is hard to deliver convincingly in practice.

What *Science and Civilisation in China* still has to effect – and why it is still so vital – is the creation of an awareness in *every* historian of science that no new idea about the origins of modern science in western Europe should have *droit de cité* until it has been tested against the Chinese case to see if it effects a discrimination that is both precise and relevant. William Eamon's wonderfully rich *Science and the Secrets of Nature* is a good example.[63] The argument is too complex for summary here, but he suggests two points, which he relates directly to the genesis of modern science, that can be

[60] Pomeranz (2000). [61] See my review in the *China Quarterly* (Sept. 2001).
[62] See Elvin (1984), especially p. 103. [63] Eamon (1994).

taken on their own without serious injustice. The first is that the conception of nature as a repository of decipherable secrets, derived from a strand of Islamic tradition, was a powerful stimulus to late medieval experimental science in the West, most notably for Roger Bacon in the 13th century. The second derives from his exploration of the early modern 'vulgarisation' (in its good sense) that led to 'a barrage of how-to-do-it manuals', the '"disenchantment" of technology', and 'a mediation between [élite] natural philosophy and the people's immediate . . . empirical experience of nature'.[64] It seems likely that Chinese alchemy, older than its Western counterpart, can broadly match the first, and that the approximately millennial tradition of Chinese printed how-to-do-it books on farming and crafts, though probably not so easily accessible to the only modestly literate, offers at least a partial parallel to the second. Thus the effect *in a context* was probably what was crucial – and most awaits definition.

Most tantalising, though, is the refinement Eamon implicitly gives to Needham's famous summing up, namely that in premodern China science and technology reached the level of da Vinci but not of Galileo. His research has brought out of the shadows the late 16th-century European concern with 'natural magic', and especially that of Della Porta who wrote the best-known contemporary book on the subject.[65] This, however, had much in common with the tastes of at least some important Chinese intellectuals at the same time. In the thousand or so pages of Hsieh Chao-Che's *Fivefold Miscellany*, published in 1608,[66] one finds an analogous mixture of rational analysis and breathtaking fantasy, a similar obsession with the marvellous and exceptional, the same interweaving of grand metaphysics with bizarre trivia, an identical relish in explaining or unmasking technical trickery (including the alleged role of demons), and a parallel alternation of sound observation and testing in practice with the blithe acceptance of the unverified and unverifiable.[67] When Eamon says that 'Della Porta's "empiricism" was not some proto-Baconian inductivism, but history and literature verified by experience',[68] he unknowingly encapsulates Hsieh's thinking with equal accuracy. Differences there were, too. Of these the most important and least remarked upon may have been the stronger European sense that all of this 'mattered'. But how the flower of science grew from one heap of intellectual compost and not from the other needs more discriminating analysis than Eamon (or anyone yet) offers. This is symbolised by the delicate but critical shift-around in 1611 on the part of the still youthful Accademia dei Lincei 'away from its esotericism and its preoccupation with natural secrets'.[69] This was caused in the immediate sense, as he recounts, by the replacement of the dominant influence of its fifth member to join by that of its sixth. The magus, Della Porta, was the fifth member, and the sixth one of the first scientists conventionally accepted as 'modern' – a certain Galileo Galilei. Thus the Chinese case shows up, as perhaps nothing else can, how our subject lives on a knife-edge.

[64] *Ibid.*, pp. 113 and 105. [65] See *ibid.*, chapter 6, 'Natural Magic and the Secrets of Nature'.
[66] *Wu Tsa Tsu* 五雜組 [or 俎], by Hsieh Chao-Che 謝肇淛 . Reprinted by Hsin Hsing Shu Chü, Taipei, 1971, 2 vols.
[67] Elvin (1993–4). [68] Eamon (1994), p. 216. [69] *Ibid.*, p. 231.

A final point: I have written 'Needham' at most places in this introduction, aware that on many occasions, it should have been at the very least 'Needham *et al.*' Joseph had a remarkable core band of collaborators. They ranged from the Chinese students who first sparked his interest in Chinese science, one of whom was Lu Gwei-Djen, the biologist, and the most important of them all over the long run, through Wang Ling, who laid the sinological foundations, and Nathan Sivin, who gave new dimensions to the discourse, down to Kenneth Robinson, whose contribution figures prominently in the present volume, as well as those who have continued to care for his legacy at the Needham Research Institute. Outstanding among these are Ho Peng-Yoke, both a major contributor in his own right and the former Director, who has long piloted the ship through difficult times, and Christopher Cullen, the former Deputy Director and now Director, and an expert in the history of Chinese cosmography and mathematics. Specific tribute should also be paid to Sir Brian Pippard FRS, who for a time chaired the executive committee, and to Sir Geoffrey Lloyd, who served as chair of the Board of Trustees. There have been many, many others, who I hope will forgive my not mentioning them individually here, especially those who authored some of the later individual volumes. All of them are scholars who are – or, for those no longer with us, were – distinguished in their own right. Perhaps the greatest tribute of all to the compelling power of Needham's vision is that people of such quality have felt honoured and enriched to spend a substantial part of their creative lives as part of the collective enterprise he created.

Mark Elvin

FOREWORD

When 'Science and Civilisation in China' was first conceived, we thought of it as a single slim volume. When in 1948 I sat down with my first chief collaborator, Wang Ling 王鈴 (Wang Ching-Ning 王靜寧), to work out the general scope of the book, we made a rapid survey of the spectrum of the sciences, and decided that the work should have seven volumes. What could not be predicted was the relative vastness of the material which we would find as we went on in the various sections. The later volumes had therefore to appear in parts, each of which was a physical volume in itself. Such is the situation in which we were compelled to acquiesce as the work went on; and here great praise is due to Cambridge University Press, which has never grudged space to the revelation of the wonderful achievements of ancient Chinese science, technology, agriculture and medicine.

It will certainly not have escaped the perspicacity of our readers that as time has gone by, the whole work has become more and more of a co-operative enterprise. About 1970 Lu Gwei-Djen 魯桂珍 and I took a weighty decision. We could have gone on pegging away single-handed, as it were, in which case the series would never have come within sight of completion in our lifetime; or we could take to ourselves suitable collaborators and entrust them with sections, sub-sections, or even whole volumes. We decided on the latter course, and in due time this policy bore wonderful fruit. Francesca Bray led off with her history of agriculture (vol. 6, pt. 2) in 1984, then came Chhien Tshun-Hsün 錢存訓 (Tsien Tsuen-Hsuin) with that of paper and printing (vol. 5, pt. 1) in 1985 and more recently in 1988 Dieter Kuhn has produced the first volume on the history of textile technology, spinning and reeling of textile fibres (vol. 5, pt. 9).[1] More and more team-work has thus been put in, Chinese and Westerners have alike contributed, more and more have specialist friends come to be relied upon. We now have about thirty collaborators scattered all over the world. And this is only natural and reasonable for, as I have often said, no one individual could summon up all the skills necessary for placing the whole development of all the sciences in their correct historical perspective, both in China and the rest of the world. It might be true to say that my own role has risen from that of a single historian of science to the conductor of a kind of orchestra. I can only hope that the resulting music will be found both delightful and fitting.

Volume 7 has for its theme 'the social and economic background' of Chinese science and technology, considered always for its relevance to the grand question which has partly inspired the whole work – why modern science originated in Europe

[1] Since this was written the following volumes have also been published: *SCC*, vol. 5, pt. 6, *Military Technology, Missiles and Sieges*, by Robin Yales (1994); *SCC*, vol. 6, pt. 3, *Agro-Industries and Forestry*, by Joseph Needham (1996); *SCC*, vol. 7, pt. 1, *Language and Logic*, by Christoph Harbsmeier (1998); *SCC*, vol. 5, pt. 13, *Mining*, by Peter J. Golas (1999); *SSC*, vol. 6, pt. 5, *Fermentations and Food Science*, by H. T. Huang (2000); *SSC*, vol. 6, pt. 6, *Medicine*, by Joseph Needham (2000).

alone. The social and economic factors considered in Volume 7 differ from the subject-matter of the preceding Volumes 3 to 6 in the way that they must be treated. Volumes 3 to 6 were all concerned with the various sciences and technologies. Sciences and technologies could be described; social and economic factors must be interpreted. As we pass from the history of science proper to its social and economic background, the discussion necessarily becomes more complex than ever. I realise very well that there must be differences of opinion, particularly as Time does not stand still. I am content to put before the reader my views as they remain today.

Volume 7, part 1, which has been written by Christoph Harbsmeier, is entirely concerned with language and logical thought, certainly one of the factors most powerfully influencing the development of scientific thinking.

'Literary Chinese as a Language for Science' by Kenneth Robinson,[2] to which I added my two pennyworth, had at one time been planned as the sequel to it, and was closely integrated with it, but is now in abbreviated form included in this volume. The close relationship between mathematics, science, logic and language hardly needs to be stressed. But it then became necessary, for reasons of size, to present it as part of the present volume. It is preceded by three articles written by me some considerable time ago, revised in recent years, and also by 'The Nature of Chinese Society: A Technical Interpretation', written by my old friend Huang Jen-Yü 黃仁宇, (Ray Huang). His modesty led him to attribute to me far more authorship than I can reasonably claim, and I have therefore insisted that his name should be given pride of place.

We are, however, most concerned in Volume 7 with the great question of why modern science did not arise in China after so many centuries of technical leadership, and, closely connected with this, why it was that modern capitalism did not develop in China. I had hoped at one time to cover a wide array of social factors which might have thrown light on this question, but there are limits to what one man can achieve. Let me now, therefore, confine myself to what I believe to be the most important: the collapse of feudalism in Europe, and the rise of the bourgeoisie. This requires us first to consider the nature of 'feudalism'.

I first began to use the term 'bureaucratic feudalism' in 1943 or thereabouts, when I was in China during the Second World War. I never felt irrevocably bound to the concept, but simply found it a useful description. Others decline to make use of the term or the concept, perhaps preferring just 'bureaucratism' – but this is an example of the differences of opinion which have to be expected in these fields. Of course, in those days 'feudal' was a pejorative word applied to all the social and economic features of imperial China; yet the term has always had meaning and significance for me. It stood over against the 'military-aristocratic' feudalism of Europe, that system where the king was always at the apex of a pyramid of nobles of different grades, each of whom by their tenure of 'fiefdoms', was bound to come to the aid

[2] See also *Comparative Criticism* 13: *Literature and Science*, ed. E. S. Shaffer (Cambridge University Press, Cambridge, 1991), pp. 3–30.

of the king, when he wanted to make war, with so many mounted knights, so many archers, so many foot soldiers, etc. This system may have seemed to be stronger, with all those knights in armour clanking about, but was actually weaker, perhaps because less rational. The heir to an earldom might be a moron, but yet by the rule of primogeniture, he would have to be next in succession – a very different situation from that of the Chinese bureaucrat, chosen by imperial examination and equipped with special expertise acquired in the job. In China individuals in each generation had to justify their promotion by their own efforts. Downward mobility for the heirs and successors of great men could only be counteracted by great effort. Thus the 'carrière ouverte aux talents' had been a Chinese principle for some 2,000 years before it became a French one.

Out of this unlikely military-aristocratic European milieu *modern* natural science could and did arise. When the merchants began to come out of their city-states in the 16th century, and capitalism arose, first mercantile and then industrial, *modern* natural science arose with it, in the time of Galileo and Torricelli. This was the 'rise of the bourgeoisie', and though other factors were involved too, such as the Protestant Reformation, it was this above all which happened in Europe, and in Europe alone.[3] The point of view which has been adopted throughout these volumes is that modern science was that form of science at which the ancient and medieval sciences of all the countries of the world were aiming,[4] but Europe alone was able to get there. Here the background of Greek logic and *mathesis universalis* was also important.

A good deal of work remains to be done on the exact nature of the tie-up between modern science and nascent capitalism. I have always pictured it as beginning with the exact specification of materials. If a merchant in the Renaissance period purchased a large quantity of oil from a Greek island he would need to know not only what its normal use was, but what it could also conceivably be used for; he would want to know its surface tension, its specific gravity, its refractive index, indeed all its properties, before he could decide who to sell it to. This would have been in the time of mercantile capitalism; in industrial capitalism there is less difficulty in imagining how intimately connected with it were science and technology. The accurate description of materials would have generated accuracy everywhere else, even in subjects like astronomy where there was no possibility of experimentation. And with exactness came the possibility of mathematisation.

Modern science has been defined elsewhere[5] as the mathematisation of hypotheses about Nature, and the testing of them rigorously by persistent experimentation. Experiment was something rather new; the Greeks had done relatively little of it, and although the Chinese had been well acquainted with it, their purposes were

[3] Since this was written Needham came to appreciate the work of Immanuel Wallerstein on 'The Four Collapses'. See 'Conclusion', p. xx. [Ed.]

[4] We have adopted the metaphor of the medieval sciences of the different civilisations being like rivers emptying into the great ocean of Modern (originally Western) Science. This has been regarded in some quarters as a Western Imperialist notion. See however 'Do the Rivers Pay Court to the Sea? The Unity of Science in East and West', *Theoria to Theory*, **5** (2) (1971), 68–77.

[5] *SCC*, vol. 3, pp. 150 ff.

primarily practical. Only the European Renaissance found out how to test mathematised hypotheses about Nature by relentless experimentation, and so to 'discover the best method of discovery'. But I hope that no one will interpret all this as meaning that I think that modern science, which grew up with capitalism, must always remain wedded to it. Events in our own time have shown that the socialist countries, notably Russia and modern China, are perfectly capable of doing successful modern science.

Of course military-aristocratic feudalism existed in other parts of the world besides Europe. I remember thinking when in Japan in 1986 how strange it was that modern science had not originated there as well. But then I reflected that the Japanese had not had the tradition of the Greek city-state, which was so important for Europe. Athens gave rise, when the Renaissance came, to Venice and Genoa, to Pisa and Florence, and these in their turn to Rotterdam and Amsterdam, the cities of the Hanseatic League and finally London. In these cities, protected by their Lord Mayor or Burgomaster and their Aldermen, the merchants could shelter from interference by the feudal nobility of the surrounding countryside, until the day when they should come forth, and, after lending money to kings, princes and nobles, run the whole show.

It is worthwhile taking a look at the idea of the town or city in China compared with Europe. In China the town was simply a node in the administrative network, held for the emperor by the civil governor, and (several bureaucratic ranks lower down) by the military commander. It was the centre of the network of outlying villages, the people of which came in to market in the city. Compare with this the well-known picture by Rembrandt of 'The Militia Company of Captain Frans Banning Cocq' (Fig. 1), a group of finely dressed citizens gathered with their weapons and immensely proud of the city they were pledged to defend. Cities in Europe were really states within states, ready in the course of time to provide governments as alternatives (however much it might be glossed over in practice) to the medieval feudal-style governments which had preceded them.

China may have been the prime example of 'bureaucratic feudalism', but all the other non-European parts of the world such as India, the South-East Asian countries and the whole Arab world may be said to have participated in it to some extent.

It was as if all the intermediate feudal lords had been abolished in China, leaving only the emperor himself, ruling 'all under heaven' by the aid of an immense bureaucracy, the like of which was never dreamed of by the feudal sovereigns of Europe.

If anyone does not like the expression 'bureaucratic feudalism' they might like to settle for 'nosphimeric bureaucratism'. Nosphimeric was a word which I invented during the war years. While on my perpetual travels I often encountered Bishop Ronald Hall of Hong Kong, visiting one of his outlying Anglican congregations, and one day we had a chance meeting at Annan 安南 in Kweichow. Talking about various things at dinner, I happened to mention to him that I needed a non-pejorative word for that squeeze, graft and corruption which has always been so characteristic of the

Figure 1. The Militia Company of Captain Frans Banning Cocq, known as 'The Night Watch' by Rembrandt. 1642, Rijksmuseum, Amsterdam.

bureaucracy in China, and which had loomed so large in the eyes of the modern Western business men who tried to buy and sell there.[6] Both our trucks were being repaired that night and his was finished earlier, so he set off first, but not before leaving me a bit of paper on which was written: 'see Acts 5:1–11'. When I got to the Bible, I found it was the story of Ananias and Sapphira, who had promised a sum of money to the church, but then kept back a portion of it, and accordingly died, blasted by St Peter. Now the word used in the Greek New Testament for 'to sequestrate' is nosphizein, and since meros means a part, we can form just the adjective required.

One can have little idea of how amazing it is for the Westerner trying to understand China to find how deeply the civil service, the mandarinate, the bureaucracy, in fact, is embedded in Chinese life. It is even in the folklore, for example the giving of bureaucratic titles to dragons, nagas, gods and spirits by conferring civil ranks upon

[6] Needham (1969b), pp. 37 ff.

them.[7] Bureaucratic feeling is everywhere. Even in the war years when I went to country places in China, I saw the inscription on red paper on the Walls: 'May the heavenly officials grant peace and plenty!' We do not have this in the West. We have never had such a civil service. On another occasion (in 1944) I was sitting in a Szechuanese teahouse with Sir Frank Eggleston, the Australian Ambassador. Seeing the common drain running down the middle of the village street, he exclaimed how medieval everything was, and said, 'You could almost expect to see a knight and men-at-arms come riding by'. I replied yes, indeed, but pointed out that it wouldn't have been a knight, but rather a civilian official, and the men-at-arms would have been represented by unarmed servitors carrying his titles and dignities on placards. It did not mean that the ultimate sanction was not force, as it has been in all human societies – but it was much better concealed by the Chinese bureaucracy.

In a word, if Chinese science, technology and medicine is to be understood, it must be related to the characteristics of Chinese civilisation. This is the point of Volume 7. Elsewhere we have explained how the bureaucratic ethos began by powerfully aiding Chinese science, while only in the later stages did it inhibit any move towards modern science. Such, at any rate, is our interpretation of the comparative developments in China and Europe.[8]

This brings up the question of where the cut-off point has been for Chinese traditional science, before the advent of modern science from the West, in all the preceding volumes. Generally we have sought to make it in the neighbourhood of 1700, about the time of the ending of the Jesuit mission; but we have often had occasion to go beyond it, and many instances present themselves. For example, the dinner-party at which the Khang-Hsi Emperor talked about the antiquity of the knowledge of gunpowder in China, in the presence of the Scottish physician John

[7] A few examples will suffice to illustrate this phenomenon. They are taken from the *Hsi Yu Chi* 西遊記 (*Journey to the West*) which draws on the resources of folklore and reflects its conventions. Chapters 10 and 11 of this novel depict the watery world of the Dragon King of the Ching River, which is staffed by 'a yaksa [*yeh chha* 夜叉 or demon] on patrol, the shrimp and crab ministers, the samli [*shih* 鰣 or shad] counsellor, the perch Subdirector of the Minor Court, and the carp President of the Board of Civil Office'. When the Thang Emperor descends to the underworld he finds a bureaucracy that mirrors his own, complete with Judge of the Underworld, King of the First Chamber, the Grand Marshall, demon messengers and soldiers, Guards of the Bridges, the Three Tribunes for trying and reviewing cases, and King Yama himself. (Anthony Yu, *Journey to the West*, vol. 1, pp. 221–2, 235–45. University of Chicago Press, Chicago, 1977.) [Ed.]

[8] Bureaucratic feudalism's initially favourable influence on science can be illustrated by the remarkable examples of organised field research that were undertaken under imperial auspices in the +8th century. These included a meridian survey directed by the Buddhist monk I-Hsing 一行 and the Astronomer Royal Nan-Kung Yüeh 南宮說 in +724–725, and at about the same time an expedition to the East Indies for the purpose of surveying the constellations of the Southern Hemisphere. See *SCC*, vol. 3, pp. 202–3, 292 ff.; Needham, Beer, Ho, *et al.* (1964); and Needham (1986), p. 7. However, bureaucratic feudalism played an inhibiting role later on. Between +1405 and +1433 the eunuch admiral Cheng Ho 鄭和 led a remarkable series of seven maritime expeditions with a fleet of sixty-three ocean-going junks that explored the south seas, collecting information on geography and sea routes, and bringing back produce, exotic animals, and luxury goods from the isles and from India, not to mention tribute from states as far west as the Persian Gulf. These expeditions encouraged advances in navigational technology, and increased China's knowledge of the outside world. However, the Confucian bureaucracy, which was never in favour of these expeditions, took advantage of a financial crisis and an over-stretched budget to call for their termination, arguing that they drained the treasury unnecessarily and that there was nothing China needed or wanted from foreign countries. Their objection was also most likely due in part to the traditional opposition between Confucian bureaucrats and power-hungry eunuchs. At any rate, the expeditions were stopped, thus putting an end to the greatest age of maritime exploration in Chinese history. See *SCC*, vol. 1, pp. 143–4, and *SCC*, vol. 4, pp. 524–5.

Bell of Antermony, took place in 1721 (*SCC*, vol. 5, pt. 7, p. 127). And in Volume 5, part 3, we recognised that if one was going to talk about the entry of modern inorganic and organic chemistry into China, one would have to come down to the closing years of the 19th century. Finally, in Lu & Needham (1980), pp. 118 ff., when discussing the development of acupuncture, it was clear that the cardinal discovery of acupuncture analgesia, permitting major surgery to be performed, was only made in the 1950s.

In writing this Foreword to the concluding part of Volume 7 I must not anticipate what I have to say in my 'Conclusion'. We have produced sixteen volumes while I have been 'on the bridge', and I am sure that our ship will arrive safely in port in the fullness of time, with another thirteen or fourteen volumes in the hold.

I am confident that none of these volumes, many of which I have seen in typescript, or have discussed with the authors, would cause me to change my views on the growth of science in China, or to come to general conclusions different from those which I have here elaborated.

This brings us to the moment for expressions of indebtedness. First, to the many authors who have joined me in the enterprise, and whose work has appeared in *Science and Civilisation in China*, and then to those firm friends and scholars of good will[9] whose writing could not be included in the published volumes, but who nevertheless did not allow disappointment to cloud our friendship. Nor can I overlook the part played by those who helped so greatly in the production of the volumes – Wang Ling, Lu Gwei-Djen, Gregory Blue (for many years my personal assistant, a constant helper and warm friend), Kenneth Robinson, Christopher Cullen and many others, librarians, secretaries, among whom I must place Diana Brodie, my private secretary for many years, gardeners, administrators, trustees and above all our friends of Cambridge University Press whose support has been unremitting since the project began, and whose names are recorded in note 9.

[9] When the Foreword was first drafted I wrote of my incalculable debt to Lu Gwei-Djen who shortly afterwards became my wife. Together we had conceived of the 'Science and Civilisation in China' project more than fifty years before. This was before I set out for China, during the Second World War, and she departed for America for the duration (as we used to say). Her death in 1991 became our incalculable loss.

We are most grateful to the following who have read and commented on distinct portions: Martin Bernal (Cornell), Francesca Bray (UCLA), Peter Burke (Cambridge), Timothy Cheek (Colorado College), Paul Connerton (Cambridge), Helen Dunstan (Bard College), Nathan Sivin (University of Pennsylvania), Chris Wickham (University of Manchester), R. Bin Wong (University of California, Irvine). It is also appropriate here to thank all of my earlier collaborators on the 'Science and Civilisation in China' project since there was hardly a conversation with any of them over the past decades that did not have some bearing on the questions we are dealing with in Volume 7, and express our thanks first to all those authors who have contributed parts or whole volumes to the series: Wang Ling, Kenneth Robinson, Lu Gwei-Djen, Tsien Tsuen-Hsuin, Ho Ping-Yü, Nathan Sivin, Robin D. S. Yates, Krzysztof Gawlikowski, Dieter Kuhn, Huang Hsing-Tsung, Francesca Bray, Christian Daniels, Nicholas Menzies, Christoph Harbsmeier, Janus Chmielewski, Peter J. Golas, Huang Jen-Yü.

Nor must we omit to mention those who made a contribution to a volume which could not ultimately be included, due to the reshaping of the series. Among these must be included Derk Bodde, Immanuel Wallerstein, Timothy Brook, Gregory Blue, Christopher Cullen, Lawrence Krader, Marianne Bastid, G. L. Hicks, S. G. Redding.

For the support of our research and writing, we must mention the National Science Foundation (USA), the Mellon and the Luce foundations, and the National Institute for Research Advancement (Japan). Without their continued support, our work would have been impossible. Most of their help was mediated through our New York Trust (Chairman, Mr John Diebold); and this is also deserving of our warmest thanks for obtaining from the Kresge foundation a grant of US$150,000 towards the building of the South Wing of our Institute.

I cannot close this Foreword without also expressing the most sincere appreciation for all those who made possible the planning, design and construction of the building in which we work, and in particular its architect Christoph Grillet. It will, I trust, be a green island of quietness in the City of Cambridge for many years to come.

Joseph Needham 1995

50 GENERAL CONCLUSIONS AND REFLECTIONS

(*a*) SCIENCE AND SOCIETY IN EAST AND WEST[1]

JOSEPH NEEDHAM

When I first formed the idea, about 1938, of writing a systematic, objective and authoritative treatise on the history of science, scientific thought, technology and medicine in the Chinese culture-area, I regarded the essential problem as that of why modern science[2] had not developed in Chinese civilisation (or Indian or Islamic) but only in that of Europe. Nevertheless, as they say in sunny France, 'Attention! Un train peut cacher un autre!' As the years went by, and as I began to find out something at last about Chinese science and society, I came to realise that there is a second question at least equally important, namely, why, between the –1st century and the 15th century, was Chinese civilisation much *more* efficient than occidental in gaining natural knowledge and in applying it to practical human needs?

The answer to all such questions lies, I now believe, primarily in the social, intellectual and economic structures of the different civilisations. The comparison between China and Europe is particularly instructive, almost a bench-test experiment one might say, because the complicating factor of climatic conditions does not enter in. Broadly speaking, the range of climate in the Chinese culture-area is not unlike that in Europe. It is therefore not possible for anyone to say of China (as has been maintained in the Indian case) that the environment of an exceptionally hot climate inhibited the rise of modern natural science.[3] Although the natural, geographical and climatic settings of the different civilisations undoubtedly played a part in the development of their specific characteristics, I am in any case not inclined to regard this suggestion as valid even for Indian culture. The point is that it cannot even be asserted of China. Nonetheless here again we see the enormous importance of geography in relation to the civilisations and the way they developed.[4] Europe, as I have often said, is like an archipelago. There is the Baltic, the North Sea, the Irish Channel, the Mediterranean, the Aegean, the Black Sea and then the great Atlantic outside the Pillars of Hercules – all inviting maritime commerce and the

[1] First published in the J. D. Bernal Presentation Volume (London, 1964), and then in *Science and Society* (1964), **28**, 385, and *Centaurus* (1964), **10**, 174; collected in *The Grand Titration* (Allen & Unwin, London, 1969), and further revised for publication here.

[2] If I were asked to define modern science, I would say that it was the combination of mathematised hypotheses about natural phenomena with relentless experimentation. On reflection, I am not sure that experimentation was not the greatest Chinese stimulus to European alchemy and so indirectly to the European Renaissance, for the Greeks did not experiment, but the Chinese did; otherwise they would never have made their fundamental discoveries such as porcelain and the magnetic compass.

[3] Cf. the writings of Huntington (1907), (1924) and especially (1945).

[4] For which see Dorn (1991).

activities of sea-captains. In contrast to all this, China is a vast land-mass, well suited for the activities of peasant-farmers in their millions. It is hardly surprising that the civilisations turned out to be so different.

From the beginning I was deeply sceptical of the validity of any of those 'physical-anthropological' or 'racial-spiritual' factors in explaining China's development. Everything I have experienced during the fifty years since I first came into close personal contact with Chinese friends and colleagues, has only confirmed me in this scepticism. They proved to be entirely, as Andreas Corsalis wrote home in a letter to Lorenzo di Medici so many centuries ago, 'di nostra qualità'. I believe that the vast historical differences between the cultures can be explained by sociological studies, and that some day they will be. The further I penetrate into the detailed history of the achievements of Chinese science and technology before the time when, like all other ethnic cultural rivers, they flowed into the ocean of modern science, the more convinced I become that the cause for the breakthrough (occurring only in Europe) was connected with the special social, intellectual and economic conditions prevailing there at the Renaissance, and can never be explained by any deficiencies either of the Chinese mind or of the Chinese intellectual and philosophical tradition. In many ways this tradition was much more congruent with modern science than was the world-outlook of Christendom. Such a point of view may or may not be a Marxist one – for me it is based on study and personal experience of life.

For the purposes of the historian of science, therefore, we have to be on the watch for some essential differences between the aristocratic-military feudalism of Europe, out of the womb of which mercantile and then industrial capitalism, together with the Renaissance and the Reformation, could be born; and those other kinds of 'feudalism' (if that was really what it was) which were characteristic of medieval Asia. From the point of view of the history of science we must have something at any rate sufficiently different from what existed in Europe to help us solve our problem. This is why I have never been sympathetic to that trend in Marxist thinking which has sought for a rigid and unitary formula of the stages of social development which all civilisations 'must have passed through'.

Primitive communalism, the earliest of these, is a concept which has evoked much debate. Though such a phase is commonly rejected by the majority of Western anthropologists and archaeologists (with, of course, some notable exceptions such as V. Gordon Childe), it has always seemed to me eminently sensible to conceive of a state of society before the differentiation of social classes, and in my studies of ancient Chinese society I have found it appearing through the mists clearly enough, time after time. Nor at the other end of the story is there any essential difficulty in the transition from feudalism to capitalism, though of course this was enormously complex in detail, and much has still to be worked out.[5] In particular the exact connections between the social and economic changes and the rise of modern science, that is to say, the successful application of mathematical hypotheses to the systematic

[5] On different aspects of feudalism see Wallerstein (1992).

experimental investigation of natural phenomena, remain elusive. All historians, no matter what their theoretical inclinations and prejudices, are necessarily constrained to admit that the rise of modern science occurred *pari passu* with the Renaissance, the Reformation and the rise of capitalism.[6] It is the intimate connections between the social and economic changes on the one hand and the success of the 'new, or experimental' science on the other which are the most difficult to pin down. A great deal can be said about this, for example the vitally important role of the 'higher artisanate' and its acceptance into the company of educated scholars at this time;[7] but the present writing is not the place for it because we are in pursuit of something else. For us the essential point is that the development of modern science occurred in Europe and nowhere else.

In comparing the position of Europe with China, one of the problems of interpretation we face is whether China ever passed through a 'slave society' analogous to that of classical Greece and Rome. The question is, of course, not merely whether the institution of slavery existed – that is quite a different matter – but whether the society was ever based on it.[8] According to my own experiences with Chinese archaeology and literature, for what they are worth, I am not very inclined to believe that Chinese society, even during the Shang and early Chou periods, was ever a slave-based society in the same sense as the Mediterranean cultures with their slave-manned galleys plying the Mediterranean and their *latifundia* spread over the fields of Italy. Here I diverge, with deep humility, from some contemporary Chinese scholars, who have been extremely impressed by the 'single-track' system of developmental stages of society prominent in Marxist thinking during the past seventy years. The subject is still under intensive debate and we cannot yet say that certainty has been achieved in any aspect of it. In the years 1956–7 at Cambridge we had a series of lectures on slavery in the different civilisations, in the course of which the participants all had to agree that the actual forms of slavery were very different in Chinese society from anything known elsewhere.[9] Owing to the dominance of clan and family obligations it was rather doubtful whether anyone in that civilisation could have been called 'free' in some of the Western senses, while on the other hand (contrary to what many believe) chattel-slavery was distinctly rare.[10] The fact is that no one fully knows what was the status of servile and semi-servile groups in the different periods in China (and there were many different kinds of such groups): neither Western sinologists

[6] The great stumbling-block here for the internalist school of historiography of science is the question of historical causation. Scenting economic determinism under every formulation, they insist that the scientific revolution, as primarily a revolution in scientific ideas, cannot have been 'derivative from' some other social movement such as the Reformation or the rise of capitalism. Perhaps for the moment we could settle for some such phrase as 'indissolubly associated with . . .'. The internalists always seem to me essentially Manichaean; they do not like to admit that scientists have bodies, eat and drink and live social lives among their fellow-men, whose practical problems cannot remain unknown to them; nor are the internalists willing to credit their scientific subjects with subconscious minds.

[7] This factor was much emphasised and elaborated by the late Edgar Zilsel (see Bibliography C). Its importance has recently been recognised by the well-known medievalist, A. C. Crombie (1963). See also his (1961), p. 13.

[8] See Wallerstein (1992) and Needham (1969a), p. 167.

[9] This symposium was organised and presided over by Professor E. G. Pulleyblank.

[10] See Pulleyblank (1958b).

nor even the Chinese scholars themselves.[11] A great amount of research remains to be done, but I think it seems already clear that neither in the economic nor in the political field was chattel-slavery ever a basis for the whole of society in China in the same way as it was at some times in the West.

Although the question of the slave basis of society has a certain importance in so far as it affects the position of science and technology among the Greeks and Romans,[12] it is of course less germane to what was originally my central point of interest, namely the origin and development of modern science in the late Renaissance in the West. It could, however, have a very important bearing on the greater success of Chinese society in the application of the sciences of Nature to human benefit during the earlier period, the first fourteen centuries of the Christian era and four or five centuries prior to that. Is it not very striking and significant that China has nothing whatever to show comparable with the use of slaves on the *latifundia* in agriculture or at sea in galleys in the Mediterranean? Sail, and a very refined use of it, was the universal method of propulsion of Chinese ships from ancient times. China has few records of the mass use of the human motor comparable with the building methods of ancient Egypt, though the building of the Great Wall is an outstanding exception. So also it is remarkable that we have never so far come across any important instance of the refusal of an invention in Chinese society due to fear of technological unemployment before the 19th century.[13] If Chinese labour-power was so vast as most people imagine, it is not easy to see why this factor should not sometimes have come into play. We have numerous examples of labour-saving devices introduced at early times in Chinese culture, very often much earlier than in Europe. A concrete case would be the wheelbarrow, not known in the West before the 13th century but common in China in the +3rd and arising there almost certainly two hundred years earlier than that. It may well be that just as the bureaucratic apparatus will explain the failure of modern science to arise spontaneously in Chinese culture, so also the absence of mass chattel-slavery may turn out to have been an important factor in the greater success of Chinese culture in fostering pure and applied science in the earlier centuries.

The far greater problem that arises when we compare the histories of Europe and China, however, is how far and in what way did Chinese feudalism (if that is the proper term for it) differ from European feudalism. In my early days, when I was still a working biochemist, I was greatly influenced by Karl A. Wittfogel's

[11] At least we have been able to show (vol. 4, p. 2, pp. 35 ff.) that servile or semi-servile rank was no barrier to official, if not very exalted, positions. This was so in the case of Hsin-Tu Fang 信都芳 (*fl.* +525) and Keng Hsün 耿詢 (*fl.* +593). The former was in charge of all the scientific apparatus at the Thopa court of Northern Wei and the latter rose to moderately high rank in the Bureau of Astronomy and Calendar under the Sui.

[12] It is always said that the Greeks and Romans were less interested in labour-saving devices than they might have been because there was always a large slave population around to carry out desired changes by muscle-power.

[13] With the mass introduction of Western technology in the 19th century, however, the situation was altered. J. Dyer Ball's article in *Things Chinese* (1904) on 'Railways', for example, quotes descriptions and sources in the Chinese press from 1891 to 1899 of the resistance to the building of railways by the Chinese government. These articles were written from the point of view of the capitalist developer and do not stress Chinese fears of unemployment and foreign domination.

book, *Wirtschaft und Gesellschaft Chinas*, written when he was a more or less orthodox Marxist in pre-Hitler Germany.[14] He was interested in developing the conception of 'Asiatic bureaucratism' or 'bureaucratic feudalism' as I found later on that some Chinese historians called it.[15] This concept arose from the works of Marx and Engels themselves who had based it partly on, or derived it from, the observations of the 17th-century Frenchman François Bernier, physician to the Mogul emperor Aurangzeb in India.[16]

Marx and Engels spoke about the 'Asiatic mode of production'. How exactly it could or should be defined has been the subject of animated discussions in recent decades.[17] Broadly speaking, it was the growth of a State apparatus fundamentally bureaucratic in character and operated by a non-hereditary élite upon the basis of a large number of relatively self-governing peasant communities, still retaining much tribal character and with little or no division of labour as between agriculture and industry. The form of exploitation here consisted essentially in the collection of taxes for the centralised State, i.e. the royal or imperial court and its regiments of bureaucratic officials. The justification of the State apparatus was, of course, twofold: on the one hand it organised the defence of the whole area (whether an ancient 'feudal' state or later the entire Chinese empire), and on the other hand it organised the construction and maintenance of public works. It is possible to say without fear of contradiction that throughout Chinese history the latter function was more important than the former, and this was one of the things that Wittfogel saw. The necessities of the country's topography and agriculture imposed from the beginning a vast series of water-works[18] directed to (a) the conservation of the great rivers, in flood-protection and the like, (b) the use of water for irrigation, especially for wet rice cultivation, and (c) the development of a far-flung canal system (Fig. 2), whereby the tax-grain could be brought to granary centres and to the capital. All this necessitated, besides tax exploitation, the organisation of corvée labour and one might say that the only duties of the self-governing peasant communities *vis-à-vis* the State apparatus were the payment of tax and the provision of labour power for public

[14] I also learnt much from a golden little book by Hellmut Wilhelm, the son of the great sinologist Richard Wilhelm, *Gesellschaft und Staat in China* (1944). It is most unfortunate that this non-Marxist work has long been quite inaccessible, and that there has never been an English translation of it. My views on 'feudalism' in China may be found in Needham (1969a), pp. 167–8.

[15] I think, looking back, that I first fixed upon the phrase 'bureaucratic feudalism' during the years when I was in China (1942–6) during the Second World War. 'Feudalism' was then becoming a term of abuse for all previous societies but that did not deter me from using it. I see now, however, that for such a term to pertain, you should have a lowest stratum of society which was bound to the land they tilled – the serfs – and if that was never the case in China, then the term should not be used. It might be better then, to say 'feudal bureaucratism'. Elsewhere in this volume I have given sufficient examples of what 'bureaucratism' meant in China, and that is the main thing.

[16] *The History of the Late Revolution of the Empire of the Great Mogul*, originally published in French (Paris, 1671); many times republished, as by Dass (Calcutta, 1909). In the course of it Bernier said, 'There is no meum and tuum among them, as there is among us'. He was one of the first to use the word 'prebendal' to mean that stage of society in which all the officials get their incomes from the imperial treasury and not from direct ownership of feudal lands. See the famous letter of Marx to Engels, 2 June 1853.

[17] See below, pp. 12 ff., for which I have had the advantage of collaboration with Gregory Blue, who originally learnt Russian in order to be able to study at first hand the debates of the 1930s on the 'Asiatic Mode of Production'. See Blue (1979).

[18] This point is developed by Dorn (1991), pp. xvi, 33 and 35 ff.

Figure 2. Illustration of a flash-lock gate (bottom-left). From *The Grand Titration*, Fig. 27, from Baylin (1929).

purposes when called upon to give it.[19] Besides this the State bureaucracy assumed the function of the general organisation of production (Fig. 3), i.e. the direction of broad agricultural policy, and for this reason the State apparatus of such a type of society may well be given the appellation of 'an economic high command'. Only in China do we find among the most ancient high officials the *Ssu Khung* 司空, the *Ssu Thu* 司徒 and the *Ssu Nung* 司農 (Director of Public Engineering Works, Director

[19] One can detect in this model similarities to the system of people's communes, by which the Chinese countryside was organised between 1958 and 1979. See, for example, Strong (1964), writing on changes introduced at that time. The principle of the rational and maximal utilisation of manpower is one which goes back more than 2,000 years in Chinese history, and its timing was one of the functions of the 'economic high command'.

Figure 3. Nominating the Right Men for Office. From *The Grand Titration*, Fig. 28.

of Public Instruction and Minister for Agriculture). Nor can we forget that the 'nationalisation' of salt and iron manufacture (the only commodities which had to travel, because not everywhere producible), suggested first in the –5th century, was thoroughly put into practice in the –2nd century. Also in the Han there was a governmental Fermented Beverages Authority; and there are many examples of similar bureaucratic industries under subsequent dynasties.[20]

Various other aspects of this situation reveal themselves as one looks further into it; for example, peasant production was not under private control, and theoretically all the land within the whole empire belonged to the Emperor and the Emperor alone. There was at first a semblance of landed property securely held by individual families, but this institution never developed in Chinese history in a way comparable with feudal fief tenures of the West, since Chinese society did not retain the system of primogeniture.[21] Hence all landed estates had to be parcelled out at each demise of the head of the family. Again, in that society, the conception of the city-state was absent; the towns were purposefully created as nodes in the administrative network, though very often no doubt they tended to grow up at spontaneous market centres. Every town was a fortified city held for the Prince or the Emperor by his civil governor and his military official. Since the economic function was so much more important in Chinese society than the military it is not surprising that the governor was usually a more highly respected person than the garrison commander.[22] Lastly, broadly speaking, slaves were not used in agricultural production, nor indeed very much in industry; slavery was primarily domestic, or as some would say, 'patriarchal' in character,[23] throughout the ages.

In its later highly developed forms such as one finds in Thang or Sung China the 'Asiatic mode of production' developed into a social system which, while fundamentally 'feudal', in the limited sense that most of the wealth was based on agricultural exploitation,[24] was essentially bureaucratic and not military-aristocratic. It is quite impossible to over-estimate the depth of the civilian *ethos* in Chinese history. Imperial power was exercised not through a hierarchy of enfeoffed barons but through

[20] Cf. Schurmann (1956).

[21] Primogeniture was, however, less widespread in Europe than is usually imagined.

[22] In a recent book on the Yangtze, van Slyke (1988) has much to say about merchants and mercantile activity in traditional China, especially in connection with rice, salt, silk and *thung* 桐 oil. He well describes the activities of the rich merchants of medieval China, and though he trembles on the verge of stating that bureaucratic feudalism meant that capitalism (as well as modern science) could not develop in China, he doesn't actually do so. But he makes interesting statements such as, 'The fact that both officials and merchants centred their activities in cities large and small meant that the merchants had no urban arena of their own in which autonomously to develop their own values and institutions, to become, that is, a genuine bourgeoisie.' Again he says that guilds in China were defensive with respect to political authority, not vehicles for the development of an alternative vision of society. Central Asia may have known some city-states, but descending from their Greek antecedents, they were overwhelmingly characteristic of Europe, where they again appeared, first in Italy, like Venice, Pisa and Genoa, then in Holland, like Rotterdam and Amsterdam, finally Antwerp and London, together with the cities of the Hanseatic League. All these were by a long tradition children of the Greek city-state.

[23] See Tokei (1959), p. 291.

[24] This must not be taken to mean that industry and trade were poorly developed in medieval China. On the contrary, especially in the Southern Sung in the 12th and 13th centuries, they were so productive and prosperous that the continuance of the typical bureaucratic form is what surprises.

an extremely elaborate civil service which Westerners know of as the 'mandarinate', enjoying no hereditary principle of succession to estates but recruited afresh in every generation. All I can say is that throughout nearly fifty years of study of Chinese culture, I have found that these conceptions have made more sense in my understanding of Chinese society than any others. I believe that it will be possible to show in some considerable detail why Asian 'bureaucratic feudalism' at first favoured the growth of natural knowledge and its application to technology for human benefit, while later on it inhibited the rise of modern capitalism and of modern science, in contrast with the other form of feudalism in Europe which favoured it – by decaying and generating the new mercantile order of society.[25] A predominantly mercantile order of society could never arise in Chinese civilisation because the basic conception of the mandarinate was opposed not only to the principles of hereditary aristocratic feudalism but also to the value-system of the wealthy merchants. Capital accumulation in Chinese society there could indeed be, but the application of it in permanently productive industrial enterprises was constantly inhibited by the scholar-bureaucrats, as indeed was any other social action which might threaten their supremacy. Thus, the merchant guilds in China never achieved anything approaching the status and power of the merchant guilds of the city-states of European civilisation.[26]

In many ways I should be prepared to say that the social and economic system of medieval China was much more rational than that of medieval Europe. The system of imperial examinations for entry into the bureaucracy, a system which had taken its origin as far back as the –2nd century, together with the age-old practice of the 'recommendation of outstanding talent', brought it about that the mandarinate recruited to itself the best brains of the nation (and the nation was a whole sub-continent) for more than 2,000 years.[27] This stands in very great contrast with the European situation where the best brains were not especially likely to arise in the families of the feudal lords, still less among the more restricted group of eldest sons of feudal lords. There were of course certain bureaucratic features of early medieval European society, for example the office of the 'Counts', the institutions which gave rise to the position of 'Lord Lieutenant', and the widely customary use of bishops and clergy as administrators under the king, but all this fell far short of the systematic utilisation of administrative talent which the Chinese system brought fully into play.

Moreover, not only was administrative talent brought forward and settled thoroughly into the right place, but so strong was the Confucian *ethos* and ideal that the chief representatives of those who were not scholar-gentry remained for the most part conscious of their lesser position in the scheme of things. When I was giving a talk to a university society on these subjects, someone asked the excellent question, 'How was it that the military men could accept their inferiority to the civil officials throughout Chinese history?' After all, 'the power of the sword'

[25] See Needham (1986). *[26] See Morse (1932).
[27] A remarkable sidelight on this will be found in the paper by Lu Gwei-Djen & Needham (1963), 'China and the Origin of (Qualifying) Examinations in Medicine'.

has been overwhelming in other civilisations. What immediately came to my mind in replying was the imperial *charisma* carried by the bureaucracy,[28] the holiness of the written character (when I first went to China the stoves for giving honourable cremation to any piece of paper with words written on it were still to be seen at every temple), and the Chinese conviction that the sword might win but only the *logos* could maintain. There is a famous story about the first Han emperor who was impatient with the elaborate ceremonies devised for the court by his attendant philosophers, till one of them said to him, 'You conquered the empire on horseback, but from horseback you will never succeed in ruling it'.[29] Thereafter the rites and ceremonies were allowed to unfold in all their liturgical majesty.[30] In ancient times the Chinese leader was often an important official and a general indiscriminately, and what is significant is that the psychology of military men themselves clearly admitted their inferiority. They were very often 'failed civilians'. Of course, force was the ultimate argument, the final sanction, as in all societies, but the question was – what force, moral or purely physical? The Chinese profoundly believed that only the former lasted, and what the latter could gain only the former could keep.

Furthermore, there may have been technical factors in the primacy of the spoken and written word in Chinese society. It has been demonstrated that in ancient times in China the progress of invention in offensive weapons, especially the efficient crossbow (Fig. 4), far outstripped progress in defensive armour.[31] There are many cases in antiquity of feudal lords being killed by commoners or peasants well armed with crossbows, the penetrating power of which made the wearing of armour useless, a situation quite unlike the favourable position of the heavily armed knight in Western medieval society. Hence, perhaps, arose the Confucian emphasis on persuasion. The Chinese were Whigs, 'for Whigs admit no force but argument'.[32] The Chinese

[28] One should add the high moral standards of Confucianism which exerted great social pressure throughout the ages upon the members of the mandarinate.

[29] These were the words of Lu Chia, in –196 (*CHS*, ch. 43, p. 6b, *TCKM*, ch. 3, p. 46b). Another court liturgiologist was Shu-Sun Thung, who reported in –201 that 'the Emperor had abolished the complex and difficult rites of Chhin . . . but the result was that when the officers drank together, they disputed about precedence, got drunk, shouted and banged their swords on the columns of the halls. The Emperor was disgusted. Shu-Sun Thung said to him, "Scholars may not be able to conquer an empire, but they can help to preserve it. I suggest that you convoke all the literati of Lu and instruct them to draw up an imperial code of rites" . . . [After the first trial of it] the Emperor said, "This day for the first time I see what imperial majesty means."' This comes from *CHS*, ch. 43, pp. 15b ff. Hence *TCKM*, ch. 3, p. 25b.

[30] See *SCC*, vol. 1, p. 103.

[31] On the early date of the crossbow, and its remarkable effects, see *SCC*, vol. 5, pt. 6.

[32] This is a quotation from an amusing battle between the Whigs of Cambridge and the Tories of Oxford in the 18th century. George I precipitated it by sending at the same time a present of books to Cambridge and a troop of cavalry to Oxford. Joseph Trapp (1679 to 1747) began it with the following verse:

> The King, observing with judicious eyes
> The State of both his universities
> To Oxford sent a troop of horse, and why?
> That learned body wanted loyalty.
> To Cambridge books, as very well discerning
> How much that loyal body wanted learning.

Figure 4. Makers of Crossbows in their Workshop. From *The Grand Titration*, Fig. 29.

peasant-farmer could not be driven into battle to defend the boundaries of his State, for instance, before the unification of the empire, since he would be quite capable of shooting his Prince first; but if he was persuaded by the philosophers, whether

To which Sir William Brown (1692 to 1774) made the following reply:

> The King to Oxford sent a troop of horse,
> For Tories own no argument but force:
> With equal skill to Cambridge books he sent,
> For Whigs admit no force but argument.

See *Oxford Dictionary of Quotations*, pp. 87, 548, citing *Nichol's Literary Anecdotes*, vol. 3, p. 330.

patriots or sophists, that it was necessary to fight for that State, as, indeed, also later for the empire, then he would march. Hence the presence of a certain amount of what one might call 'propaganda' (not necessarily in a pejorative sense) in Chinese classical and historical texts – a kind of 'personal equation' for which the historian has to make proper allowance. There was nothing peculiar to China in this. It is, of course, a world-wide phenomenon noticeable from Josephus to Gibbon, but the sinologist has always to be on the lookout for it, for it was the *défaut* of the bureaucratic civilian *qualité*.

Yet another argument is of interest in this connection, namely the fact that the Chinese were always primarily peasant-farmers, and not engaged in either animal husbandry or sea-faring.[33] These two latter occupations encourage excessive command and obedience; the cowboy or shepherd drives his animals about, the sea-captain gives orders to his crew which are neglected at the peril of everyone's life on board, but the peasant-farmer, once he has done all that is necessary for the crops, must wait for them to come up. A famous parable in Chinese philosophical literature derides a man of Sung who was discontented with the growth rate of his plants and started to pull at them to help them to come up.[34] Force, therefore, was always the wrong way of doing things, hence civil persuasion rather than military might was always the correct way of doing things. And everything that one could say for the position of the soldier *vis-à-vis* the civil official holds good, *mutatis mutandis*, for the merchant. Wealth as such was not valued. It had no spiritual power. It could give comfort, but not wisdom, and in China affluence carried comparatively little prestige. The one idea of every merchant's son was to become a scholar, to rise high in the bureaucracy.[35] Thus did the system perpetuate itself through 10,000 generations. I am not sure that it is still not alive, though raised of course to a higher plane, for is not the Party official, whose position is quite irrelevant to the accidents of his birth, expected to despise both aristocratic values on the one hand and acquisitive values on the other? In a word, perhaps socialism was the spirit of undominating justice imprisoned within the shell of Chinese medieval bureaucratism.[36] Basic Chinese traditions may perhaps be more congruent with the scientific world co-operative commonwealth than those of Europe.

Between 1920 and 1934 there were great discussions in the Soviet Union and elsewhere about what Marx had meant by the 'Asiatic mode of production'. Recent research by authors such as Giani Sofri,[37] Marian Sawer,[38] S. P. Dunn,[39] and especially V. N. Nikiforov[40] illustrates how those in the 1930s who argued for the succession: primitive communalism–slave-society–feudalism–capitalism–socialism,

[33] This contrast was, I think, first appreciated by André Haudricourt. See his article, Haudricourt (1962), p. 46.

[34] See *SCC*, vol. 2, p. 576.

[35] In Chinese opera, one of the characteristic comic characters is the *nouveau-riche* young man who thinks (wrongly) that everything can be bought for money.

[36] Of course the medieval mandarinate was part of an exploiting system, like those of Western feudalism and capitalism, but as a non-hereditary élite it did oppose both aristocratic and mercantile ways of life. Cf. the work by Brandt, Schwartz & Fairbank (1952) and Needham (1960b) in (1969a).

[37] Sofri (1973). [38] Sawer (1977). [39] Dunn (1982). [40] Nikiforov (1970).

eventually gained the day and how this became the standard order of stages in Soviet Marxist historical writing. The climate of dogmatism which prevailed in the social sciences during the personality-cult period clearly played an important role in this situation.[41] In the 1950s and especially the 1960s Marxist authors of various nationalities and persuasions expressed great embarrassment that 'feudalism', for instance, had become a meaningless term.[42] 'Obviously', they said, 'a socio-economic stage which covers both Ruanda-Urundi today and France in 1788, both China in 1900 and Norman England, is in danger of losing any kind of specific character likely to assist analysis . . .'. Subdivisions were thus desperately needed. Joan Simon posed the 'rehabilitation' of Marx's Asiatic mode or even several modes, to enable a differentiation of the nomenclature between regional variations. The use of the term 'proto-feudal' (which I believe I myself invented) was also recommended for a single basic stage which then developed in different ways. Since then the publication of Marx's *Grundrisse*[43] in other West European languages, and more recently the publication of his ethnological notebooks by Lawrence Krader,[44] have gone a long way towards making known and disseminating the original views of Marx and Engels on such subjects. However, in regard to our specific concern – the nature of traditional Chinese society – we still await an analysis, whether in terms of the 'Asiatic mode of production' or otherwise, that is completely convincing.

One great question is whether Marx and Engels regarded this as something qualitatively different from one or other of the classically distinguished types of society in the rest of the world, or only quantitatively different. It is not yet clear whether they saw it as essentially a 'transitory' situation (though in some cases it might be capable of age-long stabilisation) or whether they thought of 'bureaucratism' as a fourth fundamental type of class society. Was the 'Asiatic mode of production' simply a variation of classical feudalism? Some Chinese historians have certainly regarded it as a special type of feudalism. But sometimes Marx and Engels seemed to speak as if they did consider it as something qualitatively different from slave production or feudal production. There was also always the question how far the conceptions of 'bureaucratic feudalism' might be applicable to pre-Columbian America or other societies such as medieval Ceylon (Sri Lanka).

For many years it was usual for the name of Wittfogel to be mentioned with aversion in Marxist writings on this topic. During the Hitler period Wittfogel migrated from Germany to America, where he lived and worked until his death.[45] He was a great brandisher of tomahawks in the intellectual Cold War, and those writers who regard his book *Oriental Despotism*[46] as propaganda directed against Russia and China old and new, or against Asian cultures generally, are only too probably correct. In his American period Wittfogel sought to attribute all abuses of power, whether in totalitarian or other societies, to the principle of bureaucratism; but the fact that he

[41] See above, pp. 2 ff. [42] Simon (1962), p. 183. [43] Marx (1973). [44] Krader (1974) and (1975).
[45] See Wittfogel (1978), which is largely autobiographical.
[46] Wittfogel (1957). Among the many critiques of Wittfogel's ideas may be mentioned an interesting study from the juristic point of view by Lee (1964).

became a great opponent of the ideas regarding ancient and imperial Chinese history which I and many others favour does not alter the fact that he once set them forth quite brilliantly himself, and thus I still admire his *Wirtschaft und Gesellschaft Chinas* (1931), and his other works from this early period, while deprecating his latter-day views. Although from the 1950s he perhaps overdid it, I do not regard his theory of 'hydraulic society' as essentially erroneous, for I also believe that the spatial range of public works (river control, irrigation and the building of transport canals) in Chinese history transcended time after time the barriers between the territories of individual feudal or proto-feudal lords. It thus invariably tended to concentrate power at the centre, i.e. in the bureaucratic apparatus, arched above the granular mass of 'tribal' clan villages.[47] I think it played an important part, therefore, in making Chinese feudalism 'bureaucratic'. Of course it does not matter from the standpoint of the historian of science and technology how different Chinese feudalism was from European feudalism, but it has got to be different enough (and I firmly believe it was different enough) to account for the total inhibition of capitalism and modern science in China as against the successful development of both these features in the West.

As for bureaucracy, it is sheer nonsense to lay all social evil at its door. But was it actually productive in fostering science? I doubt it. I imagine that the invention of poundlock gates would be something typical of the hydraulic element, and the scanning of the skies in China was intimately related to the elaboration of the calendar and the mode of government. But there seems no bureaucratic reason why the Chinese should have invented the mechanical clock as they did, with the water-wheel linkwork escapement which brought round one bucket after another every twenty seconds or so on the rim of a large wheel. It will not be denied that the mechanical clock is the greatest instrument of modern science, being able to tell accurately the progress of time. Yet further use of the mechanical clock in China was not developed. The seeds of modern science were there, but their growth was inhibited. There are other examples of this which we hope to summarise below in our General Conclusions. The Chinese had the inventiveness, but lacked the social conditions for the elaboration of modern science. Now that social conditions are altered and modern science has been adopted in China, Chinese inventiveness can be expected to come into full play once again.

In a word, I cannot see that the early scientific achievement of China was due to the hydraulic bureaucratic system. But in other ways bureaucratism has been a magnificent instrument of human social organisation. Furthermore, it is going to be with us, if humanity endures, for many centuries to come. The fundamental problem before us now is the humanisation of bureaucracy, so that under socialism not only shall its organising power be used for the benefit of the ordinary man and woman; but that it shall be known and palpably felt and seen to be so used. Modern human

[47] This argument is exactly stated in the *Yen Thieh Lun* (Discourses on Salt and Iron) of about –80, ch. 13, p. 1a. See *SCC*, vol. 4, pt. 3, pp. 264 ff.

society is, and will increasingly be, based on modern science and technology, and the more this goes on the more indispensable a highly organised bureaucracy will be. The fallacy here is to compare such a system after the rise of modern science with *any* precursor systems which existed before it. For modern science has given us a vast wealth of instruments from telephones to computers which now and only now could truly implement the will to humanise bureaucracy. That will may rest on what is essentially Confucianism, Taoism, and revolutionary Christianity, as well as Marxism.[48]

The term 'oriental despotism' recalls of course the speculations of the Physiocrats in 18th-century France, who were deeply influenced by what was then known of the Chinese economic and social structure.[49] For them, of course, it was an enlightened despotism, which they much admired, not the grim and wicked system of Wittfogel's later imagination. Sinologists throughout the world were impatient with his later book[50] because it persistently selected from the facts. Thus, for example, it is impossible to say that there was no educated public opinion in medieval China. On the contrary the scholar-gentry and the scholar-bureaucrats constituted a wide and very powerful public opinion, and there were times when the Emperor might command but the bureaucracy would not obey.[51] In theory, the Emperor might be an absolute ruler, but in practice what happened was regulated by long-established precedent and convention, interpreted age after age by the Confucian exegesis of historical texts. China has always been a 'one-Party State', and for over 2,000 years the rule was that of the Confucian Party. My opinion is, therefore, that the term 'oriental despotism' is no more justified in the hands of Wittfogel than it was in those of the Physiocrats, and I never use it myself. On the other hand there are many Marxist terms, some old and some more recently gaining prominence, which I find great difficulty in adopting. For example, in some texts the 'imaginary State construct' is contrasted with the 'real substratum' of the independent peasant villages. This does not seem to me justifiable because in its way the State apparatus was quite as real as the work of the peasant-farmers. Nor do I like to apply the term 'autonomic' to the village communities because I think it was only true within very definite limitations. The truth is that we urgently need the development of some entirely new technical terms. We are dealing here with states of society far removed from anything that the West ever knew, and in coining these new technical terms I would suggest that we might make use of Chinese forms rather than continuing to insist on using Greek and Latin roots to apply to societies which were enormously different. Here the term *kuan liao* 官僚 for the bureaucracy might come in useful. If we could get a more adequate terminology it would also help us to consider certain other related problems. Here I am thinking of the remarkable fact that Japanese society was more similar to

[48] Though one of the great religions of China, Buddhism, being so unworldly, is not included here.
[49] On this see L. A. Maverick's *China: A Model for Europe* (1946) which includes a translation of F. Quesnay's (1767) *Le Despotisme de la Chine*.
[50] See, for example, the review by Pulleyblank (1958a).
[51] Cf. Liu Tzu-Chien, 'An Early Sung Reformer, Fan Chung-Yen' (1957), p. 105.

that of Western European society than that of China, and for that very reason more capable of developing modern capitalism. This has been recognised by historians for a long time past, but recent writings have pinpointed rather precisely the exact ways in which Japanese military-aristocratic feudalism could generate capitalism as Chinese bureaucratic society could not.[52]

It seems clear, at any rate, that the early superiority of Chinese science and technology through long centuries must be placed in relation to the elaborate, rationalised and conscious mechanisms of a society having the character of 'Asiatic bureaucracy'. It was a society which functioned fundamentally in a 'learned' way, the seats of power being filled by scholars, not military commanders. Central authority relied a great deal upon the 'automatic' functioning of the village communities, and in general tended to reduce to the minimum its intervention in their life. I have already written (above, pp. 1–2) of the fundamental difference between peasant-farmers on the one hand and shepherds or seamen on the other. This difference is expressed epigrammatically in the Chinese terms *wei* 爲 and *wu wei* 無爲. *Wei* meant application of the force of will-power, the determination that things, animals, or even other men, should do what they were ordered to do, but *wu wei* was the opposite of this, leaving things alone, letting Nature take her course, profiting by going with the grain of things instead of going against it, and knowing how not to interfere.[53] *Wu wei* was the great Taoist watchword throughout the ages, the untaught doctrine, the wordless edict.[54] It was summarised in that numinous group of phrases which Bertrand Russell collected from his time in China, 'production without possession, action without self-assertion, development without domination'.[55] Now *wu wei*, the lack of interference, might very well be applied to a respect for the 'automotive' capacity of the individual farmers and their peasant communities. Even when the old 'Asiatic' society had given place to 'bureaucratic feudalism' such conceptions remained very much alive in Chinese political practice and government administration that had been inherited from ancient Asian society and from the single pair of opposites, 'villages-prince'. Thus, all through Chinese history, the best magistrate was he who intervened least in society's affairs, and all through history, too, the chief aim of clans and families was to settle their affairs internally without having recourse to the courts.[56] It seems probable that a society like this would be favourable to reflection upon the world of Nature. Man should try to penetrate as far as possible into the mechanisms of the natural world and to utilise the sources of power which it contained while intervening directly as little as possible, and utilising 'action at

[52] See, for example, the monograph by Jacobs (1958), notable also for the excellence of its index. The author, a Weberian sociologist, executed the remarkable feat of making no mention of Marx and Engels.

[53] I remember that during the war I had a friend in the Foreign Office in London who had a huge Chinese scroll beside his desk, with these two characters alone on it, and when later on I became Master of Caius, I found that it was essentially a practical dictum; things worked better if you left the College Lecturers, the Dean and Chaplain and the Kitchen Office to get on with it without any interference from above.

[54] See *SCC*, vol. 2, p. 564.

[55] *SCC*, vol. 2, p. 164; from Bertrand Russell, *The Problem of China* (1922), p. 194.

[56] An aspect of the darker side of this is given in the partly autobiographical account of my old friend Kuo Yu-Shou (1963).

a distance'. Conceptions of this kind, highly intelligent, sought always to achieve effects with an economy of means,[57] and naturally encouraged the investigation of Nature for essentially Baconian reasons. Hence such early triumphs as those of the seismograph, the casting of iron, and water-power.[58]

It might thus be said that this non-interventionist conception of human activity was, to begin with, propitious for the development of the natural sciences. For example, the predilection for 'action at a distance' had great effects in early wave-theory, the discovery of the nature of the tides, the knowledge of relations between mineral bodies and plants as in geo-botanical prospecting, or again in the science of magnetism. It is often forgotten that one of the fundamental features of the great breakthrough of modern science in the time of Galileo was the knowledge of magnetic polarity, declination, etc.; and unlike Euclidean geometry and Ptolemaic astronomy, magnetical science was a totally non-European contribution.[59] Nothing had been known of it to speak of in Europe before the end of the 12th century, and its transmission from the earlier work of the Chinese is not in doubt. If the Chinese were (apart from the Babylonians) the greatest observers among all ancient peoples, was it not perhaps precisely because of the encouragement of non-interventionist principles, enshrined in the numinous poetry of the Taoists on the 'water symbol' and the 'eternal feminine'?[60]

However, if the non-interventionist character of the 'villages–prince' relationship engendered a certain conception of the world which was propitious to the progress of science, it had certain natural limitations. It was not congruent with characteristically occidental 'interventionism', so natural to a people of shepherds and sea-farers. Since it was not capable of allowing the mercantile mentality a leading place in the civilisation, it was not capable of fusing together the techniques of the higher artisanate with the methods of mathematical and logical reasoning which the scholars had worked out, so that the passage from the Vincian to the Galilean stage in the development of modern natural science was not achieved, perhaps not possible. In medieval China there had been more systematic experimentation than the Greeks had ever attempted, or medieval Europe either,[61] but so long as 'bureaucratic feudalism' remained unchanged, mathematics could not come together with empirical Nature-observation and experiment to produce something fundamentally new. The suggestion is that experiment demanded too much active intervention, and while

[57] One can see what this implies by imagining a city on the side of a hill above a river, where water was needed for the upper streets. The Confucians would have had squads of men pedalling square-pallet chain-pumps to send up the water from the river; but the Taoist way would have been quite different. They would have taken off a derivate canal from the river at a higher level and by guiding it along the contours, they would have reached the upper streets of the city on a *wu wei* principle.

[58] One might add the magnetic compass, deep borehole drilling, and the escapement of clockwork, and many other inventions listed below.

[59] See Needham (1960a). [60] Cf. *SCC*, vol. 2, p. 57.

[61] Nathan Rosenberg has suggested to us that a new attitude almost of deference to experimental results arose in Europe from the 16th century onwards, with the dominance of the bourgeoisie, which was not paralleled elsewhere. This attitude is very similar to that of merchants interested in quantitative accounting. See Rosenberg & Birdzell (1986).

this had always been accepted in the arts and trades, indeed more so than in Europe, it was perhaps more difficult in China to make it philosophically respectable.

There was another way, also, in which medieval Chinese society had been highly favourable to the growth of the natural sciences at the pre-Renaissance level. Traditional Chinese society was highly organic, highly cohesive. The State was responsible for the good functioning of the entire society, even if this responsibility was carried out with the minimum intervention. One remembers that the ancient definition of the Ideal Ruler was that he should sit simply facing the south and exert his virtue (*te* 德) in all directions so that the Ten Thousand Things would automatically be well governed. As we have been able to show over and over again, the State brought powerful aid to scientific research.[62] Astronomical observatories, for example, keeping millennial records, were part of the civil service; vast encyclopaedias, not only literary but also medical and agricultural, were published at the expense of the State, and scientific expeditions altogether remarkable for their time were successfully accomplished (one thinks of the early +8th-century geodetic survey of a meridian arc stretching from Indo-China to Mongolia, and the expedition to chart the constellations of the southern hemisphere to within twenty degrees of the south celestial pole).[63] By contrast science in Europe was generally a private enterprise. Therefore it hung back for many centuries. Yet State science and medicine in China were not capable of making, when the time came, that qualitative leap which happened in occidental science and medicine in the 16th and early 17th centuries.

Intimately associated with this qualitative leap is the idea of scientific progress. Let us now follow the conception of a progressive development of knowledge further, far beyond the level of ancient techniques. It would be quite a mistake to imagine that Chinese culture never generated this conception, for one can find textual evidence in every period showing that in spite of their veneration for the sages, Chinese scholars and scientists believed that there had been progress beyond the knowledge of their distant ancestors. Indeed the whole series of some hundred astronomical tables ('calendars') between −370 and −1742 illustrates the point, for many new emperors wanted one which would necessarily be better and more accurate than any of those that had gone before. No mathematician or astronomer in any Chinese century would have dreamed of astronomical constants. The same also may be said to be true of the pharmaceutical naturalists, whose descriptions of the kingdoms of Nature grew and grew. The number of main entries in the major pharmaceutical natural histories between +200 and +1600 show the growth of knowledge through the centuries; an unduly sharp rise just after +1100 is probably referable to increasing acquaintance with foreign, especially Arabic and Persian, minerals, plants and animals, with a synonymic multiplication which subsequently righted itself, on which see Volume 6, part 1.

[62] *SCC*, vols. 2, 3, 4, 5, 6 *passim*.

[63] See Beer, Ho Ping-Yü, Lu Gwei-Djen, Needham, Pulleyblank & Thompson, 'An Eighth-Century Meridian Line; I-Hsing's Chain of Gnomons and the Pre-History of the Metric System' (1964).

The position in China would be well worth contrasting in detail with that in Europe. In his great work Bury (1924) showed long ago that before the time of Francis Bacon only very scanty rudiments of the conception of progress are to be found in Western scholarly literature. The birth of this conception was involved in the famous 16th- and 17th-century controversy between the supporters of the 'Ancients' and those of the 'Moderns', for the studies of the humanists had made it clear that there were new things, such as gunpowder, printing and the magnetic compass, which the ancient Western world had not possessed. The fact that these (and many other innovations) had come from China or other parts of Asia was long overlooked, but the history of science and technology as we know it was born at the same time out of the perplexity which this discovery had generated.

Bury had dealt with progress in relation to the history of culture in general; Zilsel enlarged his method to deal with progress in relation to the ideal of science.[64] The 'ideal of scientific progress' included, he thought, the following ideas: (a) that scientific knowledge is built up brick by brick through the contributions of generations of workers; (b) that the building is never completed; and (c) that the scientist's aim is a disinterested contribution to this building, either for its own sake, or for the public benefit, not for fame or private personal advantage. Zilsel was able to show very clearly that expressions of these beliefs, whether in word or deed, were extremely unusual before the Renaissance, and even then they developed not among the scholars, who still sought individualistic personal glory, but among the higher artisanate, where co-operation sprang quite naturally from working conditions. Since the social situation in the era of the rise of capitalism favoured the activities of these men, their ideal was able to make headway in the world. Zilsel traces the first appearance of the idea of the continuous advancement of craftsmanship and science to Mathias Roriczer, whose book on cathedral architecture appeared in 1486.[65] Thus, 'science', said Zilsel, 'both in its theoretical and utilitarian interpretations, came to be regarded as the product of a co-operation for non-personal ends, a co-operation in which all scientists of the past, the present and the future have a part'. 'Today', he went on, 'this idea or ideal seems almost self-evident – yet no Brahmanic, Buddhist, Muslim or Latin scholastic, no Confucian scholar or Renaissance humanist, no philosopher or rhetor of classical antiquity ever achieved it'. He would have done better to leave out the reference to the Confucian scholars until Europe knew a little more about them. For in fact it would seem that the idea of cumulative, disinterested, co-operative enterprise in amassing scientific information was much more customary in medieval China than anywhere in the pre-Renaissance West.

Some Asian scholars have been suspicious of the idea of the 'Asiatic mode of production' or 'bureaucratic feudalism' because they have identified it with an assumed

[64] See Zilsel's writings between 1940 and 1945.

[65] This work is entitled *Das Büchlein von der Fialen Gerechtigkeit.* There is a facsimile reproduction published by Geldner (1965).

'stagnation' which they thought they saw in the history of their own societies.[66] In the name of the right of the Asian and African people to progress, they have projected this feeling into the past and have wished to claim for their ancestors exactly the same stages as those which the West had itself gone through, that Western world which had for a time dominated so hatefully over them. It is, I think, very important to clear up this misunderstanding, for there seems no reason at all why we should assume *a priori* that China and other ancient civilisations passed through exactly the same social stages as the European West. In fact, the word 'stagnation' was never applicable to China at all; it was purely a Western misconception. A continuing general and scientific progress manifested itself in traditional Chinese society, but this was violently overtaken by the exponential grc wth of modern science after the Renaissance in Europe. China was homoeostatic, cybernetic if you like, but never stagnant. In case after case it can be shown with overwhelming probability that fundamental discoveries and inventions made in China were transmitted to Europe, for example, magnetic science, equatorial celestial co-ordinates and the equatorial mounting of observational astronomical instruments,[67] quantitative cartography, the technology of cast iron,[68] essential components of the reciprocating steam-engine such as the double-acting principle and the standard interconversion of rotary and longitudinal motion,[69] the mechanical clock,[70] the boot stirrup and the efficient equine harnesses,[71] to say nothing of gunpowder and all that followed therefrom.[72] These many diverse discoveries and inventions had earth-shaking effects in Europe, but in China the social order of bureaucratic feudalism was very little disturbed by them. The built-in instability of European society must therefore be contrasted with a homoeostatic equilibrium in China, the product I believe of a society fundamentally more rational. What remains is an analysis of the relationships of social classes in China and Europe. The clashes between them in the West have been charted well enough, but in China the problem is much more difficult because of the non-hereditary nature of the bureaucracy.

66 An example of belief in Chinese stagnation is to be found in the small book of Chhien Wen-Yüan (1985), *The Great Inertia: Scientific Stagnation in Traditional China.*

67 Needham (1955).

68 Cf. Needham, *The Development of Iron and Steel Technology in China* (1958). This was written with the collaboration of Wang Ling, but our volume of ferrous metallurgy is now in the hands of Donald Wagner of Copenhagen whose study of the subject will be, we hope, the last word on it for many years.

69 See Needham (1961) and (1963).

70 Cf. Needham, Wang Ling & de S. Price, *Heavenly Clockwork* (1960). We still think that the idea of the escapement came over to Europe at the end of the 13th century as a kind of stimulus diffusion from China. The water-wheel link-work escapement had originated in China at least from the beginning of the +8th century onwards, and it still seems to us quite likely that somebody said in Europe at the end of the 13th century, 'In the East, men have found a way of slowing down the revolution of a wheel until it keeps pace with the diurnal revolution of the stars (mankind's primary clock)'. Whereupon the Europeans set to and developed their own escapement, which was quite a different one, namely the verge-and-foliot.

71 The latter has already been described in *SCC*, vol. 4, pt. 2, pp. 304 ff., but the former will be described in vol. 5, pt. 8, as part of the cavalry material.

72 See *SCC*, vol. 5, pt. 7. Some of the multifarious influences of Chinese inventions and discoveries of the pre-Renaissance world have been emphasised by Lynn White in his *Medieval Technology and Social Change* (1962). The book of Gimpel, *The Medieval Machine* (1976), is much less sensitive to them.

It may well be that bureaucratic feudalism lasted so long because of the characteristic methods of wet-rice cultivation. Everyone pondering it has come away with the feeling that it is a predominantly 'Gartenbau' situation (to use the German term) – rather like the art of gardening itself. Francesca Bray, in her volume on the history of agriculture in China, adumbrated this very clearly. She wrote 'From the very outset, we believe, the fundamental constraints (both technical and social) of wet-rice cultivation significantly influenced China's peculiar path of development'.[73] And again, 'Unlike dry-land agriculture, in wet-rice areas, successful farming depends less on equipment than skill, less on capital investment and economies of skill than on labour shrewdly applied'.[74] She went on to write, 'It is clear that the small size of units of production (and of individual fields) suited to efficient wet-rice production is a barrier to economies of scale and thus to technical inventiveness of the type which we owe to the European Agricultural and Industrial Revolution, that is, a tendency towards mechanisation and improved returns to labour'.[75] She concluded,

> In wet-rice societies, . . . the natural course of development appears to be not towards capitalism but towards a social formation that may conveniently be called a petty-commodity mode of production. Once this stage is reached, the relations of production in wet-rice cultivation have an internal dynamism that enables them to sustain not only significant increases in agricultural productivity, but also rapid economic diversification, without undergoing historical change.[76]

In her separate book on the rice economies Francesca Bray (1986) returns to this theme. 'European historical methodology', she writes (p. 1), 'has understandably been profoundly marked by the growth of capitalism, but it is doubtful to what extent models derived from Europe's highly specific experience are applicable to other parts of the world'. She makes a distinction (p. 7). between 'mechanical technologies', like that of European agriculture, and 'skill-oriented' technologies, such as rice cultivation, and she concludes (p. 207) that 'the transition to capitalist relations of production whereby landowners evicting their tenants in order to run large, consolidated farms using cheap wage-labour, did not occur in China'. On the last page of her volume in *Science and Civilisation in China*, she concluded, 'It seems, then, that while wet-rice cultural systems can produce rapid, sustained technical and economic *development*, it is to the harder dry-land systems that we must look for real social transformation and historical *change*'.[77]

In recent decades much interest has been aroused in the history of science and technology in the great non-European civilisations, especially China and India, interest, that is, on the part of scientists, engineers, philosophers and orientalists, but little, on the whole, among historians. Why, one may ask, has the history of Chinese and Indian science been unpopular among them? Lack of the necessary linguistic and cultural tools for approaching the original sources has naturally been an inhibition, and of course if one is primarily attracted by 18th- and 19th-century science,

[73] *SCC*, vol. 6, pt. 2, p. 613. [74] *Ibid.*, p. 611. [75] *Ibid.*, p. 613. [76] *Ibid.*, p. 616. [77] *Ibid.*

European developments will monopolise one's interest. But I believe there is a deeper reason.

The study of great civilisations in which *modern* science and technology did not spontaneously develop obviously tends to raise the causal problem of how modern science did come into being at the European end of the Old World, and it does so in acute form. Indeed, the more brilliant the achievements of the ancient and medieval Asian civilisations turn out to have been, the more discomforting the problem becomes. During the past fifty years historians of science in Western countries have tended to reject the sociological theories of the origin of modern science which had a considerable innings earlier in this century. The forms in which such hypotheses had then been presented were doubtless relatively crude,[78] but that was surely no reason why they should not have been refined. Perhaps also the hypotheses themselves were felt to be too unsettling for a period during which the history of science was establishing itself as a factual academic discipline. Many historians were prepared to see science having an influence on society, but not to admit that society influenced science, and they liked to think of the progress of science solely in terms of the internal or autonomous filiation of ideas, theories, mental or mathematical techniques handed on like torches from one great man to another. They have been essentially 'internalists' or 'autonomists'. In other words, 'there was a man sent from God, whose name was . . .' Kepler.[79]

However, the study of other civilisations places traditional historical thought in serious intellectual difficulty. For the most obvious and necessary kind of explanation which it demands is one which would demonstrate the fundamental differences in social and economic structure and mutability between Europe on the one hand and the great Asian civilisations on the other, differences which would account not only for the development of modern science in Europe alone, but also of capitalism in Europe alone, together with its typical accompaniments of Protestantism, nationalism, etc. not paralleled in any other part of the globe. They must in no way neglect the importance of a multitude of factors in the realm of ideas – language and logic, religion and philosophy, theology, music, humanitarianism, attitudes to time and change – but they will be most deeply concerned with the analysis of the society in question, its patterns, its urges, its needs, its transformations. In the internalist or

[78] Such is the adjective generally applied to B. Hessen's famous paper 'On the Social and Economic Roots of Newton's *Principia*', delivered at the International Congress of the History of Science at London in 1931 (1932). It was certainly in plain, blunt, Cromwellian style. But already half a dozen years later, R. K. Merton's remarkable monograph, 'Science, Technology and Society in Seventeenth-Century England' (1938), had achieved a considerably more refined and sophisticated presentation. Much is owing also to the works of Zilsel (1941–5).

[79] Though off the rails at various points, J. Agassi is entertaining on this topic in his monograph *Towards an Historiography of Science* (1963). The 'inductivist' historians of science, he says, are chiefly concerned with questions of whom to worship and for what reasons; but he does not like the 'conventionalists' much better. With this particular quarrel I am not here concerned, but it is surprising that Agassi did not make more use of the works of Walter Pagel (1935–68) which would have supported some of his arguments strongly. On the whole, Agassi takes his own stand for autonomism, regarding Marxism as one of the failings of inductivists, and believing that contention between different schools was the main factor in the development of science. As his monograph comes from the University of Hong Kong, it seems that he encysted himself with extraordinary success from all contact with Chinese culture.

autonomist view, such explanations are unwelcome, and do little to encourage the study of the other great civilisations.

But if you reject the validity or even the relevance of sociological accounts of the 'scientific revolution' of the late Renaissance, which brought modern science into being, if you renounce them as too revolutionary for that revolution, and if at the same time you wish to explain why Europeans were able to do what Chinese and Indians were not, then you are driven back upon an inescapable dilemma. One of its horns is called pure chance, the other is racialism however disguised. To attribute the origin of modern science entirely to chance is to declare the bankruptcy of history as a form of enlightenment of the human mind. To dwell upon geography and harp upon climate as chance factors will not save the situation, for it brings you straight into the question of city-states, maritime commerce, agriculture and the like – concrete factors with which autonomism declines to have anything to do. The 'Greek miracle', like the scientific revolution itself, is then doomed to remain miraculous. But what is the alternative to chance? Only the doctrine that one particular group of peoples, in this case the European 'race', possessed some intrinsic superiority to all other groups of peoples. Against the scientific study of human races, physical anthropology, comparative haematology and the like, there can, of course, be no objection, but the doctrine of European superiority is racialism in the political sense and has nothing in common with science. For the European autonomist, I fear, 'we are the people, and wisdom was born with us'.[80] However, since racialism (at least in its explicit forms) is neither intellectually respectable nor internationally acceptable, the autonomists are in a quandary which may be expected to become more obvious as time goes on.[81] I confidently anticipate, therefore, a great revival of interest in the relations of science and society during crucial European centuries, as well as a study ever more intense of the social structures of all the civilisations, and the delineation of how they differed in glory, one from another.

In sum, I believe that the analysable differences in social and economic pattern between China and Western Europe will in the end illuminate, as far as anything can ever throw light on it, both the earlier predominance of Chinese science and technology and also the later rise of modern science in Europe alone.

[80] Job, 12:2. He is talking to Zophar, the Na'amethite. Another translation runs, 'No doubt you are the people, and wisdom will die with you!'

[81] D. J. de S. Price, a valued collaborator of our own, knew much of the Asian contribution, but in his *Science Since Babylon* (1961) follows a 'hunch' of Einstein's and favours chance combinations of circumstances as the evocators of Greek and Renaissance science. A. R. Hall, in 'Merton Revisited' (1963), attacks anew what he calls the 'externalist' historiography of science, but significantly keeps silence about the problem posed by the Asian contributions. If he had taken a broader comparative point of view, his arguments about the European situation might have carried more conviction. A. C. Crombie (see above, p. xxviii, n. 18) alone, of the three, shows a real consciousness of the slow social changes which permitted the intellectual movements of the late Middle Ages and the Renaissance to bring modern science into being in the European culture-area, but even he pays less attention to their economic concomitants.

(*b*) THE ROLES OF EUROPE AND CHINA IN THE EVOLUTION OF OECUMENICAL SCIENCE[1]

JOSEPH NEEDHAM

Many historians of ideas and culture still blandly assume that the Asian civilisations 'had nothing that we should call science'. If slightly better informed, they are apt to say that China had humanistic but not natural sciences, or technology but not theoretical science, or even correctly that China did not generate *modern* science (as opposed to the ancient and medieval sciences). This is not the place to set such ideas to rights in any detail, but my own experience has shown that it is comparatively easy to produce a whole series of bulky volumes about the scientific and technological achievements which the Chinese are supposed not to have had. If, as is demonstrably the case, they were recording sun-spot cycles a millennium and a half before Europeans noted the existence of such blemishes upon the solar orb,[2] if every component of the parhelic system received a technical name 1,000 years before Europeans began to study them,[3] and if that key instrument of scientific revolution, the mechanical clock, began its career in early +8th-century China rather than (as usually supposed) in +14th-century Europe,[4] there must be something wrong with conventional ideas about the uniquely scientific genius of Western civilisation. Nevertheless it remains true that modern science, i.e. the testing by systematic experiment of mathematical hypotheses about natural phenomena, originated only in the West. Yet it cannot even be maintained that China contributed nothing to this great breakthrough of modern science when it occurred in the later stages of the European Renaissance, for while Euclidean geometry and Ptolemaic planetary astronomy were undeniably Greek in origin, there was a third very vital component, the knowledge of magnetic phenomena, and the foundations of this had all been laid in China.[5] There people had been worrying about the nature of magnetic declination and induction before Westerners even knew of the existence of magnetic polarity.

But from the time of Galileo (1600) onwards, the 'new, or experimental, philosophy' of the West ineluctably overtook the levels reached by the natural philosophy of China, leading in due course to the exponential rise of modern science in the 19th and 20th centuries. What metaphor then can we use to describe the way in which the medieval sciences of both West and East were subsumed in modern science? The sort of image which occurs most naturally to those who work in this field is

[1] Presidential Address delivered to Section X (General) on 31 August 1967 at the Leeds Meeting of the British Association. The present paper was given, in preliminary form, as an address at the opening of the permanent exhibition of Chinese medicine at the Wellcome Historical Medical Museum in London on 28 March 1966 (see Poynter, Barber-Lomax & Crellin (1966)).

[2] *SCC*, vol. 3, pp. 434 ff. [3] *SCC*, vol. 3, pp. 474 ff., and Ho Ping-Yü & Needham (1959a).

[4] *SCC*, vol. 4, pt. 2, pp. 435 ff., and Needham, Wang & Price (1960).

[5] *SCC*, vol. 4, pt. 1, pp. 239 ff., 334, and Needham (1964), p. 255.

that of the rivers and the sea. There is an old Chinese expression about 'the Rivers going to pay court to the Sea',[6] and indeed one can well consider the older streams of science in the different civilisations like rivers flowing into the ocean of modern science. Modern science is indeed composed of contributions from all the peoples of the Old World, and each contribution has flowed continuously into it, whether from Greek and Roman antiquity, or from the Islamic world or from the cultures of China and of India.

Here I shall confine myself to the Chinese case. In considering the situation before us there are two quite distinct questions to ask, first when in history did a particular science in its Western form fuse with its Chinese form so that all ethnic characteristics melted into the universality of modern science; and second at what point in history did the Western form decisively overtake the Chinese form? We may thus try to define the date of what may be called the 'fusion point' on the one hand, and that of the 'transcurrent point' on the other. Since by a historical coincidence the rise of modern science in Europe was closely accompanied by the activities of the Jesuit mission in China (Matteo Ricci S.J. (Li Ma-Tou 利馬竇, Fig. 5) died in Peking in 1610), there was relatively little delay in the juxtaposition of the two great traditions. Since the breakthrough occurred in the West, the transcurrent point for each of the sciences naturally preceded the fusion point, but as we shall see, the interest lies largely in the lag or delay between the two.

First let us consider the fusion points, those estuaries in time when the rivers flowed into the sea, and when full mixture took place. Here at once we find a remarkable difference between what happened in the physical sciences and in the biological sciences. On the physical side, the mathematics, astronomy and physics of West and East united very quickly after they first came together. By 1644, the end of the Ming dynasty, there was no longer any perceptible difference between the mathematics, astronomy and physics of China and Europe; they had completely fused, they had coalesced.

If at first it seemed that Western mathematics had been at a higher level than Chinese mathematics this was found to be due, as the decades went by, to a loss of the expertise which the Sung and Yuan algebraists had had, and the restoration of their techniques redressed the balance – though the lack of deductive geometry remained a debit on the Chinese side. Chinese mathematics had always been by preference algebraic rather than geometrical.[7] Astronomy differed between the civilisations in an equally fundamental way, for while Greek astronomy had always been ecliptic, planetary, angular, true and annual, Chinese astronomy had always been polar, equatorial, horary, mean and diurnal (Fig. 6).[8] The two systems were not in any way opposed or incompatible; it was just as in the mathematics, the attention of Chinese and Europeans had been concentrated upon different aspects of Nature. If the Chinese had never had the passion for geometrical models which produced

[6] Cf. *SCC*, vol. 3, p. 484.
[7] A full account of the history of Chinese mathematics is given in *SCC*, vol. 3, pp. 1–168.
[8] This epigrammatic formulation was due to Leopold de Saussure; see *SCC*, vol. 3, p. 229.

Figure 5. Portrait of Matteo Ricci S.J. and his friend Hsü Kuang-Chhi. From Kircher, *China Illustrata* (1667).

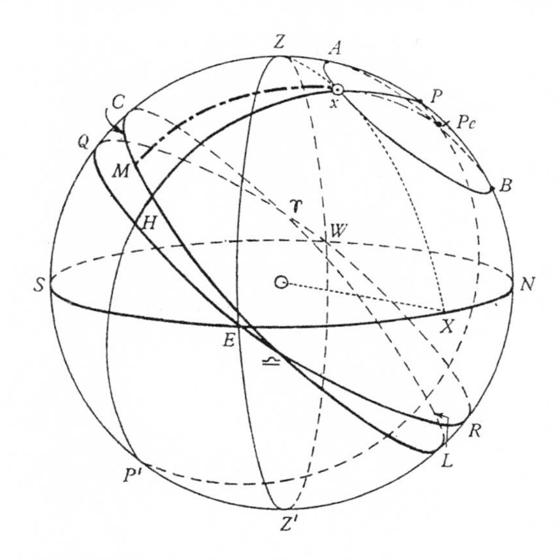

Figure 6. Diagram showing the circles of the celestial sphere. From *SCC*, vol. 3, p. 178.

the Ptolemaic epicycles and ultimately the Copernican solar system, their medieval cosmology had been far more modern than that of Europe, for instead of crystalline celestial spheres they thought in terms of infinite empty space and an almost infinitude of time.[9] By 1673 when Ferdinand Verbiest S.J. (Nan Huai-Jen 南懷仁) was reconstructing the Peking Observatory and equipping it with new instruments even more accurate than the splendid 13th-century ones of Kuo Shou-Ching 郭守敬,[10] this was perfectly understood; and the seal was set upon the matter when early in the 18th century Antoine Gaubil S.J. (Sung Chün-Jung 宋君榮) published his great works on the history and theory of Chinese astronomy.[11]

Since the breakthrough occurred in Europe first, Europe contributed rather more, giving up its crystalline celestial spheres but introducing more refined calendrical computations, giving up its Greek ecliptic co-ordinates[12] but opening the way into those undreamt-of worlds which the telescope would shortly reveal,[13] and above all introducing the new celestial mechanics and dynamics of the age of Galileo. Unified astronomy of course profited greatly by the records of celestial phenomena (eclipses, novae and supernovae, comets, etc.) which Chinese astronomers had kept, as accurately as they could, since the –5th century, and in greater abundance than any other culture.[14] Lastly one must take account of the fact that if oecumenical

[9] Cf. *SCC*, vol. 3, pp. 408, 438 ff. [10] *SCC*, vol. 3, pp. 350 ff., 367 ff., 451 ff.

[11] For the works of Gaubil see *SCC*, vol. 3, pp. 760 ff., *Histoire de l'Astronomie Chinoise, Traité de l'Astronomie Chinoise*, etc.

[12] *SCC*, vol. 3, pp. 266 ff. [13] Cf. Pasquale d'Elia; Needham & Lu Gwei-Djen (1966a).

[14] *SCC*, vol. 3, pp. 409 ff. These records are in constant use by astronomers at the present day; cf., for example, the current discussion on a nova of +1006 by Goldstein (1965), Goldstein & Ho Ping-Yü (1965), Minkowski (1965), Marsden (1965), Gardner & Milne (1965); and another on the saecular deceleration of the earth involving ancient eclipse records by Curott (1966).

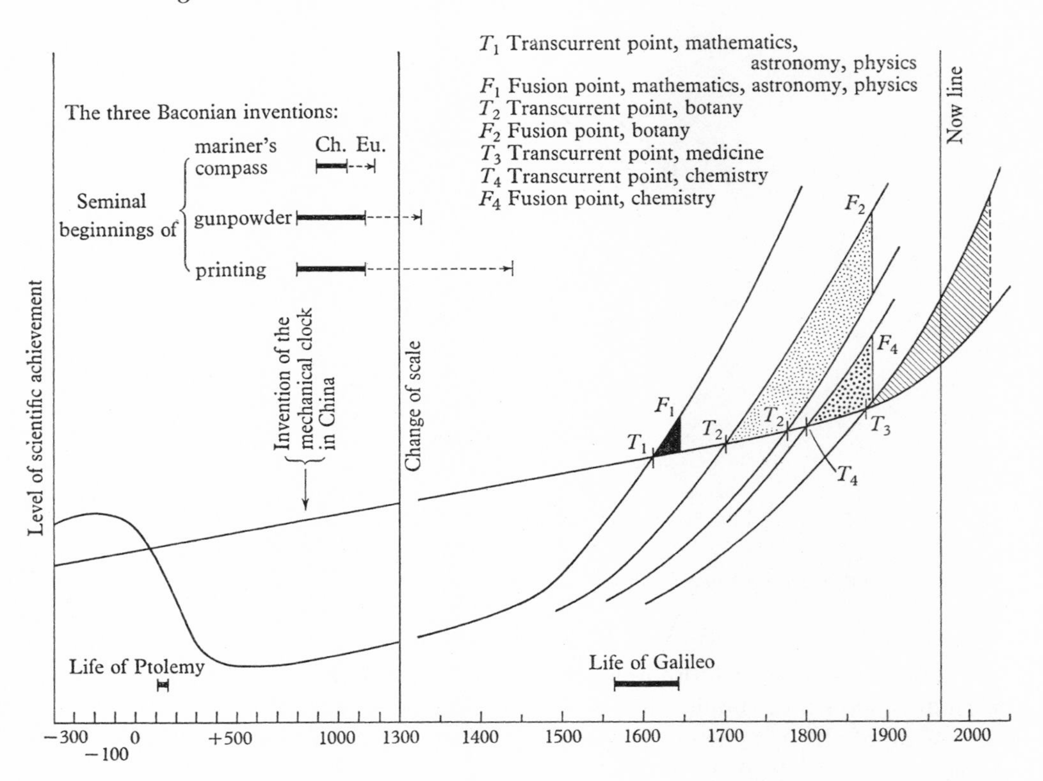

Figure 7. Graph showing transcurrent and fusion points for Chinese and Western science. From Needham (1970), *Clerks and Craftsmen*, Fig. 99.

astronomy today uses exclusively the Greek constellation patterns this is not in the slightest degree due to any inherent superiority of these over the entirely different Chinese ones; it is simply a side-effect which followed upon the meteoric rise of modern science in the West as a whole. Men like Flamsteed or Herschel would have thought it bizarre to speak of *Wei hsiu* 尾宿 or *Thien chi* 天機 instead of Scorpio,[15] but there is no intrinsic reason for this; it was an incidental result of the rise of modern science in Western civilisation first. One must always be on the look-out for such side-effects. At all events by the mid-17th century the union of the two astronomies had occurred (Fig. 7).

One can see how complete this fusion was by taking an example such as the large 18th-century Korean astronomical screen from the Yi dynasty's royal palace in Seoul, lately deposited in the Whipple Museum of the History of Science in Cambridge (Fig. 8).[16] To the right it reproduces the classical planisphere of 1395, prepared for Yi Thaejo 李太祖 by Kwon Kun 權近 and his colleagues, but though equatorial in projection this bears the Chinese names of the Western zodiacal houses round its

[15] Schlegel (1875), pp. 153 ff. [16] See Needham & Lu Gwei-Djen (1966b) (Korean Screen).

Figure 8. General view of the Korean Astronomical Screen (in the Whipple Museum of the History of Science, Cambridge University).

periphery. In the centre there are two Jesuit planispheres, ecliptically projected and using the Western 360° graduation instead of the Chinese 365¼° one, yet conserving not only the entire pattern of Chinese constellations (quite different from the Western, as we have noted) but even the age-old division of the stars into three colours based on the ancient star lists drawn up by the −4th-century astronomers, Shih Shen 石申, Kan Te 甘德 and Wu Hsien 巫咸.[17] To the left there are diagrams of the several planets, with text describing the new discovery of their moons and phases by Galileo and Cassini; and sun-spots depicted on the sun, as Chinese astronomers had noted them from the −1st century onwards. Further text described the resolution of nebulae, star clusters and the Milky Way, with the aid of the telescope. The Jesuit work preserved on the screen centres round one of the Directors of the Chinese Bureau of Astronomy, Ignatius Kögler S.J. (Tai Chin-Hsien 戴進賢), and it must have been painted about 1757, not long after his death.[18]

Another striking example can be seen in two Chinese optical virtuosi, Po Yü 薄鈺 and Sun Yün-Chhiu 孫雲球, who lived and worked in Suchow between 1620 and 1650, making apparatus such as telescopes, compound microscopes, magnifying

[17] *SCC*, vol. 3, p. 263.

[18] The circumstances of the matter have a special interest because in 1741 a Korean astronomical official on a mission to China, An Kuk-Rin 安國麟, had formed a friendship with one of Kögler's aides, André Pereira S.J. (Hsü Mou-Te 徐懋德), whose real name was Andrew Jackson, the only Englishman who is numbered among the roll of the Jesuit Mission in China.

glasses, magic lanterns, etc.[19] Po Yü, indeed, may be numbered with Leonard Digges and J.-B. della Porta, Lippershey, James Metius and Cornelius Drebbel, and all those other figures involved in the invention of the telescope and the microscope in the West. It is quite astonishing to find that within a couple of decades seekers in China were hot on the same trail. We do not yet know, and perhaps we shall never find out, exactly how independent they were, possible Jesuit intermediation being in this case obscure, but one can certainly say that at a surprisingly early date, 1635, the telescope was being applied to artillery in battle in China; and there is a strong possibility that Po Yü himself independently invented the telescope by juggling about with combinations of biconvex lenses just as a number of the inventors named above did in the West. Sun Yün-Chhiu even wrote a treatise entitled *Ching Shih* 鏡史 (History of Optick Glasses). Such was the extreme rapidity with which the mathematical and physical sciences of the Western and Chinese cultures fused after they first came in contact.

When you consider an intermediate science like botany, you find a totally different picture. It has been most interesting to observe, in work which my collaborators and I have been doing recently on the history of botany,[20] that there was a long delay after the first contacts, and one might say that the fusion point in botany did not occur until about 1880. Down to that time Chinese botany continued on its classical way. The naming, classifying and describing of plants went on along traditional lines. Even as late as 1848 the indigenous style persisted in the important work of Wu Chhi-Chün 吳其濬 called the *Chih Wu Ming Shih Thu Khao* 植物名實圖考 (Illustrated Investigation of the Names and Natures of Plants). Though written at such a recent date, this splendid and well-illustrated treatise was entirely traditional in character, and did not take any account of the advances in botany which had been made by Camerarius and Linnaeus. It is important to notice here that the Jesuit mission of the 17th century did relatively little for botanical contacts; indeed what it transmitted was westwards rather than eastwards, as witness, for example, the *Flora Sinensis* of Michael Boym S.J. (Pu Mi-Ko 卜彌格) printed in 1656. Moreover it could not have transmitted modern botany because its activity was both pre-Camerarian and pre-Linnaean in time. But when one comes to 1880, when Emil Bretschneider, the great medical officer of the Russian Ecclesiastical Exarchate in Peking, was doing his work on Chinese botany, then there began to be Chinese botanists who could speak the same language, could talk about Linnaean families and natural families, men who understood like European naturalists the function of the flower, and what the microscope could reveal of plant morphology. This was the time too at which centred the great effort of many investigators, indispensable for further development, to establish correlation as complete as possible between the Chinese traditional plant names and the Linnaean binomials.[21] Thus one might say that it was not before 1880 that the fusion point took place in botany, and a decade or so later might be a better guess.

Beyond this, when one passes on to medicine, one finds a situation in which the fusion of the sciences, pure and applied, in East and West, has not taken place

[19] See Needham & Lu Gwei-Djen (1966b).
[20] *SCC*, vol. 6, pt. 1.
[21] Cf. the works of Bretschneider, which are still quite indispensable.

even yet. I dare say that this is the case because, although physicists don't quite like you to say so, and astronomers equally may demur, nevertheless the phenomena of these sciences are surely much simpler than those with which biologists have to deal, and *a fortiori* physiologists, pathologists and medical men. Wherever the living cell is concerned, and *a fortiori* the living cell in its metazoan forms of high organisation, the puzzles are profounder, the tools both practical and conceptual more inadequate, the room for doubt greater. However optimistic one may feel as a young biologist or biochemist, the secret of life is still not yet just round the next bend. I speak from experience. Thus the coming together of the two cultural traditions, the fusion of them into a unitary modern medical science, has not even now been effected.

Many people, of course, when they think of Chinese medicine today, imagine it as some kind of 'folk-medicine', something bizarre and quite outdated, some sort of meaningless curiosity, but in truth these are all entirely wrong ways of reacting to it. It is, one must say, the product of a very great culture, a civilisation equal in complexity and subtlety to that of Europe.[22] While conserving a medieval body of theory, it contains a wealth of empirical experience which has got to be taken account of. Just as in other science, we can find many Chinese priorities; for example the compilation of a great classificatory description of disease entities without therapeutic material, the *Chu Ping Yüan Hou Lun* 諸病源候論 (Systematic Treatise on Diseases and their Aetiology), by Chhao Yüan-Fang 巢元方 in +610,[23] a whole millennium before Felix Platter[24] and Thomas Sydenham.[25] Or again the first handbook of forensic medicine in any civilisation, the *Hsi Yüan Lu* 洗冤錄 (Washing Away of Wrongs) by Sung Tzhu 宋慈 (1247),[26] appeared a considerable time before the European foundation-stones of the subject, the books of Fortunato Fedele[27] and Paolo Zacchia.[28] However, the rationale of some of the most important Chinese therapeutic practices, such as acupuncture, to which I will return in a moment, is not yet clearly understood; and obviously not all the drugs of the very rich traditional Chinese pharmacopoeia have yet been thoroughly examined from the biochemical and pharmacological point of view.

Equally important, there has until now been little unification of concepts. The original stimulus for this study arose indeed from problems of translation and technical terminology. In all the inorganic sciences Dr Wang Ling (Wang Ching-Ning) and I found long ago that once you know exactly what the ancient or medieval Chinese writer is talking about you can find the occidental equivalent for his words

[22] Palos (1963) rightly emphasises this.

[23] There is no adequate translation as yet of any part of this remarkable work, but in the meantime a paper by Rall (1962) may be referred to with due reserve.

[24] 1536 to 1614 Garrison (1929), pp. 271 ff.; Castiglioni (1947), pp. 429, 441, 452; his *Praxis Medica* (1608) has been described as the first attempt at a systematic classification of diseases.

[25] 1624 to 1689 Garrison (1929), pp. 269 ff.; Castiglioni (1947), pp. 546 ff.; his *Observationes Medicae* (1676), *De Podagra et Hydrope* (1683) and *Opera Universa* (1685) were all outstanding for their pathognostic descriptions.

[26] Partial translation by Giles (1924).

[27] 1550 to 1630 Castiglioni (1947), p. 557; his *De Relationes Medicorum* (1602) defined the subject.

[28] 1584 to 1659 Castiglioni (1947), p. 557; his *Quaestiones Medico-Legales* (1635) is the great landmark in the European history of forensic medicine.

without too much difficulty, and everything makes good sense. So for example one can converse with him across the centuries about solstices (*chih* 至) or equinoxes (*fen* 分), square roots (*khai fang* 開方) or comets (*hui hsing* 慧星), Lowitz arcs (*thi* 提), rock-salt and brachiopods (*shih yen* 石鹽, 石燕, written in two different ways); equally in the technological world about norias (*thung chhe* 筒車), water-powered reciprocating engines (*shui phai* 水排) or chain-and-link work (*thieh ho hsi* 鐵鶴膝). This is true only to a slightly lesser extent in fields such as botany and zoology, where *sui* 穗 is a spike or raceme and *thai* 苔 a capitulum or flower-head, while the corolla (*pa* 葩) is clearly distinguished from the calyx (*o* 萼). Similarly *wei* 胃 cannot mean anything but the stomach of an animal nor *fan chhu wei* 反芻胃 anything but a rumen. Alchemy and early chemistry have their own special problems exactly as they do in the West and for the same reasons, of purposive concealment and the like, but even amidst the spagyrical flights of poetic fancy there is far greater regularity than might be supposed, so that the 'river chariot' (*ho chhe* 河車) always means metallic lead,[29] and the 'food left behind by Yü the Great' (*Yü yü liang* 禹餘糧) always means brown nodular masses of haematite (ferric oxide). China, moreover, had her Martin Ruhland, but her *Lexicon Alchemiae* was nearly 1,000 years older than his, the *Shih Yao Erh Ya* 石藥爾雅 (Synonymic Dictionary of Minerals and Drugs) by Mei Piao 梅彪, *c.* +806, and it is extremely useful still today. Similarly, on the technological side *fan* 礬 is always alum, *shih tan* 石膽 always copper sulphate, and *huo yao* 火藥, the 'fire chemical', never means anything else than a gunpowder composition. On the whole the terminology of medieval Chinese chemistry, though far from unravelled as yet, presents no fundamental difficulties. Dr Ho Ping-Yü and I have found that it is quite possible to make medieval Chinese alchemical writing intelligible, though many years must pass before all its secrets become known.

But it is when one comes to the medical sciences that the translator finds himself in a really embarrassing position. Medical texts bristle with technical terms for which no equivalents in Western languages exist, some being ordinary words like *han* 寒 (cold) used in a highly technical sense, others specially constructed ideographs (often using the 'disease' radical), for example, *i* 疫 (infectious epidemic illness), *nüeh* or *yao* 瘧 (malaria-like fevers) or *li* 痢 (dysenteries of various origin). The key words of the highly systematised medical philosophy are the most difficult, for even Chinese lexicographic works do not dare to define them, since it was always expected that physicians would acquire their correct usage during long apprenticeship. Nowadays, to be sure, there are many works produced by the schools of traditional medicine in China which help to expound the terms, though not of course to translate them. Since there can be no exact equivalents in the Western world, where the evolution of physiological and medical thought followed very different courses, my collaborators, especially Dr Lu Gwei-Djen, and I, are adopting a new technique in translation, i.e. constructing an entirely fresh series of 'words of art' from Greek and Latin roots designed to express the innermost senses of the Chinese medical technical terms,

[29] But *caveat emptor*, of course, for in medico-physio-logical terminology the same words mean the placenta.

and then using them systematically.[30] All other possible procedures are open to the gravest objections; to leave the terms untranslated makes the result unreadable; to translate them by mechanical use of the dictionary makes it archaic, quaint and ridiculous as well as incorrect; while to seek to replace them by modern technical terms in current use is liable to distort the traditional ideas in a very dangerous way. The effect of our method is to make the medieval Chinese physicians talk like 16th- and 17th-century European medical writers, Ambrose Paré or Thomas Willis, in the same genre as it were, but obviously in a completely different tradition, and this is just the effect desired.

Now what all this essentially means is that the medical philosophies and theories of China and the West are by no means as yet mutually expressible, so that we have a situation quite different from what pertains in the inorganic and the simpler organic sciences. Indeed there are important technical terms in Chinese medicine, such as *hsü* 虛 and *shih* 實, *piao* 表 and *li* 裏, which almost defy any rendering in a Western language – almost but, we believe, not quite.[31] Such considerations came very prominently to mind during the discussions which Dr Lu and I had with Professor Chhen Pang-Hsien 陳邦賢, the eminent historian of medicine, and other colleagues, in Peking in 1958; and it was out of them that the ideas expressed in the present paper first arose.

One finds oneself obliged to speak (as herein I do) of 'modern-Western' medicine. It is not fair to call it just 'Western' as if it was wholly on a par with 'Chinese' or 'Indian' medicine, because it is palpably based on modern science in a way which the medicine of the non-European civilisations is not; but it is equally unfair to call it blandly 'modern' medicine, because that implies that no non-European civilisation has anything to contribute to it. On the contrary, truly modern and oecumenical medicine will not come into being until all these contributions are gathered in. Therefore I contrast 'modern-Western' with 'Chinese-traditional' medicine.

Let us now examine more closely the transcurrent points rather than the fusion points. One can hope, I believe, to define a certain number of moments in history when modern science, as we know it in the West since the time of Galileo, took clearly and decisively the lead over against Chinese science. One has to remember of course the earlier situation, pertaining in the Middle Ages, when nearly every science and every technique, from cartography[32] to chemical explosives,[33] was much more developed in China than in the West. From the beginning of our era down almost to the time of Columbus, Chinese science and technology had very often been far ahead

[30] For example, we composed the word *enchymoma*. See *SCC*, vol. 5, pt. 5, pp. 27 ff.

[31] Thus we propose for their equivalents respectively 'eremotic, plerotic, patefact and subdite'. Needless to say, the systematic use of such language in the translation of medical texts will be preceded by a deep analysis of the content of the Chinese terms and a semantic justification of the linguistic components chosen in the construction of their equivalents. Obviously the whole task is one of extreme difficulty, not only because of the conceptual discordance between the cultures, but also because naturally Chinese medical terminology changed slowly through the centuries. Nevertheless the ideas can, we believe, be harmonised if sufficiently understood; and there is a sufficient consensus in the language of Chinese medicine to allow of first-approximation Graeco-Latin equivalents to be very widely valid.

[32] See *SCC*, vol. 3, pp. 525 ff.

[33] *SCC*, vol. 5, pt. 7.

of anything that Europeans knew. Just to take only one or two examples, seismology was cultivated in China generations before the West, Chang Heng 張衡 in the +2nd century devising apparatus for locating the azimuth direction of the epicentre and recording the force of the shock.[34] So also while the Roman agriculturists despaired of classifying and describing soils, their Chinese equivalents before the +2nd century had brought into use more than fifty definable pedological terms, at the same time laying the foundations of all oecology and plant geography.[35] Or again, nobody in Europe (that Europe of the later-boasted 'iron horse' and irresistible 'ironclads') could reliably obtain a single pig of cast iron until about +1380, while the Chinese had been great masters of the art of iron-casting ever since the −1st century.[36] The standard method of interconversion of rotary and longitudinal motion, the eccentric, connecting-rod and piston-rod assembly, was not known in Europe before about +1450 but it had been fully at work in China since +970 and the combination of the first two components of it goes back there to +600 at least.[37] This is why I have elsewhere said that Chinese physical science attained a Vincian but not a Galilean level.[38] When then can we find these turning points of transition, these transcurrent points, as I call them, when modern science and technology originating in the West decisively took over from the Chinese level?

In the case of mathematics, astronomy and physics, I think one can say that it happened almost at the same time as the fusion point, or only a very short time before. What the Jesuits brought to China included of course Euclidean geometry and Ptolemaic planetary astronomy, both of which were very ancient, and certainly not part of modern science. But they also brought the algebraic notation of François Viète, which had only just been developed in the middle of the 16th century,[39] and later the logarithms of John Napier; and above all they brought the new dynamics, mechanics and optics of Kepler and Galileo. It is interesting to reflect that Tycho Brahe, the observational father of modern astronomy, filling his ledgers of data on the island of Hveen, was a patently Chinese figure, not in technique or conception much more advanced than Shen Kua 沈括 or Su Sung 蘇頌; it was only the next generations that begin to overpass the Chinese levels. Although the Jesuits played down the Copernican theory itself they gave full publicity to the results which Galileo obtained with the telescope from 1610 onwards.[40] It will be remembered that he got the idea of the telescope from Holland[41] and then made his own, after which things began to happen very quickly. In mathematics, astronomy and physics, therefore, it

[34] *SCC*, vol. 3, pp. 626 ff.

[35] There are good studies in Chinese of this birth time of pedology, oecology and plant geography, but little in Western languages. See, however, *SCC*, vol. 6, pt. 1, pp. 56 ff.

[36] See Needham (1958b).

[37] See *SCC*, vol. 4, pt. 2, pp. 369 ff., 380 ff.; Needham (1963). In *SCC* about two centuries of priority could be substantiated, but the discovery by Cheng Wei 鄭為 of a scroll-painting of *c.* +970 by Wei Hsien 衛賢, depicting a large water-mill which includes a reciprocating bolter, gives about five centuries. As for the connecting-rod and eccentric only, see *SCC*, vol. 4, pt. 2, p. 759.

[38] *SCC*, vol. 3, p. 160. [39] *SCC*, vol. 3, p. 438. [40] Needham & Lu Gwei-Djen (1966b).

[41] The 'optic tube' or reflecting telescope had, however, apparently been known and used in England since a date between 1540 and 1559. See Ronan (1991).

would seem that the transcurrent point came only a few decades before the fusion point.

In botany, on the other hand, there was, as we have seen (Fig. 7), a great time-lag, because the fusion point did not occur until after 1880. The transcurrent point we should have to put as occurring some time between 1695, when Camerarius first demonstrated the nature of the flower, Linnaeus' prime in 1735, and the restorative work of the great Adanson in 1780. Were it not for the fact that the Linnaean sexual system of classification was a sort of branch-line or siding, and not in the main line of advance, one might be tempted to say that Chinese botany attained a Magnolian or Tournefortian, but not a Linnaean level. But it would perhaps be fairer to take Adanson as the turning-point about 1780 and to say that then it was that botany in the West began to be decisively ahead of Chinese botany. Yet after that there was in China a lag of some hundred years, between 1780 and 1880, hence the feelings of superiority of the early 19th-century plant collectors, admirers though they were of the Chinese horticulturists whose gardens they delightedly pillaged.[42]

Next then we come to consider the question when did Western medicine decisively draw ahead of Chinese medicine? I confess that the more we think about it the later we are inclined to put this moment. I am beginning to doubt whether the transcurrent point was really much earlier than about 1900, perhaps 1850 or 1870. There are many things to be considered.[43] One has to weigh, for example, the clinical discoveries (Morgagni and Auenbrugger, 1761; Corvisart, 1808; Laënnec, 1819);[44] the rise of pharmaceutical chemistry (Pelletier & Caventou, 1820) with the study of alkaloids as its centre;[45] the new understanding of neurophysiology (Bell, 1811; Magendie, 1822); the development of bacteriology after Pasteur (1857);[46] the growth of immunology from Jenner *c.* 1798 (itself originating from a Chinese technique, variolation); the development of antiseptic surgery (Lister, 1865) and anaesthesia (1846); radiology (Röntgen, 1896), radiotherapy (the Curies, 1901) and radio-isotopes (Joliot-Curie, 1931); then parasitology with the discovery of the malaria plasmodium and its life-cycle (Laveran, 1880; Ross, 1898); eventually the coming of vitamins (Hopkins, 1912),[47] sulpha-drugs (Domagk, 1932), antibiotics (1940) and so on.[48] All this requires more thought, but if therapeutic success rather than diagnostic understanding is taken as the criterion, I suspect that it was not much before 1900 that medicine in

[42] See the monograph by Cox (1945).

[43] Here a valuable help is the monograph of Keele (1963) on the evolution of clinical diagnostic methods.

[44] Pathological anatomy, percussion, auscultation, the invention of the stethoscope. The thermometer came somewhat earlier, the sphygmograph much later.

[45] Isolation of strychnine and quinine; there is a fine statue commemorating these two chemists on the Boulevard St Michel in Paris.

[46] Necessarily involving microscopy. A landmark in this was Beale's *The Microscope in Medicine and its Application to Clinical Medicine* (1854).

[47] Biochemistry was to contribute fundamentally of course to clinical diagnosis as well as treatment but not much before the 20th century. Qualitative and quantitative urine analysis was how it began. This was first systematised by Neubauer & Vogel in 1860, in a book to which I remember referring myself in my young days.

[48] The other side of the medal is how long medieval practices continued in Europe. Galenic pulse-lore was still being systematised in 1828, and the Manchester Royal Infirmary ceased its bulk purchasing of leeches only in 1882.

the West drew decisively ahead of medicine in China.[49] Naturally, terms need careful definition. The work of Vesalius was not done in vain, and surgery and morbid anatomy were therefore correspondingly far ahead of China already by 1800. It may well be that all the sciences basic to medicine were much more advanced throughout the 19th century than what was known in China, and this must certainly be true of physiology as well as anatomy; yet from the point of view of the patient these branches of knowledge were rather slow in application, so that if we judge strictly clinically, the patient may not have been much better off in Europe than in China before the beginning of the 20th century.[50] In one single day in 1890 my father, himself a physician, lost both his first wife and beloved teenage daughter of diphtheria, no antitoxin being then available, though in later years he could constantly use it. I mention this family tragedy in order to emphasise that it may be an entire illusion to suppose, as so many do, that European medicine enjoyed a serene superiority over Chinese medicine throughout the 18th and 19th centuries. The 'enteric fever' of the Boer War could be another striking example. So when one finds a traveller like Dr Dinwiddie, who accompanied the Macartney Embassy to China in 1793, putting on airs of great superiority about Chinese science and medicine, one realises today that he had very little reason for being so pleased with himself.[51] By 1900, of course, perhaps even by 1870 or 1885, there were very good and solid reasons. But if the change-over or transcurrent point came so late, what is certain is that the fusion point has not even yet been attained. Indeed many decades must doubtless pass before this is achieved. Today the traditional medical doctors in China are working side by side with what we may call the 'modern-Western' physicians in full co-operation.[52] This is a very remarkable fact, which my collaborators and I have ourselves seen, in 1952, 1958 and 1964. It has been brought about in China by national renaissance, social conditions, and the paucity of medical doctors trained in modern style, during the past fifteen years. The two types of physicians and surgeons have joint observations, joint clinical examinations, and there is the possibility for patients to choose whether they will have their treatment in the traditional or the modern way; in other cases the physicians themselves decide which is best and proceed to apply it. And if one reads the *Chinese Medical Journal*, for example, carefully, one will find certain fields, as for instance the treatment of fractures,[53] where prolonged consideration has decided that in fact there were many valuable features in the traditional methods, and what

[49] To take two last examples, the electro-cardiogram dates only from 1903 and the electro-encephalogram only from 1929.

[50] In 1826, the year of Laënnec's death and the appearance of D. M. P. Martinet's significant *Manuel de Pathologie*, the basic sciences, says Keele, received more lip-service than application in the practice of medicine, even in France where there was most awareness of their value.

[51] *A fortiori* his even more self-satisfied colleague, the surgeon Dr Gillan; see the papers of the McCartney Embassy recently published by Cranmer-Byng (1962).

[52] This raises questions about traditional Chinese conceptions of clinical diagnostic method, regimen and therapy, as well as the nature of the medical philosophy itself. We cannot go into these here but must refer the reader to *SCC*, vol. 6, pt. 6. In the meantime we may cite as the least misleading among Western books about Chinese medicine, the following publications: Hume (1940); Morse (1934); Beau (1965); Palos (1963); Chamfrault & Ung Kang-Sam (1954).

[53] See the series of papers by Fang Hsien-Chih 方先之 *et al.* (1963–4).

is in use now is a combination of the two, the Chinese and the Western. Such fusion is going to happen more and more, giving rise to a medical science which is truly modern and oecumenical and not qualifiedly modern-Western. Here is only one example of it.

Now I shall say briefly something more on the question of acupuncture. As is generally known, this is a method of therapy, developed some 2,000 years ago, which involves the implantation of very thin needles into the body in different places according to a scheme or chart based on traditional physiological ideas and thoroughly systematised at an early date, certainly by the Thang and Sung periods.[54] We ourselves have seen the way in which this implantation of needles is done, attending acupuncture clinics in several Chinese cities. The method is still used very widely indeed in China at the present day. The problem arises of how its action can come about, and one may say without fear of contradiction that there are dozens of laboratories in China and Japan at the present day which are actively working with modern methods of a physiological and biochemical character to elucidate what happens. There are many possibilities; for example, that the stimulation of the autonomic nervous system by this method may increase the antibody titre in the blood, or increase the cortisone production by the suprarenal cortex, or it may exert a neurosecretory influence upon the pituitary gland. A wealth of experimental approaches lies open.[55] Moreover, it must be recognised that the acupuncture system connects in many ways with assured facts in neurophysiology, notably the Head Zones of the skin in mammals, which are related with specific viscera, and the remarkable phenomena of referred pain.

No one will ever really know the effectiveness of acupuncture or the other special Chinese treatments until accurate clinical statistics have been kept for several decades. The Chinese are not getting around to this at the present time [1967, Ed.] because the practical job of looking after the health of 700,000,000 people does not readily permit it, but I have no doubt that within a century accurate clinical statistics will

[54] Though there are many more recent books in Western languages, those of Soulié de Morant (1939) have not been superseded. From 1901 he studied directly under two eminent physicians named Yang and Chang, respectively, at Peking and Shanghai; and thirty years later, on returning to France, he set forth at length the classical system of acupuncture. Among the writings derivative from this tradition are those of Baratoux (1942) and the Lavergnes (1947). Since then several different strains of transmission have led to Europe. From Formosa the influence of Wu Hui-Phing 吳惠平 has generated the books of Lavier, Moss and the Lawson-Woods. From Vietnam that of Nguyen van Nha has affected those of Mann. Japanese studies have also exerted influence in Europe (Nakayama and Sakurazawa). In approaching acupuncture through the works of representatives of the present-day European practitioners some reserve should be exercised, for (a) very few of them have had linguistic access to the voluminous Chinese sources of many different periods, (b) it is often not quite clear how far their training has given them direct continuity with the living Chinese clinical traditions, (c) the history in their works is generally quite unscholarly, and accounts of theory very inadequate, and (d) their works are naturally much influenced by Western concepts of disease aetiology and semeiography so that they seem not to practise the classical Chinese methods of holistic classification and diagnosis. Nevertheless, pending the historical and theoretical account which we ourselves hope to give in *SCC*, vol. 6, pt. 6, this literature has its value. A brief and anonymous but authoritative statement issued by the National Academy and Research Institute of Chinese Traditional Medicine at Peking a few years ago is an important document.

[55] It is hard to give any adequate references here since so much of the literature is published in Chinese and Japanese, and in journals both difficult of access and not sought after by Western medical libraries. Much time must yet, I fear, elapse before all this is digested into a form available to the world scientific and medical public.

be kept, and this will be a fundamental contribution to our knowledge of traditional Chinese medicine.

A view commonly expressed (mostly by Westerners) is that acupuncture acts purely by suggestion, like many other things in what they often call 'fringe' medicine. This is, I believe, a question of what one might call relative credibility (or perhaps the calculus of credulity), a choice of what is the most difficult thing to believe. I must say that to my mind it is more difficult to believe that a treatment which has been engaged in and accepted by so many millions of people for something like twenty centuries has no basis in physiology and pathology, than to believe that it has been of purely psychological value. Of course it is true that the practices of phlebotomy and urinoscopy in the West had exceedingly little physiological and pathological basis on which to sustain their extraordinary and long-enduring popularity, but none of these had the subtlety of the acupuncture system. Possibly blood-letting had some slight value in hypertension, and extremely abnormal urines could tell their story, but neither contributed much to modern practice.[56] I can only say that for my own part I find the purely psychological explanation of acupuncture much more hard to credit than an explanation couched in terms of physiology and pathology. Animal experiments, where the psychological factor is ruled out, support this opinion. My own view is that in due course the scientific rationale of the method will be found. But until it is, Chinese and modern-Western medicine will not have fused.

Something more remains to be said about the theoretical setting of acupuncture and other traditional methods, such as the medical gymnastics for example which originated very early in China.[57] I have in mind the relative value placed in Chinese and Western medicine on aid to the healing power of the body on the one side, and a direct attack on the invading influences on the other. Now in Western medicine and in Chinese medicine alike these conceptions are both to be found. On the one hand, in the West, besides the seemingly dominant idea of direct attack on the pathogen, we have also the conception of the *vis medicatrix naturae*, which my father was always telling me about when I was a boy; for resistance and the strengthening of resistance to disease is an idea strongly embedded in Western medicine from Hippocrates and Galen onwards. On the other hand, one can also affirm that in China, where the holistic approach might be thought to have dominated, there was the idea of combating external disease agents, whether these were sinister *pneumata*, the *hsieh chhi* 邪氣 from outside, of unknown nature, or whether they were distinct venoms or toxins left behind for example when insects had been crawling over food – this is a very old conception in China – so that the combating of external agents was certainly present in Chinese medical thought too.[58] This may be called the *i liao* 醫療 aspect (or, in the ordinary parlance, *chih ping* 治病); and the other one, the *vis medicatrix naturae*, was what

[56] On these see the books of Keele (1963) and Brockbank (1954). Keele rightly congratulates the ancient Chinese on their freedom from 'magico-religious concepts of disease', though he calls their traditional medical philosophy 'metaphysical', but he might have added that they were also always free from that individual genethliacal astrology which played so painfully prominent a part in medieval European medicine, as he himself shows.

[57] Cf. Dudgeon (1895). [58] See Needham & Lu Gwei-Djen (1962).

was meant in China by *yang sheng* 養生, the strengthening of resistance. Nevertheless I think it is clear that whatever the acupuncture procedure does, it must be along the lines of strengthening the patient's resistance (e.g., by increasing the antibody or cortisone production), and not along the line of fighting the invading *pneumata* or organisms, venoms or toxins, i.e. not the characteristic 'antiseptic' attack which naturally has dominated in the West since the time of origin of modern bacteriology. This is shown by the significant fact that while Westerners are often prepared to grant value to acupuncture in affections such as sciatica or lumbago (for which modern-Western medicine can do very little anyway), Chinese physicians have never been prepared to limit either acupuncture or the related moxa (mild cautery and heat-treatment) to such fields; on the contrary they have recommended and practised it in many diseases for which we believe we know clearly the invading organisms, for example, typhoid, cholera or appendicitis, and they claim at least remission if not radical cure. The effect is thus cortisone-like. It is surely very interesting that both these conceptions (the exhibition of hostile drugs and the strengthening of the body's resistance) have developed in both civilisations, in the medicine of both cultures; and one of the things which any adequate history of medicine in China will have to do will be to elucidate the extent to which these two opposite ideas dominated in the systems of East and West at different times.

These remarks would not be complete without a reference to the importance of the traditional Chinese pharmacopoeia. I do not think that anyone today is inclined to despise pharmacopoeias of a traditional or empirical character developed among non-European peoples.[59] Since the recognition of the use of ephedrine from *Ephedra sinica* in the Chinese pharmacopoeia, one of its greatest triumphs, there have been many more shocks administered to pharmacologists in the West, as, for example, the famous case of *Rauwolfia* with its numerous powerful and highly peculiar alkaloids. I suppose that the whole modern science of chemotherapy has been closely connected with, if not directly dependent on, the investigation of naturally occurring drugs of an alkaloidal or otherwise highly complex organic character. The pharmacopoeia in China is in fact full of things which are of great interest from this point of view. When I was in China during the war running a scientific liaison mission between the Chinese and the Western allies, as Director of the British Scientific Mission in China, I had a good deal to do with *Dichroa febrifuga* (in Chinese *chhang-shan* 常山). People were looking about very urgently for anti-malarials other than quinine, and *chhang-shan* was therefore studied a good deal. Various pharmaceutical laboratories in China worked on it, gaining positive results early on in the war; these were doubted in the West, but eventually at the National Institute of Medical Research Dr Thomas Work made a study of it and it has turned out to be a quite powerful anti-malarial.[60] Its interest is indeed somewhat impaired by a variety of side-effects, presumably due

[59] See, for example, the symposium recently edited by Chhen Kho-Khuei, Mukerji & Volicer (1965). Or the monograph of Mosig & Schramm (1955).

[60] See Chang Chhang-Shao (1945); Fu Feng-Yüng & Chang Chhang-Shao (1948); Chang, Fu, Huang & Wang (1948); Tonkin & Work (1945); Duggar & Singleton (1953).

to other substances present, but if the active principle could be purified (and I am not quite sure how far this has been done) it could be a valuable medicament.

The naming of plants in the Linnaean system after personal names is often regarded as very modern, but sometimes the names of particular people were given to drug plants in China too. There is one, for example, called *shih chün tzu* 使君子 named after a physician Kuo Shih-Chün 郭使君 who studied and used it in the Wu Tai or Thang period about the +10th century. This is *Quisqualis indica* (or Rangoon creeper), a really valuable anthelmintic, especially in paediatric use, and still employed today on a wide scale.

If we may now sum up the results of our meditation so far we can make a very simple table to assemble certain figures that have been mentioned, and to accompany Figure 7.

Science	Transcurrent point	Fusion point	Lag in years
Mathematics, Astronomy, Physics	1610	1640	30
Botany	1700	1880	180
	or 1780	1880	100
Medicine	1800, 1870 or 1900	not yet	?

From this one might be tempted to deduce quite tentatively a 'law of oecumenogenesis' which would state that the more organic the subject-matter of a science, the higher the integrative level of the phenomena with which it deals, the longer will be the interval elapsing between the transcurrent point and the fusion point, as between Europe and an Asian civilisation. If this were in general principle right one might try to test it by looking into the history of chemistry in East and West, for which one would expect a figure intermediate between those for the physical sciences and for botany.

This subject bristles with difficulties partly because of the contingent historical trends which intervene. Chemistry as we know it is of course a science like that of the branch of physics which deals with electricity – wholly post-Renaissance, indeed 18th century in character. The pre-history of chemistry goes far back into antiquity and the Middle Ages, and it does so in China at least as much as the West.[61] In the West there were first the mystical aurificers of Alexandria, and in the East at the same time the pharmaceutical alchemists of China.[62] There is overwhelmingly

[61] See *SCC*, vol. 5, parts 1, 2, 3, 4, 5, 6, 7 and 9 of which are already published. I may also refer to the pioneer book of Li Chhiao-Phing 李喬苹 (1948). Studies of particular fields reach, as usual, a more scholarly level; see, for example, Dubs (1947); Sivin (1965); Ho Ping-Yü 何丙郁 & Needham (1959c); Tshao Thien-Chhin 曹天欽, Ho Ping-Yü & Needham (1959). The brief chapter in Leicester's (1965) handbook is balanced and perceptive. See also the monograph of Eliade (1956), full of insights.

[62] These terms are 'words of art'. The Greek aurificers were aurifictors, imitating gold but not believing that they could make it from other substances. But the essence of alchemy is macrobiotics, the chemical search for means of longevity and material immortality, and the combination of this with belief in aurifaction was fundamentally Chinese.

strong ground for believing that Arabic alchemy was influenced from China (even the very name is probably Chinese in origin),[63] and that it handed on its alchemical afflatus, coupling the art of making gold with that of finding the elixir of immortality, to the European alchemists of the +10th to the +15th centuries whose triumphs included the discovery of alcohol.[64] Vast treasuries of Chinese alchemy from the +3rd to the +14th centuries are contained in the *Tao Tsang* 道藏 or Taoist patrology, and besides all this there are rich texts of other genres which tell of metallurgy and the chemical industries. When Paracelsus in the +16th century inaugurated iatro-chemistry he was only copying unknowingly in Europe just what had come about in China somewhat earlier, with the difference that in that culture there had never been any prejudice against mineral remedies. So brilliant was the iatro-chemical period in China (+11th to +17th centuries) that it has been possible to show how the adepts of those times were able to prepare mixtures of crystalline steroid sex hormones and use them for therapy in cases for which they are normally prescribed today.[65]

But all this was not theoretical modern chemistry. The foundations of this were laid, as everyone knows, during the later 18th and early 19th centuries, with the exploration of the nature of gases by Priestley and others (1760 to 1780), the 'revolution in chemistry' effected by Lavoisier (1789), and then the atomic theory of Dalton (1810) followed by the far-reaching insights of the founder of organic chemistry Justus von Liebig (1830 to 1840).[66] This was already the beginning of the Opium Wars and the Thai-Phing 太平 Rebellion, but as soon as quiet recurred in China and modern science was able to strike roots again, chemistry in its new forms was introduced. There were no obstacles to fusion, because there had been no competing Chinese theories in the past;[67] the basic facts of chemical change which alchemists, industrial workers and medical doctors had long known fitted simply into the new explanations, the superiority of which over the traditional Yin and Yang and Five-Element theories was much more obvious here than in physiology or medicine. Modern chemistry was taught at all the Chinese universities after 1896, and books on it had been published by the Translation Department of the famous Kiangnan Arsenal from the time of its foundation by Ting Jih-Chhang 丁日昌 in 1865 onwards. Private institutions such as the Ko Chih Shu Yüan 格致書院 at Shanghai, which opened in 1874, also propagated chemical knowledge.[68] It may therefore be quite fair to set a period of some eighty years, say between 1800 (the approximate transcurrent point) and 1880 (the approximate fusion point), as the time elapsing. This evidently fits in adequately

[63] Etymologies of 'chem-' from Greek or Egyptian roots having long been notoriously unconvincing, its derivation from Chinese *chin* 金 (gold) or *chin i* 金液 (gold juice) was suggested independently in 1946 by the writer (Needham (1946) and Mahdihassan (1962)), who since then has elaborated the equation in many papers. This view has now gained general acceptance (see, e.g., Dubs (1961), and Schneider (1959)).

[64] Cf. the masterly little books of Sherwood Taylor (1951) and Holmyard (1957). See also Partington (1961).

[65] See Lu Gwei-Djen & Needham (1964).

[66] Cf. the usual accounts: Thorpe (1921); Lowry (1936); Partington (1957).

[67] This does not mean that there were no theories in Chinese alchemy and chemical industry, but they always remained essentially medieval in type (cf. Ho Ping-Yü & Needham (1959c)).

[68] This was headed for a time by Wang Thao 王韜, the Chinese collaborator of the great sinologist James Legge.

with the general picture, but I should not like to put too much emphasis on it, partly because of the adventitious historical circumstances, and partly because modern chemistry, as a relative late-comer in modern science, had no alternative system to meet with when it reached the Chinese culture-area, as all the other sciences considered had.

In conclusion, then, what we have done here is to examine the time elapsing between the first sprouts of particular natural sciences *in their modern forms* in European culture, and their fusion with the traditional forms as Chinese culture had known them, to form the universal oecumenical body of the natural sciences at the present day. The more 'biological' the science, the more organic its subject-matter, the longer the process seems to take; and in the most difficult field of all, the study of the human and animal body in health and disease, the process is as yet far from accomplished. Needless to say, the standpoint here adopted assumes that in the investigation of natural phenomena all men are potentially equal, that the oecumenism of modern science embodies a universal language that they can all comprehensibly speak, that the ancient and medieval sciences (though bearing an obvious ethnic stamp) were concerned with the same natural world and could therefore be subsumed into the same oecumenical natural philosophy, and that this has grown, and will continue to grow among men, *pari passu* with the vast growth of organisation and integration in human society, until the coming of the world co-operative commonwealth which will include all peoples as the waters cover the sea.

(c) THE NATURE OF CHINESE SOCIETY: A TECHNICAL INTERPRETATION[1]

HUANG JEN-YÜ 黃仁宇 AND JOSEPH NEEDHAM

(1) INTRODUCTION

When the People's Republic of China was established in 1949, many external observers believed that the new leadership would break away completely from old traditions. Today, nearly half a century later, the epoch-making effect of the revolution is clearly seen, yet it has also become increasingly evident that China's social and cultural heritage has been subsumed in the new order. It will undoubtedly exert great influence in determining the destiny of mankind in the future, and not only that of the Chinese themselves.

The historical continuity of the most populous country in the world has sometimes been regarded as a triumph of intellectual guidance. Many Western scholars have found theoretical points in common between Chinese communism, Confucianism, Taoism and even Buddhism. Such connections can be substantiated in a number of ways, though points of difference are also many; but it is easy to over-stress the influence of philosophers and to minimise the effect of concrete environmental and economic factors.

The formation of the Chinese ethos is of course a topic lending itself to a variety of interpretations. But as we see it, the importance of ideology, however great, can never obscure the basic fact that underneath lie the material forces of climate, geography and social integration. Chinese history differs from that of all other civilisations in the world because during the centuries before our era China had already developed a centralised political system, and since this could necessarily command only quite primitive technical support, it constituted a very bold advance indeed. The high degree of centralisation attained did not, however, grow out of the imagination of political thinkers; it was imposed by circumstances, geography being a leading factor. During subsequent centuries this centralised system had to be continually refined and sustained until China's political and ethical maturity ran far ahead of the development of other institutions, such as diversified economy, codified civil contractual law, and a system of jurisprudence protective for the individual. As time went on, these latter elements, still relatively rudimentary, were positively restrained lest they should disturb the uniform administration of the empire. One undesirable effect was that lacking what one might call technical sophistication, Chinese bureaucratic government always appeared impressive in breadth while remaining shallow in depth.

[1] This text was previously published, Needham (1974b), and is reprinted here with minor revisions. We take pleasure in acknowledging the clarification and help derived from discussions with Dr J. J. Broeze, Dr C. Feinstein, Dr Iain Macpherson, the late Prof. Joan Robinson, and Dr Mikulas Teich.

Basically its support was derived from social institutions and social values. Individuals were encouraged to adhere in primary groups, and if an enlightened code of ethics laid emphasis on man's duty to his fellow-men, it also inhibited all claim for natural rights. These conditions changed little during the past 2,000 years.

One can at once descry a number of factors in traditional Chinese society that would be inimical to the growth of a capitalism but favourable to a transition to socialism. The concept of a single united national culture, the unity in part arising from an ideographic[2] non-agglutinative language not easily learnt by outsiders, the reign of morality, order and organisation rather than law, the absence (as we shall see) of the development of any middle class, and the consistent restraint of mercantile interests and initiatives – all these were features advantageous for socialism once the time had come. In order to evaluate where China stands today, therefore, our account must first survey some of the major developments in the past.

(2) The Early Unification and Centralisation

It is surely significant that Chinese civilisation started along the Yellow River.[3] That great stream runs into and out of a large area of loess, a deposit of compacted wind-blown sand often more than 300 feet deep. Consequently the river's silt-content is phenomenal.[4] While in most major flows in the world a silt content of over 4 or 5 per cent can be regarded as significantly high, that in the current of the Yellow River has established observed records of 46 per cent.

One of its tributaries has produced an unheard-of 63 per cent. With the last 500 miles of its course receiving not a single tributary, the Yellow River has therefore had a constant tendency to fill up its own channel and over-run its dykes, so that today its bed is well above the level of the surrounding North China plain.

Before the imperial period, the execution of works of water-control by the divided feudal States remained a constant cause of controversy between them, as such devices usually relegated the problem to the neighbouring States, with intensified harm. As early as −651 a league convention was held by the principalities concerned, and during its sessions the participants solemnly pledged not to modify the course of the Yellow River so as to cause difficulty to the other princedoms. The promise was ineffective. As China entered the period of the Warring States, from the early −5th century to the late −3rd century, contesting princedoms even broke the dykes deliberately in order to flood the territories of their enemies. Not until the emergence

[2] This term has often been criticised, for example, Sampson (1985), pp. 34 ff.: 'the term [ideographic] . . . seems to be used in a way that blurs the crucial distinction between semasiographic and logographic systems'. Nevertheless the idea of ideographic as opposed to alphabetical languages is so widely understood that it may stand here. [Glen Dudbridge points out that 'ideographic' is an obsolete term in relation to post-Shang Chinese script. Joseph Needham was aware of this, but when I discussed it with him in 1993 he was unwilling to sacrifice the term which he had used from Volume 1 of *Science and Civilisation in China* onwards, and which had been used before him by such revered masters as Bernard Karlgren. Ed.]

[3] This remains true in spite of the wide variety of neolithic sites discovered in very recent times.

[4] Cf. sub-sections on hydraulic engineering in Section 28 (*SCC*, vol. 4, pt. 3).

of a united China in −221 could the problem reach a settlement. Thus there is undeniable truth, over-emphasised by some,[5] that the need for hydraulic works, not only in flood protection but also for irrigation and later for bulk transport, did favour unification of the empire. And not only so, but this circumstance equally remains one of the best explanations of why China was traditionally bureaucratic, not aristocratic.[6] Water control and management always tended to transcend the boundaries of feudal domains, since only an emperor and his hierarchy of officials were competent to master it.[7]

Of course China was not the only civilisation that arose in a great river valley; this was also true of Egypt based on the Nile, Babylonia based on the Tigris and the Euphrates, and the Mohenjodaro culture based on the Indus. But all these were relatively near the heartland of the Old World, exposed rather prominently to mutual influences and conquests, not isolated beyond that great Tibetan mountain massif which generated so naturally and appropriately the Indian and Chinese concepts of the central super-mountain, Mt. Meru or Khun-Lun Shan 昆侖山. This must be part of the answer to any enquiry why all the other ancient river-valley civilisations disappeared as recognisable entities, replaced by later cultures, Gupta, Mogul, Roman, Byzantine, Islamic; while Chinese continuity has endured without break from the neolithic Yang-shao 仰韶 and Lung-shan 龍山 people right down to the present day.

Rivers and silt posed great problems, but climate was not kind to China either.[8] Early Chinese writers, without the benefit of modern meteorology, took it for granted that every six years there would be a serious crop failure and every twelve years a general famine. The official dynastic histories down to 1911 report that in those 2,117 years no fewer than 1,621 floods and 1,392 droughts were recorded, thus on an average more than one disaster in each year. Only recently has the phenomenon found an adequate explanation.

Rainfall in China has a highly seasonal distribution, some 80 per cent of it coming in three summer months; and during the same period the prevailing wind direction changes. The fluctuation of the annual rainfall can also be drastic, because the monsoon in China is more cyclonic than orographic. In other words, the humid monsoon air relies not so much on high mountains for its cooling agent as on the northerly and north-eastern cold air currents, which lift it to a sufficient height to yield its moisture. While the regular pattern of crossing currents is unchanging, the actual result varies widely from one year to another, since it depends upon the synchronisation of two sets of variables. The points of convergence could be more numerous than expected, or else rather fewer. This variation produced the high

[5] The name of Wittfogel has become overwhelmingly associated with this but there are some strange exceptions; Sri Lanka, for example, had many wonderful hydraulic engineering works from the Middle Ages onwards yet never developed a bureaucracy.

[6] See again Section 28 of *SCC*, vol. 4, pt. 3.

[7] This is stated in so many words in the *Yen Thieh Lun* (*c.* −80). See *SCC*, vol. 4, pt. 3, p. 264.

[8] See Section 21 in *SCC*, vol. 3, and Section 47 in *SCC*, vol. 7, pt. 1.

frequency of floods and droughts in China – a meteorographical phenomenon well known in history, yet still seen yearly today – and sometimes with the two kinds of natural disaster coming at the same time but in different localities.

The centuries prior to −221 saw a series of internecine wars. The extant records create an impression that the fighting enabled the princes to fulfil their schemes of aggrandisement and settle their private feuds. But sufficient clues are also left in the source materials to suggest that natural disasters were at least partially responsible for the disturbances.[9] Peasant-farmers participated in the campaigns because of hunger and food shortages. Crops were seized by the engaging armies; stoppage of food supplies in time of famine usually constituted a *casus belli*. Moreover, large states capable of distributing famine relief inevitably emerged victorious; they gathered larger followings. Thus the sequence of events worked out almost automatically a rational solution to the problem, pointing to a unified China and a centralised bureaucratic administration, which alone would be able to command the necessary resources for relieving local distress.

But Nature imposed on China yet another problem the solution of which also hinged upon unification and centralisation, and that was the security of the northern frontier, manifested spectacularly by the emergence of the Great Wall.[10] In meteorological terms, that barrier coincides in general with the fifteen-inch isohyet line. This simply means that south of the Wall an annual rainfall of fifteen inches can be expected, the minimum for agricultural production. Beyond the Wall rain is more sparse, most rivers dry up before they can reach the sea, and nomadic pastoralism provided the only means of livelihood. Called by Lattimore 'one of the most absolute frontiers in the world',[11] the Great Wall has for more than 2,000 years been a demarcation line separating cultural groups, social customs, languages and religions. Since the peaceful assimilation of minority groups on the other side of the Wall was never permanently possible, the northern frontier remained a perennial skirmishing ground between Chinese farmers and nomadic tribesmen. There were occasions when the Chinese launched offensive wars into the steppeland and desert in the north and north-west, but as often as not they were put on the defensive by the invading tribesmen whose mobility commanded a great advantage. And it is a commonplace of history that large parts of China were sometimes ruled by dynastic houses springing from nomadic tribal origins; indeed the united empire itself fell to the Mongols in the Yuan period and to the Manchus in the Chhing. There is an obvious parallel here with the Roman Empire, but in the end Rome was more or less 'barbarised', and China never. At all events, and in every case, mobilisation and military logistics called for a united China. And once again, bureaucrats, even in uniform, could mobilise forces beyond the dreams of any aristocrats.

Besides these factors, the compactness of the Chinese *oikoumene*, its general suitability for agriculture, its network of inland waterways, and the linguistic homogeneity of its inhabitants all favoured unification.

[9] See Section 5 in *SCC*, vol. 1. [10] See *SCC*, vol. 4, pt. 3. [11] Lattimore (1940), p. 21.

Nor was there any insurmountable obstacle in the way of establishing a centralised system. But unlike the defence against the floods of the Yellow River, not to mention the relief of recurring natural disasters, and the control of the nomadic menace on the frontier, the favourable conditions of human and social geography hardly added a sense of urgency to the issue of unification. Left to themselves, they would in all probability have worked towards China's political centralisation, but at a slower pace and possibly on a more solid foundation. In the course of history this made a great deal of difference. As it happened, unification and centralisation came to China, one could almost say, soon after the end of the Bronze Age. Pressed by matters of survival, the process allowed no time for local institutions and customary practices to mature. Aristocratic idiosyncrasies might have fostered them, not a far-flung bureaucracy with its mind on efficiency and uniformity. Taoist technology and *laisser-faire*, Mohist science and religion, Legalist uniformity and Hedonist self-cultivation, were all in a sense casualties of the paramount necessity for the centralised Confucian bureaucratic State. The authority of imperial China came as an inheritance from two major sources: the constitutional structure of the proto-bureaucratic feudal principalities, and the corresponding thought of the political philosophers, especially Confucians, who grew up within them.

Mention was made a few paragraphs above of the other great ancient river-valley civilisations, and any discussion of the characteristics of Chinese feudal bureaucratism does always raise the question of why the resemblances to it in India are so very partial. India also suffered from floods and droughts; India also was menaced, in her case chiefly from the north-western frontier, by the incursions of alien invaders many times indeed in her long history; India also needed important works of hydraulic engineering, though apart from the Vijayanagar kingdom in the south these were never forthcoming to the extent that would have been desirable. What was the X-factor that made the Chinese experience so different from that of most other cultures? Here perhaps is where those 'less urgent' factors just mentioned are seen to have outstanding importance. In India linguistic, religious and social (caste) diversification dominated the sub-continent from very early times; in striking contrast assuredly with the homogeneity of the Chinese people, who absorbed peripheral tribes into their culture (e.g., the Yüeh 越, the Man 蠻, the Chhi-tan 契丹, the Jurchen 女真) without in any way diluting it, and could even assimilate their conquerors, Mongols or Manchus, till hardly anything recognisable of them was left. Not only this, but China radiated her culture over all East Asia, so that Korea, Japan and Vietnam are all in one way or another her children. Surely the 'monolithic' ideographic language had a great part to play in this, as also the deeply characteristic styles of Chinese agriculture and its associated technical arts, so well described by Francesca Bray,[12] and the specific procedures of Chinese bureaucratic administration which through the centuries had been moulded to suit them. It must also be of great importance that no institution of a caste character developed in China, and that

[12] See Section 41 in *SCC*, vol. 6, pt. 2.

none of the three religions (*san chiao* 三教) claimed over-riding obedience or temporal power.

Perhaps it is no coincidence that the only other civilisation comparable to that of China in its long enduringness was that of Ancient Egypt strung along the Nile, and that there also a homogeneous ideographic language prevailed. There were of course differences. Egypt's life-span was set at an earlier time in history than China's for the Pharaonic institution was already more than two millennia old when the Shang kingdom in China started the ball rolling there, while the unitary Chinese State is still a reality today a millennium and a half after the triumph of Arabic and Islamic culture. Egypt was also very narrowly bounded by the desert on each side of the strip, quite unlike the spaciousness of China's rivers, mountains and plains stretching over an area two or three times the size of Europe. In sum, then, it may not be too difficult to understand why the Chinese experience was not paralleled in any of the other river-valley civilisations.

(i) *The bureaucratic management of an agrarian society*

The generation of a well-knit bureaucratic apparatus by an agrarian nation of China's size has no parallel in world history. Paid in salaries (when paid at all),[13] the professional managerial group regarded administrative positions as fully interchangeable within a hierarchical order; this required a great deal of standardisation and precision. Two consequences followed, first the bulk transport of taxation goods meant that salaries and local expenses had generally to be found by deduction at the source, not paid from the capital city, hence a perpetual tendency towards peculation at all levels; and secondly the principle of inter-changeability of general-purpose district officers meant that specialists (hydraulic engineers, mathematicians, astronomers, medical experts, etc.) were always shunted aside and could rarely reach high office. Since the functionaries did not represent the financial interests of the provinces or districts that they managed, the government could never gain stability through a balance of those interests. Its usual approach, indeed, was to restrain the more advanced sectors of the national economy in order not to outstrip the more backward sectors. The relationship between the governing and the governed was contingent solely upon their mutual subjection to the imperial rule. The central government appeared to be, and was, the only source of power, yet on account of the vastness of the country this power could not always be exercised effectively.[14]

These requirements and conditions generated various solutions in the early phases of China's unification. For instance, codes of law were written to facilitate the work

[13] This it was which led to a mass of peculation, 'squeeze, graft and corruption' at all levels which always upset the businessmen from the West who were accustomed to quantitative accounting. One of the first phrases I learnt in Chinese was *Ta kuan fa tshai* 大官發財, 'Become a high official and get very wealthy'. Since the government never paid all the outlying provincial officials adequately, they had naturally to supplement their incomes by 'taxing the Emperor at source', i.e. by 'taking a cut off the joint at all levels'.

[14] The bureaucratic character of the administration runs like a thread through the recent book of Chinese history by Huang Jen-Yü (1988), see pp. 15, 42, 60, 80, 89, 111, 143, 170, 200 and 203, but this meant throughout the ages that China was not in Huang Jen-Yü's words 'mathematically manageable'.

of the administrators, though in general such systems avoided that legal abstraction characteristic of Roman law, and every case was judged as a totality of circumstances, not in isolation from factors that might seem irrelevant. Since law was always thought of primarily as criminal law, the institution of advocates did not develop. Yet such was the passion for justice among the best Chinese officials that the oldest books on forensic medicine in any civilisation are due to China, for example, the *Hsi Yüan Lu* (The Washing Away of Wrongs), compiled in 1247 by Sung Tzhu 宋慈.[15] Administrative discipline was quickly re-established during changes of dynasty, when military régimes were transformed into civil government with a minimum relaxation of control. By and large it became the custom of the scholar-gentry to transfer allegiance to a new ruling house as soon as it had established by arms its claim to have received the 'mandate of Heaven' (*thien ming* 天命). The government itself always proclaimed moral conduct as its aim, and public education was mobilised to strengthen its position.

All these features, however, important as they were, could never compensate for the lack of adequate technical support in terms of transport, communication, banking techniques, accounting methods, information-gathering, and data-processing expertise – the means that any modern bureaucratic management would regard as indispensable. This is not to say that archives were not kept and censuses not taken: they were; but in spite of the Bureau of Historiography and the Chancellor's Secretariats, Chinese government organisation far outran the technique on which it had to be based. One remembers Chhin Shih Huang Ti 秦始皇帝, the First Emperor, reading through cartloads of bamboo and wooden slips daily, and the extreme delays to which official despatches were subject century after century.

Thus in the history of developing bureaucratic administration, China followed a pattern of 'forward three steps, but backward two'. The dilemma was that while an over-tight centralised system would create too much stress and strain, any decentralisation could start a liquidating process in which the entire organisation was liable to disintegrate. One sees this at work in every period of political fragmentation and particularly clearly when independent theocracies sprang up at the end of the Han, and when the provincial governors (*chien tu shih* 監督使) took over at the end of the Thang. But given time, reunion invariably followed, and even the foreign and colonialist attacks of the past century and a half could not distort this pattern, as we see in the centralised strength of the present day after the provincialist warlordism of the Kuo-min-tang 國民黨 period.[16] Here the ideographic nature of the language had always been of cardinal importance, for it did much to prevent the definitive

[15] See Section 44 in *SCC*, vol. 6, pt. 6.

[16] This can be well seen from the Table of Chinese Dynasties (Table 5 on p. 78 of *SCC*, vol. 1) repeated immediately after the index at the end of all the subsequent volumes of the series. Of the twenty-two centuries which have elapsed since the first unification of the empire by Chhin Shih Huang Ti 秦始皇帝, a period of no more than 250 to 337 years (according to how one interprets the word dynasty) has seen China not unified and under the sway of a centralised administration. This amounts to between 11.3 per cent and 15.3 per cent of the time that has elapsed since the first unification of the empire in −221. It is true that Chinese scholars have often gone about to prove that one or other of the contending dynasties in the periods when China was divided maintained the imperial rule and so handed down the succession, but that was a largely theoretical pursuit and may be disregarded here. The fact is that there were at most only 337 years when China was radically fragmented.

centrifugal separatism.[17] China had civil wars often enough, internal national conflicts never. And correspondingly there was to some extent an absence of the stimuli that war gave to science and technology in Europe.

Of course in dealing with economic problems different dynasties in China devised different schemes. Yet inasmuch as certain background factors underwent little change, a number of common features in the administration became perennial. One such feature was a vigorous and persistent promotion of agriculture by the State, virtually uninterrupted for 2,000 years. At the founding of each major dynasty the imperial government always busied itself with farm rehabilitation. Land, seeds and draught animals were distributed to displaced persons; reclamation of new land was encouraged, with tax relief as a premium; and improved forms of agricultural tools and rural machinery were popularised. At different times the government also promoted new food plants, disseminated information about improved farm methods, and conducted surveys of agricultural production. The execution of water-control and irrigation projects was always recognised as a vital function of the State. Moreover, even alien dynasties were equally aware that the economic foundation of the empire rested on agriculture. It was Khubilai Khan who authorised the compilation of the *Nung Sang Chi Yao* (Basic Elements of Agriculture and Sericulture), a handbook reprinted many times during the Yuan period. Its edition of 1315 was printed in 10,000 copies.[18] Here again there was a general contrast with the Roman Empire, which came to grief in part precisely because it did not give adequate attention to the promotion of agriculture.

Since taxes were collected direct from the general population, the State naturally regarded intermediate groups capable of intercepting the income of the primary producers with great disfavour. Energetic emperors in early Chinese history dealt with aristocratic households as vigorously as if inspired by socialist ideas. Both the Northern Wei and the Sui delivered fatal blows to the élite clans which dominated the period of disunity from the +4th to the +6th century. The Thang dynasty confiscated Buddhist monastic properties on a mass scale from time to time. In the last two decades of its life, the Southern Sung proceeded to purchase from owners on the eastern seaboard one-third of their landholdings in excess of 100 *mou* 畝 per household, but as the purchase prices were almost nominal the transaction differed little from confiscation. The founder of the Ming dynasty, a man of peasant origin himself, staged a series of expropriations of the powerful gentry families, during which no less than 100,000 persons seem to have lost their lives. After that, in 1397, the Ministry of Revenue submitted to the sovereign a list of remaining substantial landowners across the empire. The 14,341 names on it represented those who owned 700 *mou* of land or more. Their total holdings are not disclosed, but

[17] Latin was once the universal language of all Western Europe, but as soon as French, English, Spanish, Portuguese, Italian and Romanian began to be written down in the way that they were pronounced, the nation-states could arise, as they duly did. Cantonese and Fukienese were indeed different dialects but the ideographic written language was a unity, partaking of the nature of an algebraic script.

[18] This is the kind of thing which has led American agriculturalists to say that 'agricultural extension programmes' have been more characteristic of China for a couple of thousand years than anywhere else in the world.

both the minimum possession of those listed, and the number of households, give the impression that the entire 'upper' or 'middle' class could not have been a very formidable group.

Unfortunately, this is an area in which a great deal of misunderstanding has yet to be cleared up. The elimination of plutocratic influence by the imperial government was not always successful. During the later Han, in the early centuries of our era, the élite clans and regional landholding interests got out of control, and this resulted in a dissolution of the imperial order which ushered in the longest period of turbulence in Chinese history. At other times there were similar situations but on a lesser scale. There were occasions when indulgent emperors, acting against common sense, lavished land grants upon their favourites, as happened in the late Ming.[19] Contemporary writers, deploring such practices because they were contrary to their sense of sound government, protested vociferously. But it is rather too simplistic to interpret all this, as some modern scholars have done, as evidences of the re-emergence of a feudal age or a manorial system.

Of course class antagonism and class oppression existed in a sense in China, for the whole apparatus of court and gentry-bureaucracy hung like a yoke round the neck of the peasant-farmers, even though the mandarinate was not hereditary. Yet the differences between imperial China and medieval Europe were drastic and profound. Private landholdings of the size of baronial estates appeared in China as rare exceptions rather than a general rule. The gentry-landlords were too numerous and too scattered to act as an organised group; never did they openly and collectively claim their rights and demand the advancement of their common interests in the style of *Magna Charta*.[20] At times outstanding merchants commanded sufficient financial resources to influence the court, to put themselves on inside terms with the bureaucratic apparatus, and to evade the law; but they were never influential enough to press for legislative concessions from the government which could have facilitated their business transactions. From time to time the peasant-farmers rose in rebellions to redress their grievances; but they in their turn could never provide an organisational alternative to the imperial order. Technically it would have been impossible for the body politic to maintain its high degree of centralisation and yet at the same time promote particular economic interests. The practice of admitting socio-economic groups to the government as 'estates' was never attempted in China; in fact, the traditional Chinese State often operated clean, contrary to the principles laid down by Harrington in his *Oceana*.[21] It demonstrated its strength precisely by holding dominant economic groups at bay; the moment private parties were able to

[19] See Huang Jen-Yü (1988), pp. 149 ff., 164 ff.

[20] This charter is better known today as Magna Carta, but *Carta* is not a classical Latin word; it was an old English word in use, before the Norman Conquest, later believed to be Latin, and dropped from English. The word is derived from the Greek *chartes* meaning a leaf of papyrus. The Latin form was *charta*. The *Oxford English Dictionary* refers to the *Magna Charta* and the Encyclopaedia Britannica to *Magna Carta*.

[21] *The Common-wealth of Oceana* (England) was a political romance by James Harrington published in 1656, perhaps as a reply to the *Leviathan* of Thomas Hobbes (1651). At the head of the State is a prince or Archon, elected like all the other ministers by the people, who live in a condition of freedom and equality, detesting war. Property and land is limited 'so that no one man or a number of men . . . can come to overpower the whole people'. Hobbes' book,

convert their economic power into political power it was on the verge of collapse. So one could say that neither Western aristocratic-military feudalism nor Western city-state mercantile independence found any counterpart in traditional Chinese society.

Thus while feudal Europe with practically no disguise converted public affairs into a mosaic of private domains, the Chinese system was permeated homogeneously by a sort of public spirit. Its drawback was that this public spirit was sustained only at the vigilance of the Emperor. When this failed, the system permitted a horde of persons associated with the government to exploit the poor and inarticulate. The structural weakness of the Chinese system lay in the fact that between the imperial authority at the top and the mass of tax-paying population below the mid-echelon level, there was an administrative vacuum. Local government was always under-staffed. In order to prevent sub-systems from arising within the imperial order, regional autonomy was ruled out. With but few exceptions in Chinese history, the State also persistently refused to enlist the assistance of mercantile groups which might have helped to bridge the logistic gaps. Services provided by the merchants, as a rule commandeered rather than accepted in the form of partnership co-operation, were never influential enough to alter the style of the administration. In a word, therefore, there arose no middle class, or, if you like, no bourgeoisie, however much wealth individual merchants were sometimes able to acquire. And they were never encouraged, indeed most of the time positively discouraged, in projects of industrial capital investment – always the fortunes were swallowed up by the land market and the purchase of entry by one means or another into the ranks of the scholar-gentry bureaucrats.

On all these grounds the precision and standardisation characteristic of a modern bureaucracy were unattainable. The entire establishment lacked structural firmness. As the centralised financial system continued to take hold, fiscal authority was retained at the top, yet fiscal responsibility resided with the lower echelons, so that irrational features in the system were driven down towards the operational level, causing grave discrepancies between theories and practice.[22] Undoubtedly this was also one of the basic reasons for the authoritarian tone of government in traditional China. Officials often found that directives from above could not be questioned, nor could they see their own orders challenged by the populace, because if enquiries had been constantly conducted in the manner of the judiciary reviews of the Western world, the entire mechanism of the imperial government would have been immobilised. It is clear that such a system precluded the performance of complicated afferent and efferent functions.

The Leviathan of the Matter, Forme and Power of a Commonwealth, Ecclesiastical and Civil, 1657, was a treatise of political philosophy, much more *fa-chia* 法家 (legalist) in character. Hobbes believed that man is not, as Aristotle held, naturally a social being, but a purely selfish creature, seeking only his own advantage, and resisting the competing claims of others. The life of man was 'solitary, poor, nasty, brutish and short'. To escape from such intolerable conditions, mankind has adopted certain 'articles of peace'. All men must enter into a contract 'to confer all their power and strength upon one man or upon an assembly of men'. This union is called a commonwealth. See further in the *Oxford Companion to English Literature.*

[22] On this cf. Huang Jen-Yü (1988), pp. 149 ff.

As time went on there grew up increasing resistance to any diversification of the national economy. In carrying out its vital functions such as providing frontier defence, suppressing internal rebellions, executing major public works, and providing famine relief, the imperial government concerned itself mainly with the fostering of man-power and food supplies. Quantity rather than quality tended to be the key factor.[23] Surrounded by an enormous agrarian economy, it was assured of thorough mobilisation, and there was little to be gained by pushing the more advanced sectors, such as mining or sea trade. Moreover, the expansion of industry and commerce, which necessarily tended to develop in particular geographical areas, could only too easily produce those regional imbalances which the bureaucracy, educated and trained to manage a society of agrarian simplicity, would find itself unable to handle. One can never forget the rungs of the traditional social order, *shih* 士, *nung* 農, *kung* 工, *shang* 商, the farmers second only to the scholar-gentry, the artisans next and the merchants taking the lowest place. It repeatedly happened in Chinese history that dynasties arose with robust energy out of a crude and backward economy, only to lose their vitality and turn decadent after guiding that economy to a higher stage of development.

The nub of the matter is that here was a vast agrarian society which found itself, in a sense prematurely, under the centralised administration of bureaucracy; and the size of the economy made it non-competitive.[24] Stability was always valued more than change or progress. And yet industrial and technical development often went ahead of the West. The strange thing is that China was able to absorb these earth-shaking discoveries and inventions while Europe was gravely affected by them, as has been shown in detail in previous volumes. Gunpowder weapons made relatively little difference to the fighting in and around China, while in Europe they ruined the feudal castle and the armoured knight. Stirrups, also Chinese, had brought him into existence in the first place, but mounted archery in East Asia continued as before. The magnetic compass and the axial rudder permitted Europeans to discover the Americas, but Chinese sea-captains pursued their peaceful ways in the Indian Ocean and the Pacific as of old. Printing helped to launch the Reformation and the revival of learning in the West; all it did in China (apart from preserving a host of books which would otherwise have perished) was to open civil service recruiting to a wider range of society.[25] Perhaps there has hardly ever been so cybernetic and homeostatic a culture as that of China, but to say that is by no means to speak of 'stagnation', as so many Westerners have done; the rate of advance in China simply continued at its characteristic rate, while after the scientific revolution in Europe change entered an exponential phase.

[23] This can be traced back to the time of Chhin Shih Huang Ti and the first unification of the empire. See Huang Jen-Yü (1988), ch. 2.

[24] Moreover, competitiveness was always one of the values placed low down in the scale by every Chinese philosopher.

[25] For the chapter and verse of all these, see on gunpowder, *SCC*, vol. 5, pt. 7; on cavalry techniques, *SCC*, vol. 5, pt. 8; on the compass and the rudder, *SCC*, vol. 4, pt. 3; for printing, see *SCC*, vol. 5, pt. 1.

(3) Science, Technology and the Under-development of a Money Economy

China's early advancement in social organisation had as its counterpart a retarded Europe. Conversely, her lack of substantial social progress after 1450 contrasts with the great movements in the West that brought the modern world into being, the Reformation, the rise of capitalism, and the scientific revolution. But in considering the birth of distinctively modern science it is always necessary to remember that apart from the heights of Greek achievement, science and technology had consistently reached a level higher, and often much higher, in China than in Europe, during the previous twenty centuries. Of course, science and technology must always have been affected by the society in which they flourished. Thus the suggestion has been made that the greatest Chinese inventions, including paper and printing, hydro-mechanical clockwork, seismographs and advanced astronomical and meteorological instruments (such as rain-gauges), together with segmental-arch bridges and suspension-bridges, were all in one way or another useful to a centralised bureaucratic State.[26] On the other hand, gunpowder, the magnetic compass, the axial rudder, fore-and-aft sails, and paddle-wheel propulsion would have been even more likely to arise, one might have thought, in maritime mercantile cultures than in the milieu where they actually did.[27] Yet there certainly was some brake on technical developments in this society which was itself so intrinsically fertile in creating them. The gear-wheel, the crank, the piston-rod, the blast-furnace, and the standard method of interconversion of rotary and longitudinal motion, all existing earlier, in some cases far earlier, in China than in Europe[28] were doubtless utilised less than they might have been because they lacked application to the needs of that agrarian society which the bureaucracy was determined to protect and stabilise. In other words Chinese society did not always succeed in moving from invention to innovation, Schumpeter's term for the wide application of an invention.[29] There are even not a few instances where discoveries and inventions were allowed to die out, for example, in seismology, horology and some developments in iatro-chemistry.[30]

The organisational ethos may also have affected scientific thought. Chinese thinkers preferred to see the universe as an organic whole, recoiling from the analysis of the inner mechanisms of its parts, and steadily refusing to draw a clear distinction between the spiritual and the material. There were great strengths in this, as modern science is only now coming to recognise, but there were serious weaknesses too, especially on the heuristic side. Furthermore, there were particular vices of

[26] On these, see Section 32 in *SCC*, vol. 5, pt. 1; Section 27 in *SCC*, vol. 4, pt. 2; Section 24 in *SCC*, vol. 3; Section 25 in *SCC*, vol. 3; and Section 28 in *SCC*, vol. 4, pt. 3.

[27] On the first of these see Section 30 in *SCC*, vol. 5; on the second see Section 26 in *SCC*, vol. 4, pt. 1; on the third, fourth and fifth, see Section 29 in *SCC*, vol. 4, pt. 3.

[28] On the first, second and third of these, see Section 27 in *SCC*, vol. 4, pt. 2; on the fourth see Section 36 in *SCC*, vol. 5, pt. 12; and on the fifth see Section 27 in *SCC*, vol. 4, pt. 2.

[29] See Schumpeter (1939), vol. 1, p. 84.

[30] See Section 24 in *SCC*, vol. 3; Section 27 in *SCC*, vol. 4, pt. 2; and Section 33 in *SCC*, vol. 5, pt. 4.

a bureaucratic character in Chinese scientific thought,[31] notably that pigeon-hole repository of abstract concepts furnished by the *I Ching* (Book of Changes), only too readily hypostatised into real and active working forces.[32] Nevertheless, the Chinese record was a glorious one, whether in mathematics, astronomy, acoustics, magnetical science, proto-chemistry, or botany and pharmacology.

But may we not accept here a clue from the great Nicholas Copernicus, Canon-Treasurer of the Cathedral of Frombork on the Baltic, who wrote not only his world-shaking *De Revolutionibus Orbium Coelestium* but also, at the invitation of King Sigismond I of Poland in 1526, an important treatise on money, the *Monete Cudende Ratio*, in which he anticipated the statement of Gresham's Law that bad money drives out good? Apart from all speculation about the possible influences of bureaucratic society upon the shape of science and technology, let us see whether in concrete terms the lack of sustained progress could have been connected with the under-development of a money economy. Could this be one of our better measures of that inhibition which prevented the harnessing of China's multifarious inventions and brilliant scientific insights for the public good? Of course we recognise that money is in itself a secondary phenomenon, a yardstick, as it were, because if the entrepreneurial urge had been permitted in Chinese society, a true money economy must have arisen. But although money reflects the other forces at work in the economy, it may in turn exercise profound influence upon them. The fundamental cause of the under-development of a money economy in China was that the imperial government was never so efficient as to be able to invest agrarian surplus in the modern sectors of the economy, yet it was quite efficient enough to prevent private parties from doing so.

The assertion that money was a great limiting factor may at first seem strange, since China has a history of bronze coins in circulation covering more than 2,000 years. The so-called five-*shu* 銖 money was designed to serve the general public, and at the high point of its production under the Northern Sung, more than five billion such coins were minted every year. China was also the first nation in the world to have a national paper currency, or *fei chhien* 飛錢; 'flying money', the earliest letters of credit in any culture, appeared even under the Thang.

But the difference is that a modern monetary system is neither absolutely real nor entirely abstract. Its formation depends both upon governmental regulation and public participation. Inseparable from credit, its general acceptance increases the alienability of property, and enables wider borrowing which puts idle capital to work. Accelerated economic activities first use up under-employed labour, then cause a demand for labour-saving devices. The modern monetary system which grew up under capitalism has certainly proved a useful instrument for the advancement of technology. To be sure, we have encountered hardly any instances of labour-saving inventions being rejected in ancient and medieval China for fear of technological unemployment. But money in traditional China certainly did not fulfil all the above-mentioned functions.

[31] See Section 38 in *SCC*, vol. 6, pt. 1, pp. 10–11 and 307–8.

[32] See Section 13 in *SCC*, vol. 2, pp. 304 ff.

Despite its circulation of bronze coins, the Chinese government never ceased collecting taxes in kind, and requisitioning materials and labour services (*corvée*) from the people. It usually also paid officials and soldiers in grain, while rentals of tenant farmers were paid almost exclusively in kind. Large business transactions, furthermore, used bales of cloth, silk fabrics and precious metals as media of exchange. Copper coins therefore had only a limited usage, never backed by public or private assets, and their value was largely determined by their actual metal content.[33] The bullion theory of money became so firmly established in traditional China that it was virtually impossible to deviate the face value of the bronze coins from their net worth. When the former was higher than the latter, counterfeiting became impossible to stop, and the public refused or discounted the coins. When the former was lower than the latter, then users remelted the coins for profit.

The circulation of paper currency involved the other extreme; it was, one might say, too abstract. The *chiao tzu* 交子 under the Northern Sung and the *chung thung chhao* 中統鈔 under the Yuan are said to have been backed by adequate reserves; but the total amount of the former and the initial issue of the latter were so small that they could hardly qualify as national currencies. When the circulation of the paper notes involved a substantial sum, there was no reserve, as no such treasury deposit was available. These cases also illustrate the utter inadequacy of relying on governmental income alone to back a nationwide currency. The tool simply did not match the job.

When the paper money was issued without backing, the theory of money seems to have been close to that advocated by some modern economists, who hold that the real source of the value of currency is an 'acceptance' which can be developed into a 'pattern of behaviour'. Yet in the Western world paper currency traces its origin to the goldsmiths' receipts and demand deposits at the banks, which in the 17th and 18th centuries facilitated the organisation of private capital. In the last stage of development the funding of a central bank, the regularisation of the national debt, and the establishment of business law, aligned both public finance and the nation's wealth behind a unified currency, until all transactions were under its jurisdiction. Tax revenue, in turn, grew in proportion with money supply and national wealth. The process, characteristic of developing capitalism, became powerful and irreversible. There was probably no way for China to omit all these steps and yet develop before the 20th century a successful semi-abstract money. Bronze coins relying for value on metal content alone could not suffice, paper notes backed solely by imperial order could not fit the bill either; for there was no public participation in forming a currency of the necessary degree of abstraction. To claim, therefore, as many scholars do, both Japanese and Western, that medieval China saw 'revolutions in money and credit'

[33] Those accustomed to the history of Western European countries are often impressed by the tenacity with which the Western governments kept control over the minting process, while in China it was 'privatised' as early as the Shang. Moreover with few exceptions coins in China were generally cast and never 'struck'. See *SCC*, vol. 1, Section 6, pp. 107, 109, 139, 144; Section 7 pp. 243, 246–7.

is quite misleading, since they never explain why these were not followed by basic institutional changes. Only European capitalism brought these to birth.

When the last issue of paper currency failed under the Ming in the 15th century, China adopted unminted silver as a common medium of exchange and a general means of payment.[34] But this was to subject the population to the tyranny of the hardest currency of all, similar to the situation which would pertain nowadays if consumers were required to buy their petrol and pay their grocery bills with chunks and bits of gold. The government had no knowledge of the amount of money in circulation, still less the ability to readjust its volume. Wealthy families buried the bullion underground for safe-keeping, while the well-to-do lavished gold ornaments and silver utensils on themselves, knowing that the precious metals could be converted to cash whenever needed. Interest rates were high; the most developed credit institutions were the pawn shops (*tang phu* 當鋪). In the late 16th century China had 20,000 of these, and even in the late 19th century 7,000 of them were still in operation.

Certainly the under-development of credit was linked to the absence of rigorous commercial law.[35] But such laws on the Western model could hardly have been enacted in traditional China. Their enforcement, through an independent judiciary, would have involved the acceptance of property rights as absolute, contrary to China's social values and the principles of her social organisations. These values and principles supported the political structure of traditional China, which had been founded on the premise that public well-being must always take precedence over private interest. If that had been abandoned the bureaucratic administration of the far-flung empire would have broken down completely. Of course honesty is better than legal sanctions, but whether modern society could have grown up on that basis alone is very debatable.

(4) The Social Consequences

At least three times in Chinese history trends towards a money economy made upsurging movements, during the Han, under the Sung, and then under the late Ming. Chinese historians often like to call the great Han industrialists 'capitalist *entrepreneurs manqués*', and to identify the late Ming movement with a 'budding capitalism'; Japanese scholars, followed by some Western sinologists, have called the turning-point under the Sung a 'renaissance' and a 'commercial revolution'. Such terms cannot be accepted without misgivings. And in any case the movements were all abortive. Our view simply is that there was a fundamental institutional incompatibility between the centralised bureaucratic administration of an agrarian society

[34] This was the tael or shoe-shaped silver ingot. It was still in action in Shanghai as a medium of exchange after the First World War.

[35] As is generally recognised Chinese law was overwhelmingly criminal law. See Lu Gwei-Djen and Needham (1988) on the history of forensic medicine in China. Civil and commercial issues were 'settled out of court' within family conclaves.

and the development of a money economy. Without massive commerce and industry it could never be born.[36]

There is no argument that the development of rice-producing areas in the south and the improvement of water-transportation during the previous centuries brought forth phenomenal prosperity in the Sung. Similarly, the importation of Spanish-American bullion and the spread of cotton-weaving boosted the economy on China's seaboard before the end of the Ming dynasty. But there were no basic organisational revisions in the system of inter-provincial commerce which could have produced a qualitative change throughout the country.

Modern commercial practices feature a wide extension of credit, impersonal management, and co-ordination of service facilities. None of those conditions existed under the Sung or under the Ming. Contemporary literature rarely gives any direct description of business practices, but a large volume of fictional writing provides a rather complete picture which can be checked for consistency with other sources. Until the 17th century the most important person on the trade route was the travelling merchant, who usually kept his entire capital liquid. He normally employed one or more accountant assistants, but his own presence in business transactions was indispensable. Hence the importance of the *hui kuan* 會館 or commodious rest-houses maintained in one province by the merchant guild of another.[37] Buying and selling in a city or market town was handled by a residential agent, who also provided lodging and storage space. The implication was that there were no regular suppliers and outlets, and manufacturing was thoroughly scattered. Cotton-weaving (until the rise of factories in the treaty ports towards the end of the 19th century) remained a cottage industry, providing extra income to peasant households. Commodities such as tea and lacquer had to be collected from the village producers with the travelling merchants waiting, and extension of credit was a personal favour under unusual circumstances rather than a general practice. Although the long-term association between travelling merchants and local agents, sometimes continuing through several successive generations, developed into a fraternal bond, their operations never merged nor could the relationship be institutionalised. Purchase by correspondence was not practised; standing orders and automatic deliveries of goods were unheard-of.

The picture suggests that without a credit system it was impossible for any individual to extend his business operations beyond his personal presence. Financing of manufacture was fragmented, and the cash-and-carry policy covered merchandise of every sort by the piece. Yet despite the relatively crude commercial framework, the capitalisation of some of the travelling merchants was impressive; before the end of the 16th century, the carrying of something like 30,000 ounces of silver by an individual on a business trip appears to have been not altogether unusual. Because

[36] Commerce and industry there certainly were in medieval China but they never reached the volume and intensity that they had in early capitalist Europe.

[37] I can speak from personal experience of the commodious resources maintained by the merchants in provincial towns where they were not at home, because my party and I were often accommodated at one of these during our travels all over China during the Second World War.

of the size of China's national economy and the concentration of consumer markets at political centres, descriptions of major cities, whether by Chinese or Westerners, always create an impression of richness; but their favourable observations could not alter the fundamental fact that the operational characteristics gave no indication of a major breakthrough in inland trade.

Another inhibiting factor for internal trade in old China was the lack of public safety corresponding to conceptions of 'law and order.'[38] Apart from the activities of bodies of organised bandits along the trade routes, carriers and watermen were often bandits thinly disguised, while the cities and towns always harboured large numbers of *liu mang* 流氓 individuals, unemployed or semi-employed, certainly under-employed, and not averse from turning a dishonest string of cash if opportunity offered. In times of civil war, naturally, all this became confusion worse confounded, and even when peace was supposed to be reigning, the *yamen* 衙門 runners, heralds, clerks and gaolers were far from sufficient as a police force either in the town or the surrounding country.

Besides, not content with other taxes, the government levied a transport tax or inland customs duty (the notorious *likin* 釐金 of the 19th century) on all commodities passing along the trade-routes whether by road or water-way. This institution never matured into a form compatible with modern commercial practice – the operations of the customs posts remained un-coordinated, no director-general was ever appointed to integrate the collections, the accounts were never audited, and peculation at all levels of the bureaucratic management was normal. Mindful only of the income, governments never attempted to facilitate the circulation of nationally important daily necessities while taxing luxuries more heavily, no distinction being made between staple goods transport in large volumes, and minor items. Although the published duty rates were universally low, the collection was repeated at unreasonably short distances along the chains of inland way-stations. So the universal transit tax never developed into an instrument for the control of commercial transactions.[39]

Thus business risk was high; and the possibility for expansion severely restricted. The temptation for exceptionally successful merchants and their heirs to withdraw from trade was therefore great, all the more so because social prestige pertained alone to the educated scholar, able to join the charmed circle of the 'gentlemen' of the civil service. Who would want to be a wealthy parvenu if classical studies could give him gentility? It depended on what one's aims in life were; going into business and making money was just not the way of life classically most admired in China. Hence mercantile profits always tended to be used for acquiring land and real estate property, or in buying luxury goods, or even for the purchase of positions in the civil service, whenever this was possible; not for industrial investment. The situation improved little in later times. The famous draft banks of Shansi merchants became prominent only in the 18th century, and the users were limited to a rather closed

[38] This went so far that it was the custom in well-to-do families to sew gold bars into the clothes of children so that if they got lost or carried away, they might be able to turn it into money and come home again safely.

[39] On 'squeeze' at all levels see above, p. 48, n. 13.

group. Business commitment, as in the collection of debts, was still always honoured on ethical principle, not enforced by law; while the financing of manufactures was still shunned by men of substantial means. All this emphasises yet once again that China was a society which for 2,000 years denied to merchants a leading role, whether as bankers or *entrepreneurs*, in the affairs of State; and if it remains true that the rise of modern science and technology in the West was connected with the rise of mercantile and then industrial capitalism, then once again we may be nearing an explanation of why that phenomenon happened only in Europe. There was a built-in ideological anti-commercialism in Chinese life,[40] and it seems that if a Fugger or a Gresham could not have arisen in it, a Galileo or a Harvey could not have come into being either.

Perhaps the most fruitful line of development that ought to be taken by Chinese history would concentrate attention upon the more economic aspects of Chinese inventions and contrivances. Why did the simple invention of the wheelbarrow[41] for inland transport flood through the empire without hindrance almost as soon as it appeared in the San Kuo period (+3rd century), nearly a millennium before Europe had it? Why did the enormous quantitative upsurge of iron and steel production[42] in the two Sung dynasties (+10th to +13th centuries) not lead to some kind of industrial capitalism? When one knows that water-power was very widely applied to textile machinery in the 13th and 14th centuries (challenging comparison with what happened in Europe in the 18th), and that those same spinning, doubling and twisting machines must have inspired the Italian silk industry of very little later, one is irresistibly impelled to ask why factory production did not quickly ensue?[43] We really do not have as yet all the full answers. In the past the magnetic mesmerism of the bureaucratic ethos on the successful merchant families has occupied a good deal of attention; now, besides this, we must think of factors like the failure of a free surplus labour market to develop, the failure to open up mass overseas outlets for Chinese goods and to integrate them with domestic production, the failure to widen capital expenditure with the formation of a factory system, and all reflected in the inhibition of a money economy. Much closer analysis is needed of why social innovation only sometimes followed upon invention. The square-pallet chain-pump, for example,[44] like the wheelbarrow and the co-fusion method of steel-making, spread everywhere within a few decades, but on the other hand the invention of hydro-mechanical clockwork, so brilliant an anticipation in the +8th century of the European invention of the +14th, never found application on a mass scale, and mechanical time-keeping remained confined to the imperial governors.[45] In truth the bureaucratic State had its own impetus, and if social stability was more desired than economic gain or

[40] This may be thought to have been connected with the ancient Chinese dislike for competition.

[41] See Section 27 in *SCC*, vol. 4, pt. 2.

[42] See Needham (1958b), *The Development of Iron and Steel Technology in China*, first published by the Newcomen Society in 1958 and a second edition by Heffers, Cambridge, in 1964. This monograph was the result of collaboration between my first collaborator, Dr Wang Ching-Ning (Wang Ling) and myself.

[43] See *SCC*, vol. 5, pt. 9, esp. pp. 225 ff. and 404 ff.

[44] See Section 27 in *SCC*, vol. 4, pt. 2.

[45] See Needham, Wang & Price (1960), reprinted by Cambridge University Press in 1985.

general prosperity this might be only another way of saying that it was more to the advantage of the court and the bureaucratic scholar-gentry to maintain the basic agrarian social structure than to engage in, or even permit, any forms of commercial or industrial development whatsoever.[46]

One could perhaps say that Chinese inventions were often labour-saving in an agrarian society which needed labour to produce food in order to satisfy the fundamental requirements of the very sizable population, but they could hardly become the starting-point of radical changes in production because while farm labour was often seasonally under-employed, the surplus man-power was hardly transportable. The Chinese bureaucratic State organisation generally collected taxes and paid its apparatus in kind, and that proved to be incompatible with the accumulation of surplus capital in private hands. This then reinforced the immobility of the agrarian society, hindered the development of a diversified social stratification, and retarded the stimuli for the commercialisation of agriculture and the development of productive forces, despite the existence of a tremendous reservoir of technical skill and inventive genius.

Furthermore, lack of investment impeded the exploitation of the nation's natural resources. Though little mentioned in view of the seeming prosperity of the late 16th century, numerous mines throughout China were at that time abandoned. Previously a number of mines under both private and public ownership had failed; and the miners, receiving no compensation, had been obliged to take up banditry. In 1567 it actually became a national policy to seal off mines in the provinces. From Chekiang to Shantung, the sites were guarded by soldiers, the population in the adjacent areas was relocated, and even roads leading to the mines were destroyed. Inadequate financing also affected salt production, an industry long under state monopoly. At the beginning of the Ming, salt producers evaporated the brine with huge iron pans measuring about twenty feet in diameter, but before the end of the dynasty most of those had disappeared, and bamboo baskets lined with alkali-treated paper were used in their place. Since the government itself was under financial pressure, it over-sold the future stock of its salt production to the merchants as a means of deficit financing, sometimes over a ten-year period. The accrued interests over the pre-payments were so high that the stock was no longer profitable. Now and then the sale came to a complete halt, and in the inland provinces acute salt shortages were felt. Unlike Europe and Japan, where banking consortia made advances when the government was under a credit pinch, the government of imperial China had no regular source to turn to, and often the result was that public services were curtailed.

The overwhelming volume of evidence concerning official abuses and governmental corruption in the Ming should not distract us from observing that the normal income of the government around 1600, at less than thirty million ounces of silver, was a rather small budget. Down to 1900 the annual revenue of China under the Chhing was close to one hundred million ounces of silver, comparable to the level

[46] For the ancient objection of the Chinese to 'competition' see Hallpike (1986), pp. 319 ff.

in England in 1700. But Great Britain in 1700 had a population less than 2 per cent of China's in 1900. The different approaches to the monetary problem may account in part for the difference.

The most sombre aspect of the story is that the proceeds from agricultural surplus, given no proper outlet, tended to return to landholding and agrarian exploitation. It is an over-simplification, reflecting even a degree of naïveté, to say that the exploitation only took the form of excessive rents demanded from tenant farmers by a few landlords. If that were the case, the revolution in China would not have become such a painful and prolonged struggle, nor could Western and Japanese imperialism have made inroads in China to the extent that they did. The exploitation was in fact carried out both on large and small scales, it spread over all levels in village life, and very often by those who had economic and social backgrounds quite similar to those of the exploited. Rents represented only one form of exploitation; other forms were usury, mortgage and share-cropping. Extant contracts testify that at least in the 16th century such practices were already quite common, sometimes cousins holding liens on parcels of land belonging to cousins, with interests on loans as high as 5 per cent monthly. Documentation during land reforms in the 1940s disclosed that the same practices were still being carried on then.

The form of borrowing and lending was consistent with small-scale financing in traditional China. The contracts, drawn up in age-honoured tradition, became binding when witnessed by the village intermediaries, without governmental sanction. Dispossession of the agricultural land in question took place as the fortunes of individuals changed, providing a high degree of social mobility in an agrarian setting. In the past 500 years both China's population and her cultivated area have increased tremendously, and there can be little doubt that the expansion was in part financed by this type of inter-household loan. But in the light of the national economy, all these channelled the agricultural surplus to support the smallest possible scale of farm operation, applying the greatest pressure to the cultivators who were least equipped to make technological improvements. The wages of the primary producers remained at subsistence level, and the idling of a large number of exploiters and partial-exploiters presented a more substantial loss to the economy than did the landlords. In other words, the whole continued process worked towards a large population, a general lower standard of living, even more scarcity of capital, and the grave economic and social problems which, exacerbated by colonialist incursions, agonised modern China.

In a way, ironically, it was just because the agricultural technology of the Chinese countryside was so successful that the path lay open for increases in productivity to be swamped by growth of numbers of the labour force. There was therefore no incentive for further mechanisation, as there would have been if a scarcity of labour relative to capital accumulation had been present to stimulate technical progress. But even so, the advances in technology which would have brought Chinese agriculture into the modern world are not thinkable without the appearance of modern science, and thus we are brought back again to the problem of the origin of that in Europe.

(5) Europe's Past

Capitalism in Europe was a phenomenon undoubtedly the product of many factors. Military-aristocratic feudalism, in spite of all appearances, was weak and indefensible just where Chinese feudal bureaucratism was strong. Geographically Europe was an archipelago, studded with city-states in maritime trade and not averse from foreign conquest; moreover it existed in close proximity to another region of mercantile interests, the lands of the Arabic-speaking peoples. Greek science, reaching the Latin world through them, afforded a large part of the basis on which modern science was to be built (not all, because there were Chinese contributions); and inspired thereby, a manageable mathematics developed, with all its implications for accounting, banking and navigation.[47] The success of the Reformation involved a decisive break with tradition, and Europeans were not slow to reach the conclusion that there could in fact be real change in history, and that the Lord would truly make all things new. Protestantism, with its direct access to God, meant literacy, and what began with the necessity of reading the Scriptures ended in the hitherto unheard-of phenomenon of a really literate labour force, so that the class barrier of the written word was swept away, and managers, engineers, artisans and workmen could shade into one another without sharp distinctions. Europe after the Renaissance could almost be thought of as a heap of tinder, and an 'industrial revolution' was bound to follow.

It is interesting that the establishing of modern banks followed the path and timetable of the Renaissance, starting in the Italian cities, extending to northern Europe, and then spreading to England. The rising national States, unhampered by the Chinese type of bureaucratic centralisation, quickly shifted their emphasis from agricultural wealth to the development of revenue from industrial and commercial sources. The organisation of national debts and joint-stock companies ensued.

Once the movement of modern science had started, a train of circumstances and events also followed, and it is possible to trace how from one discovery and invention to another the modern world came into being in Europe and North America. The people of those regions were given extraordinary cues from Nature, and the chances were accepted. Before the end of the 13th century the British Isles generated important trading cities, especially London, and owing to their latitude there was great need for fuel. Open-seam coal was won at least as early as that, but it could not suffice. Then came the discovery of atmospheric pressure in Renaissance times,[48] and by the 18th century the draining of mines, previously an insuperable limiting factor, could be accomplished thanks to the inventions of Savery and Newcomen (Fig. 9). Completed by Watt, the steam-engine brought the industrial revolution into being, spear-headed by the textile industries, especially cotton, and gave rise to the steam-ship and the railway. The new cotton industry in particular was assured of an almost infinite market. The steam-engine then reacted back on science with the

[47] The importance of the conception of the quantitative cannot be minimised here. It must certainly be regarded as one of the roots of modern science. 'How much?' and 'how many?' are vital.

[48] The realisation that everything was subject to a pressure of 14 lbs per square inch led the way to the vacuum pump and the steam-engine. See Section 26 in *SCC*, vol. 4, pt. 1.

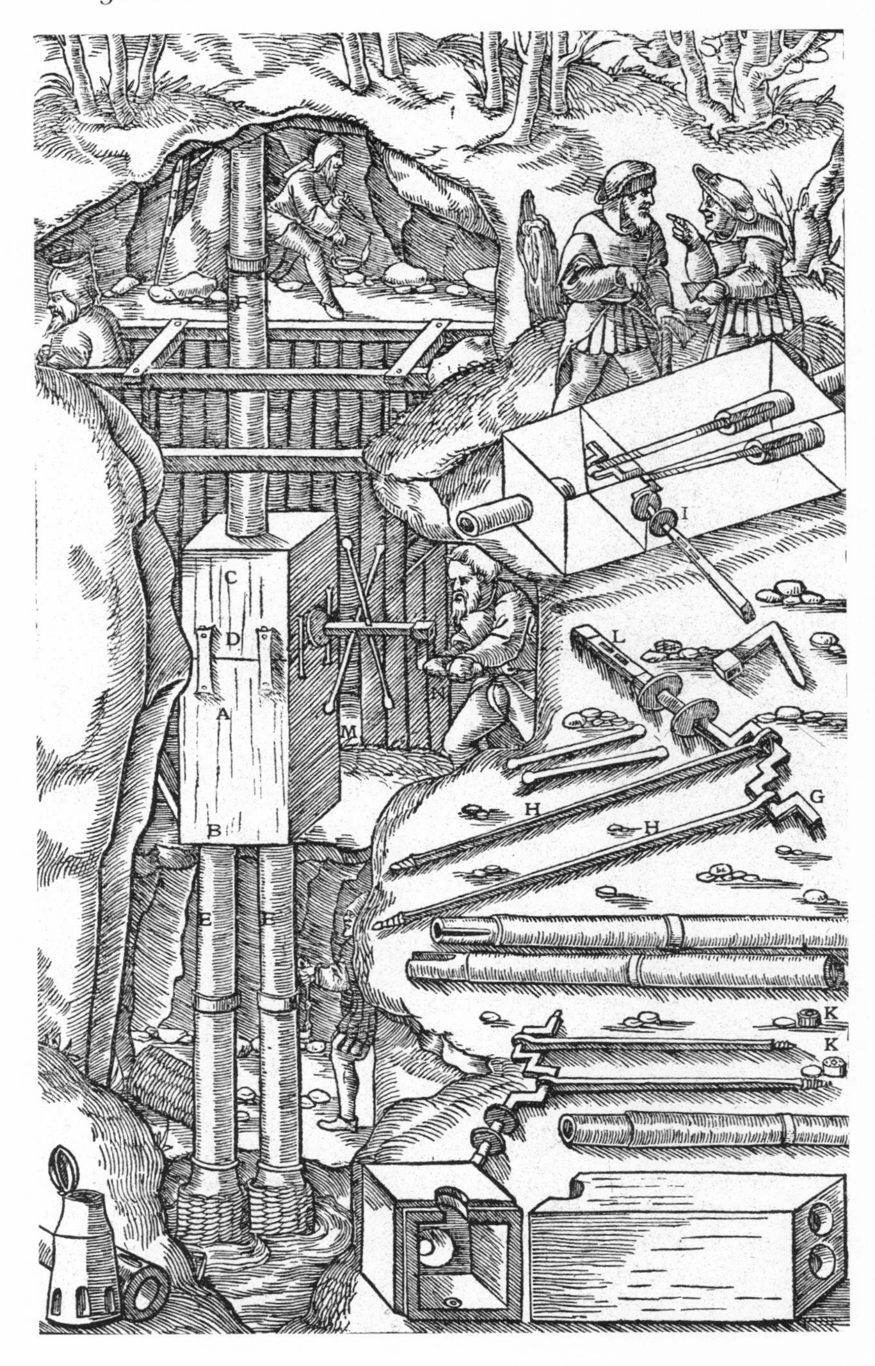

Figure 9. Miners fitting a force pump. From Agricola's *De Re Metallica*, p. 180. See *SCC*, vol. 4, pt. 2, pp. 135 ff.

development of energetics and thermodynamics. In turn the 18th-century science of electricity found employment as electrical engineering, able to provide cheap and convenient artificial light and tractive power from central generating stations.[49] Finally came petroleum engineering, also giving light at first, and lubricants too, but then with the internal combustion engine affording simple prime movers that could work unattended and drive in all directions, the small carriages arrived with which we are so familiar. Coal, iron and oil were thus the real wealth of the Western world, far more so than gold and silver, and the peoples of the Western world were fortunate in that vast supplies of these natural endowments were to be found under their territories. Without them, modern science would have been nugatory. With them, but lacking modern science, as in China, new forms of society could not come to birth.

(6) China's Future

With the above exposé as a background, it might seem that the world ought to have advised China to separate herself from her past and copy the West in earnest. In fact such proposals gained currency towards the end of the 19th century and were even accepted by a handful of Chinese intellectuals. History however has shown them to be fraught with problems.

China did not enshrine the concept of the ownership of property as an inalienable right, not because her philosophers were incapable of conceiving the notion, but because the ideal showed itself as unworkable within China's physical environment. China abstained from developing an independent judiciary, not because the Chinese were by nature contemptuous of law, but because in their history no stalemate between equal citizens of city-states, or between king and feudal barons, calling for the arbitration of the jurist, ever developed. The lack of capitalist enterprise among the Chinese in late medieval and modern China was perhaps due to the conviction that political stability was a much greater good than economic gain. To be sure, Chinese merchants never lacked the virtues of initiative, honesty, thrift, capable accountancy and ingenuity, as has been abundantly shown by their success as businessmen, outstanding beyond their neighbours, in all the countries of the 19th-century *hai-wai* 海外 (overseas) diaspora; but there the government and its difficulties were the care of others. What this background really indicates is that the solutions to China's problems are far more likely to be found from readjustments of classical motifs within rather than imitations of the world outside. A socialist form of society would seem to be more congruent with China's past than any capitalist one could be, hence in part the reason for the victory of the *Kung-chhan-tang* 共產黨 (Communists) over the *Kuo-min-tang* 國民黨 (Nationalists) in our century; but now it will be likely to generate

[49] It is difficult nowadays to think back to the time before there were power sockets on the walls of our living-rooms. An early science-fiction novel by Bulwer Lytton pictured a universal source of energy named vril which could be taken off from such sockets as a source of light and power.

managers on a totally different and higher plane. Once China finds the solutions to her problems, her experience should be invaluable to the rest of the world.

It would be unwise indeed for them to abandon their assets and adopt modes of life which have always been foreign to them, but a realistic appraisal of the situation deters us from concluding that all problems have been solved. The technique of advancing a modern economy is to take advantage of situations of imbalance, not to build up an artificial balance. Transportation and labour costs, material availability, and market conditions determine the feasibility of an undertaking. Such considerations can be put off when necessary in a socialist country, but disregard of short-term profit usually arises because of emphasis on long-term aims, or to serve non-economic purposes. How to manipulate the priorities and serve divergent needs is a problem that is taxing China's best brains today. There is no existing model to follow either, as a problem exactly matching China's in nature and dimension has not arisen before. The next step to stimulate speedy economic growth may well be to generate greater inter-regional and inter-communal trade. This is of particular importance as China is abandoning the traditional approach of maintaining a large, undiversified and non-competitive economy. Local all-round development, with much decentralisation of industry, is felt to be more appropriate.

The greatest risk that China may be exposed to is the creation of a generation of technocrats. The enormous problems of planning and co-ordinating economic activities inevitably lead to a certain degree of exclusive or jealous professionalism, which could only too easily pave the way for a new managerial class unwilling to take responsibility for the people's livelihood. With the background of so large a working force, this could turn socialism into revival of the bureaucratic rule of imperial China, the very thought of which makes Chinese hair stand on end. The great preventive measure against this is the reform of education. Political and ethical education has therefore been given from time to time at least as much emphasis as technical knowledge, while qualification for higher education has been judged on the individual's enthusiasm and dedication in serving society. This emphasis may sometimes have been pushed too far, but there is little sign of any retreat from it. All these, we may conclude, are relative rather than absolute matters. When any revolution begins to ripen, it always finds more flexibility in decision-making. Today the problem confronting the Chinese is the same as that which confronts the rest of the world; how to find a reconciliation of economic rationality with other qualities of life. What will make the Chinese solution different is China's unique historical background, and everyone will have something to learn from it.

(*d*) HISTORY AND HUMAN VALUES: A CHINESE PERSPECTIVE FOR WORLD SCIENCE AND TECHNOLOGY

JOSEPH NEEDHAM

The first conception of this paper was that it should discuss the relationships between human values and science in the contemporary world, especially in China. But the more I thought about it, the more I came to feel that we needed not 'a perspective for Chinese science' but a 'Chinese perspective for world science'. It seemed to me that certain Chinese values might be vitally helpful to man face to face with his own embarrassing knowledge.[1]

First then we might consider: what really have been the relations between Chinese culture, and science, technology and medicine? What has the work of the past decades done, since the Second World War, to elucidate their relationship? And, thirdly, most important of all, what clues can we get about the help which the Chinese tradition and contemporary China could perhaps give for the *ethos* and operation of the World Co-operative Commonwealth[2] of the future?

At the outset, of course, I have to declare, as is done in the House of Commons, a private interest; yet at the same time a certain authorisation for tackling these subjects. I am, then, a scientist by profession and training, a biochemist and embryologist, whose life, nevertheless, at sundry times and places, came by chance into close contact with the fields both of engineering and medicine. For four years during the Second World War I was Scientific Counsellor at the British Embassy in Chungking, directing the Sino-British Science Co-operation Office (*Chung Ying Kho Hsüeh Ho Tso Kuan* 中英科學合作館), a liaison bureau which kept the scientists, engineers and doctors in beleaguered China in close touch with the free world of the Western Allies.[3] I had been 'converted' (if the expression is permissible) to an understanding of the Chinese world-outlook by friends who had come to work some years before in my own and neighbouring Cambridge laboratories for their doctorates, especially Lu Gwei-Djen, today my chief collaborator. This was the origin of the 'Science and Civilisation in China' project,[4] which has taken me away from experimental researches for some thirty years past. Seven volumes have appeared, and the eighth is now going through the press; so one can calculate that my collaborators and I have succeeded in producing one in each 3.25 years. Since seven or eight more volumes are in active preparation I have little difficulty in computing

[1] Earlier presentations of some rather similar thoughts have appeared in Needham (1974a) and in Needham (1975), pp. 45, 49. This chapter, in somewhat abbreviated form, was given as a paper before the Canadian Association of Asian Studies in Montreal, May 1975.

[2] I use this phrase to designate what I believe to be the immediate goal of human social evolution. It was a favourite one on the lips of Conrad Noel, parish priest of Thaxted in Essex, among the greatest of my teachers.

[3] An account of those days is contained in Needham & Needham (1948), and Needham (1945).

[4] Cambridge, 1954.

the probabilities of my being able to correct the proofs of the last volume *in propria persona*.

Since the series began to come out, two very great changes have taken place. When I first decided to become a full-time historian of science, I had not the slightest idea how topical the subject of Chinese science would turn out to be. A new 'Chinoiserie' period has almost come about. People simply do not know enough about Chinese culture, and everywhere they are avid for information. We never thought that we of all people would be called upon to supply it. Our original question was: why had modern science originated only in Western Europe soon after the Renaissance? *Mais faites attention; un train peut cacher un autre*! We soon came to realise that there was an even more intriguing question behind that, namely, why had China been more successful than Europe in gaining scientific knowledge and applying it for human benefit for fourteen previous centuries?

We wanted to know many things. For example, how far did the Chinese get in the various particular sciences before the era of oecumenical world science began; and again, what, if anything, did they contribute to the origins of modern science itself? These are not questions I can answer in a few words here. All I want to say is that while we were sapping and mining to discover vast stores of information which had never been appreciated by the world as a whole, because still buried within the bosom of the ideographic language, historical events occurred in China leading indubitably to her elevation to great power status. Hence the feeling shared so widely all over the world that we must know more about Chinese culture and the Chinese.

I would like to return in a moment to these historical aspects, but before that I want to mention the second great change that has come about during the past thirty years – I mean that powerful movement away from science and all its works which characterises what has been called the 'counter-culture', and which is widely prevalent among the youth of the Western world at the present day. It is found not only in the West but also to some extent in the underdeveloped parts of the world, and I should be inclined to call it a deep psychological aversion from 'big technology' and the science which has given rise to it.[5] Moreover, the 'disenchantment' with science cuts across all political boundaries, because the inhumaneness of science-based technology is felt to a certain degree in the socialist countries as well as the capitalist ones. This is something on which there is much more to be said, but meanwhile my collaborators and I would like to declare for ourselves that we have in no way lost faith in science as a component of the highest civilisation, and we believe that it has done incalculably more good than harm to human beings. Indeed its development from the first discovery of fire onwards has been a single epic story involving the whole of mankind, and by no means separated out into incommensurable Spenglerian cultural entities.[6] At the same time it is only too clear

[5] This phrase echoes the 'big science' and 'little science' antithesis and the 'background noise of low technology', discussed in Price (1961) and other works.

[6] In Spengler (1926), a book famous in its time, Oswald Spengler pictured the successive cultural entities, for example, Ancient Egyptian, Islamic, Indian, Magian, Chinese, Faustian, etc. as separate organisms having very little interaction between each other, and arising, flourishing, decaying and dying like incommensurable plant or animal forms. A revaluation has been done recently by Northrop Frye in Frye (1974).

that modern science and technology, whether in the physical, chemical or biological realms, are now daily making discoveries of enormous potential danger to man and his society. Their control must be essentially ethical and political, and I shall suggest that this perhaps is where the special genius of the Chinese people could affect the entire human world.

(1) The Historical Background of Science in China

May I now return for a short time to the historical development of science and technology in China. I think it would be fair to say that the history of science, technology and medicine in Chinese culture has now really obtained *droit de cité*; it has become accepted and established (I hope not in any bad sense). It is worth looking briefly at a few details of the Chinese achievements. The mechanical clock has by long tradition been accepted as one of the most outstanding achievements of European mechanical genius at the close of the Middle Ages, and something without which modern science would never have been possible – but it remained to be shown that some six centuries of hydro-mechanical clockwork had anticipated the first clocks of Europe. Indeed one can say that the soul of the mechanical clock, the escapement, was not the invention of an unknown artisan about 1280 in Europe, but rather that of a Tantric monk and mathematician, I-Hsing 一行 and his collaborator the civil official Liang Ling-Tsan 梁令瓚, in China in +725. (Fig. 10.) Here there was a close connection with mechanical models of the heavens continuously rotated, a development which arose as early as the +2nd century in China as a result of the fact that there astronomy was polar and equatorial rather than ecliptic.[7]

For nearly 2,000 years before the beginning of modern science there were no people in the world who observed the changes in the heavens more persistently, and recorded them with greater amplitude and accuracy than the Chinese.[8] I read recently an essay by the physicist Victor Weisskopf in which he spoke about the termination of the evolution of certain stars in those tremendous explosions which men observe and describe as supernovas.[9]

> One of these occurred in the year +1054 and left behind the famous Crab Nebula [Fig. 11] in which we see the expanding remnants of the explosion with a pulsar in the centre. The explosion must have been a very conspicuous phenomenon, in its first days surpassing the planet Venus in brightness. So different from today's attitudes was the mental horizon in Europe at that time that nobody found this phenomenon worth recording. No records whatever are found in contemporary European chronicles, whereas the Chinese have left us meticulous quantitative descriptions of the apparition and its steady decline. What a telling demonstration of the tremendous change in European thinking that took place at the Renaissance![10]

[7] The story has been told in detail in Needham, Wang & Price (1960). On the astronomical background see *SCC*, vol. 3, pp. 229 ff., 339 ff.

[8] *SCC*, vol. 3, pp. 409 ff.

[9] Weisskopf (1972), p. 94.

[10] Quite recently there was some correspondence in the London *Times* about the pulsar mentioned here, where it was said that the Chinese had described the bright new star in 1954, so of course I was unable to resist the temptation of writing to point out that this misprint had knocked off 900 years from their achievement.

Figure 10. An early mechanical clock embodying the escapement principle. See *SCC*, vol. 3, p. 458, Fig. 657.

This is perfectly true, and radio-astronomers today habitually make use of the Chinese records, at least as far back as the beginning of our era. Many other examples could be taken from these branches of science. For one thing only, the complete description of the phenomena of parhelia (mock suns and Lowitz arcs) was given in China in the +7th century, but not in Europe until the +17th.[11]

Again, turning to technology, one could look at the morphology of the reciprocating steam-engine. It is, I am sure, not widely appreciated that this was in all respects completed in China some 500 years before the time of Newcomen and Watt.[12] The great difference was that the direction of energy transfer by interconversion of

[11] Cf. *SCC*, vol. 3, pp. 473 ff. More fully, Ho & Needham (1959a). [12] Needham (1963).

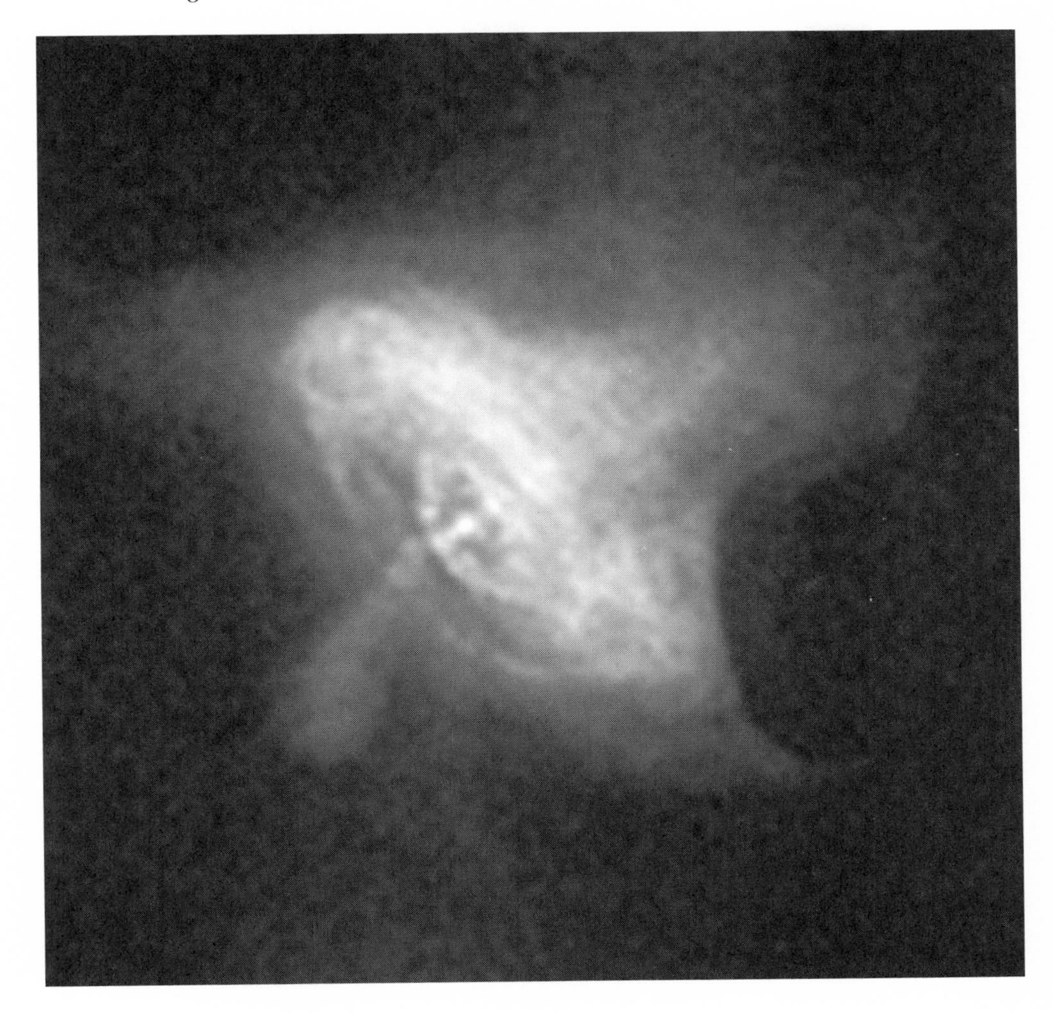

Figure 11. The Crab Nebula in Taurus.

longitudinal and rotary motion was the opposite to the steam-engine, since the old Chinese machine was driven by its water-wheel while its piston-rod worked upon metallurgical furnace bellows. Conversely, in the steam-engine, first steam and vacuum, then steam alone, acted on the piston-rod, which accordingly by an identical conversion, produced useful rotary motion (Fig. 12).

In China one can find a step-by-step evolution of all the components.[13] The eccentric came first of all, in the −4th century, for the rotation of querns; the crank-handle in the +2nd century; the eccentric and connecting-rod, so arranged that several people could exert power on the quern's rotation at one time, in the +5th century; and then the addition of the piston-rod in the +6th century, allowing the

[13] All details will be found in *SCC*, vol. 4, pt. 2, and Needham (1970) but new discoveries are still being made.

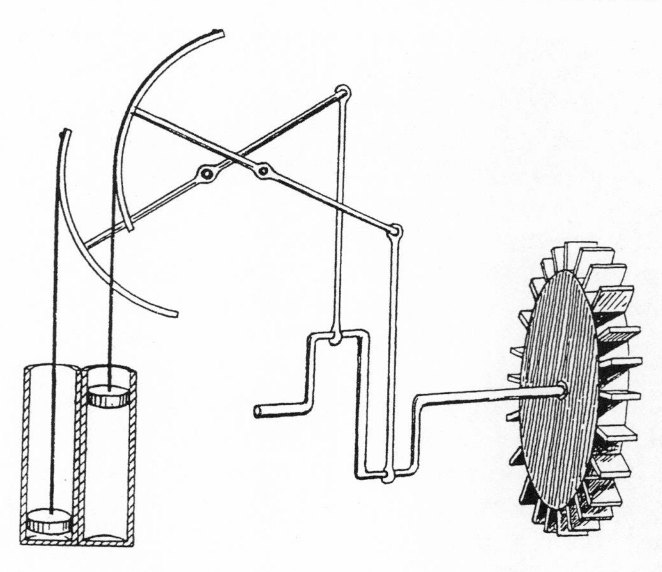

Figure 12. Diagram of interconversion of longitudinal and rotary motion. See *SCC*, vol. 4, pt. 2, Fig. 607.

application of water-power to bolting machines for cereal sifting.[14] By the +13th century, and probably a good deal earlier, the assembly was employed for the blowing of furnace bellows, a fact which may well be one of the reasons for the astonishing development of the iron and steel industry in the Sung period.[15]

I must say that the achievements of traditional China in science, technology and medicine continue to surprise even ourselves all the time. Just recently, for example, we have resumed our work on the medical volume and have been drafting the sections on acupuncture and other medical techniques. In the course of this we have naturally had to say a good deal about the ancient Chinese theories of the circulation of the *chhi* 氣 (*pneuma*) and the blood in man. I myself was at one time inclined to believe that those interpreters were right who had seen the ancient Chinese circulation as being a very slow one of twenty-four hours duration; but much closer study of the essential passages in the *Huang Ti Nei Ching* convinced us that as far back as the −2nd century the Chinese physicians had a conception of the circulation of the blood only sixty times slower than the Harveian circulation which we understand today. Many have long appreciated that Chinese civilisation was far more circulation-minded than any of the other ancient cultures, such as that of the Greeks,[16] but I was not prepared to find them so close on the heels of modern physiology.

Secondly, we have been deeply impressed with the brilliance of the tie-up between the acupuncture tracts and points on the surface of the body and the state of

[14] This last development is attested for this date by a passage in the *Loyang Chhieh Lan Chi* (Description of the Buddhist Temples and Monasteries in Loyang), written by Yang Hsüan-Chih in +547. It was first noticed by Dr William Jenner.

[15] See Robert Hartwell's important papers, for example, Hartwell (1962), (1966) and (1967).

[16] Cf. Huard & Huang (1956).

affairs, pathological or otherwise, in the viscera and other internal organs. A full two millennia before modern physiology the ancient Chinese physicians were aware of phenomena which today we would describe as viscero-cutaneous reflexes, referred pain, and those areas of nerve distribution on the surface of the body which we call Head Zones and dermatomes.[17] Since the ancient and medieval Chinese scientists, engineers and doctors, have been consistently surprising us ever since we started on this work at the end of the Second World War, I must say I am now expecting them to continue to do so for the rest of my life – and for very much longer if Westerners are still willing to look into the work of their Chinese predecessors.

I hope truly that they will be large-minded enough to do this because after all, as the Roman said, humanity is one, and nothing human should be regarded as alien from us. I dare say it will take a certain high-mindedness, because I have had many opportunities of observing that our work has been irritating for conventional Western minds; the achievements of science and technology were often what they were most proud of racially, and to show that others had done as much or more was hurtful to pride.[18] But pride is a thing we can always do without. It weighs little in the balance of Anubis compared with love and friendship.

Such was the grudging attitude of the establishment, of course, in the West, but now as time has gone on, we are faced with an entirely different and opposite reaction, that of the younger generation and the counter-culture, who would be inclined to say that if Chinese civilisation had done so much for science and technology, that would give it a bad mark in their estimation. So you cannot win. But that does not bother us, because we believe in science (though not in science alone), and we hold fast to what we regard as true historical perspective, without which no understanding of the past is possible.

In the light of all this, some estimates of China with relation to science need complete revision, in fact relegation to the scrap-heap. The idea, for example, that Chinese scientific development has been 'just a transplant', something out of keeping with the cultural inheritance, is sheer nonsense. So also is the idea that China (or Japan, or Africa, or anywhere else for that matter) will be necessitated to adopt the ways of life of the Western capitalist world as a whole in adopting science and technology, is also nonsense, whatever may have been the superficial situation so far. Look at the almost total failure of the Church as a human institution in Asia. I found when I was in China during the war that people often used to refer to modern science as Western science (*Hsi-yang kho-hsüeh* 西洋科學), and I always used

[17] The literature of acupuncture is inchoate, difficult of assembly and full of pitfalls, especially for those who have no linguistic access to Chinese, but Mann (1962b), often reprinted, is a good beginning. The latest edition should be used.

[18] It has given us no small amusement to collect examples of these reactions. Obliged at last to acknowledge a non-European achievement, Western historians of science and technology retire in good order by redefining (to their own advantage) what the achievement really ought to be considered to be. The first of all clock escapements may be Chinese, but the only true, the only important, escapement, was the verge-and-foliot of course (*SCC*, vol. 4, pt. 2, p. 545). The magnetic needle may have been used first by Chinese mariners, but of course only the wind-rose attached to the magnet is the *véritable boussole* (*SCC*, vol. 4, pt. 3, p. 564). Chinese seamen first mounted axial rudders, no doubt, but of course there could be no true rudder without a sternpost (*SCC*, vol. 4, pt. 3, p. 651). This is what we call the Department of Face-Saving Redefinitions.

to say in speeches that on the contrary it should be regarded as *Hsien-tai ti kho-hsüeh* 現代的科學, distinctively modern science, open for the participation of all trained men and women, totally irrespective of race, class, colour, creed and so on.[19] Long afterwards I found that this attitude had been exactly the same as that adopted by the Khang-Hsi emperor in 1669, when he had insisted that the title of the work on mathematics and calendrical science prepared by the Jesuits should be *Hsin Fa Suan Shu* 新法算術, and not *Hsi-Yang Hsin Fa Li Shu* 西洋新法歷書, as it had been twenty years earlier.[20] Though the emperor probably knew very little of the Royal Society, founded just at this time, he recognised with great acumen that the 'new, or experimental, science' was essentially new and not essentially Western.

Of course the rate of growth in science and technology differed greatly between China and the Western world. The Chinese rise was slow and steady; it never had anything corresponding to the 'Dark Ages' which Europe went through.[21] Some of the Greek contributions (perhaps Aristotle and Euclid) rose above the Chinese level, but between the +4th and the +14th centuries the level in Europe was abysmally below it; only with the Renaissance and the unleashing of modern science did the European curve rise quickly above the Chinese one and pursue its exponential increase to bring about the world we know. There was thus what might be called a *décalage* in the development, but there is one striking thing to note, namely that some of the greatest Chinese inventions took place precisely during the period when Europe was at its lowest ebb. For example, block-printing was going by the +8th century and moveable-type printing by the +11th. The formula for the first chemical explosive known to man, gunpowder, was established in the +9th century; and the magnetic compass, which had existed long before as a geomantic instrument, was adapted to navigation in the +10th. These were the three inventions singled out by Francis Bacon.[22] Clockwork too, as I have already mentioned, took its rise in the +8th century, about the time when scholars like Isidore of Seville and Beatus Libaniensis were trying to save what remained of ancient knowledge, and when the number of plant species known – quite a good index of scientific capabilities – sank to its absolute minimum in Europe.[23]

(2) The Counter-culture and the Anti-science Movement

Let us now return to the anti-science movement and the counter-culture.[24] Some of the most telling formulations of the disenchantment of the young with science have been presented by Theodore Roszak in his books, *The Making of a Counter-Culture*,[25] and *Where the Wasteland Ends*.[26] He and the young are against modern science – because they feel that it has had evil, totalitarian and inhuman social consequences. They are not content to put this down merely to misapplied technology; their criticism of

[19] Cf. Needham & Needham (1948), pp. 257–8.
[20] *SCC*, vol. 3, pp. 448 ff. Needham (1958a). The change was from 'The West's New Calendrical Science' to 'The New Mathematics and Calendrical Science'.
[21] See above, Needham, 'The Roles of Europe and China in the Evolution of Oecumenical Science'.
[22] *SCC*, vol. 1, p. 19; Needham (1969a), pp. 62–3.
[23] Cf. Arber (1912), repr. 1953 and Jessen (1864). [24] Cf. *Ibid.* [25] Roszak (1969). [26] Roszak (1972).

science itself goes deep. They attack 'the myth of objective consciousness',[27] detesting that 'alienative dichotomy' which separates the observing self from the phenomena in Nature,[28] and sets up what they call an 'invidious hierarchy' which raises the observer to an inquisitorial level, free to torment Nature, living or dead, in whatever way will bring intellectual light.[29] They feel too that science encourages a 'mechanistic imperative', that is to say, an urge to apply every piece of knowledge, in every possible way, whether or not its application is health-giving for human beings, or preservative of the non-human world in which they have to live.[30] The scientific world-view is thus accused of a cerebral and ego-centric mode of consciousness, completely heartless in its activity. It is not as if scientific methods of control were applied only to non-human nature; the 'scientisation of culture' is calculated to enslave man himself.[31] There are many techniques of human control, such as the behavioural and management sciences, systems analysis, control of information, administration of personnel, market and motivational research, and the mathematisation of human persons and human society.[32] In a word, technocracy is rampant, and the more complete the domination of Nature, the more fully does it become possible for ruling élites to increase their control of individual human behaviour.[33]

Since Francis Bacon's time the essence of the scientific method has been alienation, in the sense of an absolute distinction between the observer and the external world, with which he can have 'no sense of fellowship nor personal intimacy, nor any strong belonging'.[34] Nothing inhibits the ability to understand, and after understanding to manipulate and exploit to the full. Hence many of the cruelties which have undeniably been perpetrated in the name of modern biology and physiology. As I write these words there is news of a demonstration at the annual meeting of a great British chemical industry, protesting against the use of dogs as experimental animals in studies of the cancer-producing properties of tobacco.[35] There is much that can be said against the 'callousness' of science, yet without what one might call clinical detachment no scientific medicine would ever have come into being. In the same way the opponents of science cannot deny its pragmatic value in that pharmacological knowledge does lead to the relief, or cure, of disease, and that flight would be impossible without a knowledge of aerodynamics and thermodynamics. The anti-science movement is in rather a quandary here for it can hardly wish mankind to return to the infancy of pre-scientific ignorance, while at the same time it is justifiably uneasy, indeed outraged, at the uses which are constantly made of scientific knowledge;[36] and filled with fears for the future.[37]

[27] Roszak (1969), pp. 205 ff. [28] *Ibid.*, pp. 217 ff. [29] *Ibid.*, p. 222.

[30] *Ibid.*, pp. 227 ff. This is the opposite of what I quote below from the *Kuan Yin Tzu* book and the Gospels.

[31] Roszak (1972), p. 31. [32] Cf. Ellul (1964). [33] Cf. Leiss (1972). [34] Roszak (1972), p. 168.

[35] For a recent study, by a biologist, of the ethics of animal experimentation, see Roberts (1974). Cf. also Pappworth (1967).

[36] One has only to mention the Vietnam War in this connection, with its fragmentation bombs, napalm, computer analysis of terrain, vacuum asphyxia weapons, defoliants, etc.

[37] This whole subject has been the theme of the late C. H. Waddington's Bernal Lecture at the Royal Society: 'The New Atlantis Revisited' (*PRSB* (1975a), **190**, p. 301). He drew attention to the remarkable recent book by Pirsig (1974).

(3) The Forms of Human Experience

I am inclined to think that the real meaning behind the anti-science movement is the conviction that science should not be taken as the only valid form of human experience. Actually philosophers have been calling this in question for many years past, and the forms of human experience – religion, aesthetics, history and philosophy, as well as science – have been delineated in many integrated surveys.[38] Roszak himself hints this, as when he denies that scientific objectivity can be 'the only authentic source of truth', or when he says that 'we must be prepared to see the truth as a multi-dimensional experience'.[39]

Some of his most interesting insights arise from an exegesis of William Blake, where he demonstrates that Blake's hatred of what he called 'single-vision' was really a criticism of the estranged monocular scientific experience considered as the only possible manner of apprehending the universe.[40] Many other expositors of Blake have shown that his image of the 'Four Zoas' was really a recognition of the diversity of the forms of experience.[41] For example, he wrote,

> Four mighty ones are in every man; a perfect unity Cannot exist but from the universal brotherhood of Eden, The Universal Man. . . .

In this peculiar psychology of mythical entities Urizen stands for the cold scientific reason, Luvah for energy, passion, feeling and love, Urthona for prophecy (aesthetic?) and Tharmas for the power of the spirit (history and philosophy?). Some of these have consorts, *shaktis* or *alter egos*. Thus most penetratingly Blake named the 'fallen state' of Urizen as Satan, the god of brute unfeeling power, the spirit of the dark Satanic mills with their 'cogs tyrannic'. So also the fallen state of Luvah was Orc, the chained beast of thwarted desire, religious persecution and repressive or revolutionary terror. Urthona contains in itself (hermaphrodite) a Yin/Yang pair, Los and Enitharmon, and all these war with one another, or become reconciled with one another, on the battlefield of Albion, the individual man.

Among the most interesting recent treatments of the diverse forms of human experience is that of Victor Weisskopf, who suggests that they stand to one another in a relationship something like the Heisenberg uncertainty principle in physics, the dilemma of indeterminacy or complementarity.[42] If one tries to investigate the details of the quantum state by some sharp instrument of observation, one can only do so by pouring much energy into it, and so destroying the quantum state. The necessary coarseness of our means of observation makes 'exact' observation in the old sense

[38] I call to mind especially here a book which had great influence on me when I was young, Collingwood (1924) (*Speculum Mentis*). Current discussions are numerous – for example Roszak (1974) together with the article by Weinberg (1974), p. 47.

[39] Roszak (1972), pp. 106, 189. [40] *Ibid.*, pp. 74 ff.

[41] For example, Raine (1968) and Beer (1968). In *Vala, or the Four Zoas*, G. E. Bentley has reconstructed the poem from Blake's chaotic notes, and given a facsimile reproduction of it (Bentley (1963)).

[42] Weisskopf (1972), p. 58.

impossible. One can either know the speed or the position of a sub-atomic particle, but never both at the same time. Weisskopf writes:[43]

> The claim of the 'completeness' of science is that every experience is potentially amenable to scientific analysis and understanding. Of course many experiences, in particular in the social and psychological realm, are far from being understood today by science, but it is claimed that there is no limit in principle to such scientific insights. I believe that both the defenders and the attackers of this view could be correct, because we are facing here a typical 'complementary' situation. A system of description can be complete in the sense that there is no experience that doesn't have a logical place in it, but it could still leave out important aspects which in principle have no place within its system. . . . Classical physics is 'complete' in the sense that it could never be proved false within its own framework of concepts, but it does not encompass the all-important quantum effects. There is thus a difference between 'complete' and 'all-encompassing'.
>
> The well-known claim of science for universal validity of its insights may also have its complementary aspects. There is a scientific way to understand every phenomenon, but this does not exclude the existence of human experience that remains outside science . . . Such complementary aspects are found in every human situation. . . .

Weisskopf goes on to say that whenever in the history of human thought one way of thinking has developed with force, other ways of thinking or apprehending have become unduly neglected, and subjected to an overriding philosophy claiming to encompass all human experience. This was obviously the case with religion and theology in medieval Europe, and it is certainly the case with natural science today. And he goes on to draw the very existentialist conclusion that

> the nature of most human problems is such that universally valid answers do not exist, because there is more than one aspect to each of them. In both these two examples great creative forces were released and great human suffering resulted from abuses, exaggerations and neglect of complementary ways of thinking.[44]

Indeed this is most true, and the only way forward is the existentialist realisation that the forms of experience, which have a habit of contradicting each other flatly, are all basically inadequate ways of apprehending reality, and can only be synthesised within the individual life as lived. Here mutual control and mutual balance are of the essence. Couldn't we hope to learn something of this from our Chinese friends? As Weisskopf has said in another place: 'human existence depends upon compassion, and curiosity leading to knowledge; but curiosity-and-knowledge without compassion is inhuman, and compassion without curiosity-and-knowledge is ineffectual'.[45] This is a *philosophia perennis*. We find it again in Julian of Norwich, writing soon after 1373: 'By reason alone we cannot advance, but only if there is also insight and love'.[46]

The whole anti-science movement has arisen because of two characteristics of our Western civilisation: on the one hand the conviction that the scientific method is the only valid way of understanding and apprehending the universe, and on the

[43] *Ibid.*, p. 349. Similar ideas have been expressed by T. R. Blackburn in Blackburn (1971).
[44] Weisskopf (1972), p. 351. [45] *Ibid.*, p. 364. [46] Juliana (*c.* 1373), ch. 56, p. 162.

other hand the belief that it is quite proper for the results of this science to be applied in a rapacious technology often at the service of private capitalist profit. The first of these convictions is held as a semi-conscious assumption by a great many working scientists, though formulated clearly only by a small number;[47] at the same time it spreads widely through the populations, often leading to great callousness and insensitivity in personal relationships, quite beyond the power of the traditional codes of religion and ethics to modify. Similarly, the mass-production technology of the capitalist world, so freely paralleled and imitated in the Soviet Union and the Socialist Republics of Eastern Europe, has indeed supplied the peoples of the developed world with a vast wealth of material goods,[48] but only at the cost of debauching their aspirations, limiting their freedoms, and imposing controls every day more insidious and unhealthy.

(4) China, Europe, 'Scientism'

This brings me to the first of the fundamental points which I want to make. The contribution of China to the world may be particularly valuable because Chinese culture has never had this post-Renaissance 'scientism'. Since modern science did not develop in China originally, but only in Europe, there was never any necessity for a Chinese William Blake to oppose the 'single-vision' of Isaac Newton's conception of the universe, nor did a Chinese Drummond or Kropotkin have to do what he could to oppose a Chinese Huxley. In spite of the debate about scientism in China in the first decades of this century,[49] Chinese culture has never been really tempted to regard the natural sciences as the sole vehicle of human understanding. This was all the more the case because for twenty centuries an ethical system not based on supernatural sanctions had been completely dominant in Chinese society; and it was history, not theology and not physics which had been the *regina scientiarum*. This dominance of morality is perfectly represented in China at the present day by the slogan 'put politics in command', for this essentially means human moral values, and that in practice means the health and well-being of your brother and sister at the bench, in the field, on the shop-floor, and next to you in the office or at the council table. In China you will not be tempted to regard him or her as 'nothing more than' a behaviourist automaton, or a walking flask of amino-acids and enzymes, because by the wider wisdom of Chinese organic humanism, you know that he or she is a lot of other things as well, for the full reality of which mechanistic or reductionist scientific explanations, however successful for limited ends, can never suffice.[50] This is my first fundamental point. It may be that scientism, the idea that scientific truth alone gives understanding of the world, is nothing but a Euro-American disease, and

[47] Cf. Crick (1966), Skinner (1971) and Monod (1971). See also the critical enquiry on this by eleven writers edited by Lewis (1974). To the same tradition of 'scientific positivism' and reductionism belong Morris (1967), *The Naked Ape*, and Medawar (1967), *The Art of the Soluble*.

[48] Described as affluence without altruism – the failure of the 'Welfare State'.

[49] See Kwok (1965).

[50] The Chinese talk about the love and service of 'class brothers and sisters' but this is what they really mean.

that the great contribution of China may be to save us from the body of this death by restoring humanistic values based on all the forms of human experience.

It may be that the roots of the matter, why Europe has suffered so much from this, go back very far. Roszak, in the second of his books, has traced the desacralisation of Nature,[51] which has been so important for modern science, to the ancient Jewish anti-idolatry complex inherited by Christianity.[52] This led, he suggests, to what one might call 'nothing-but-ism', i.e. (in its first form) the conviction that the statue of a Greek or Roman god was nothing but a statue, totally devoid of all magical or worshipful properties. Islam of course inherited this too; 'thou shalt have no other god but God, and Muhammad is his prophet'. And Christendom and Islam both inherited the aggressive intolerance which Hebrew monotheism had from the beginning infused into this issue. But to modern minds, especially those acquainted with Buddhism or Taoism, it would seem that far too much fuss was made about idolatry by the Christian theologians, rising to a crescendo with the Protestant Reformation (so significant always for the birth of modern science); because in a way it never really existed at all – for every intelligent person the statues must always have been thought of as symbols and temporary residences of the god or spirit invoked. Even quite simple people must have understood this. Certainly it was the attitude of the Neo-Platonists.[53] But the point is that this theological 'nothing-but-ism' was a pre-figuration of the reductionism of modern science. It was a mental predisposition, a *preparatio evangelica scientifica damnata*.

First, for example, there was the suppression of secondary qualities from Galileo onwards, then after the coming of modern chemistry there was the belief that all the phenomena of life and mind could be explained without residue by the properties of atoms and molecules, ultimately of the sub-atomic particles themselves. We are not complaining that all this banned magic entirely from the world, or denied to Nature its authentic enchantment, because that is just the sort of thing which science has to do – within its own realm – but we are saying that the anti-idolatry complex paved the way through the centuries for the 'nothing-but-ness' of mechanical materialism and scientism. It attacked all the lesser sacralities in the name of the God of Abraham and Jacob, and this was all right as long as the creator deity of the Hebrews remained secure on his celestial throne, but when belief in the supernatural collapsed under the attacks of mechanical materialism, nothing holy was left anywhere.

How different was the situation in East Asia. Since the Chinese were never among the 'People of the Book',[54] like Jews, Christians and Muslims, they never had the anti-idolatry obsession. The holy was never identified with the supernatural because

[51] It is of much interest that a Persian philosopher of science, Said Husain Nasr, has recently also inveighed against this, from the standpoint of Islam. A critique of his interesting books, Said Husain Nasr (1968a) and Said Husain Nasr (1968b) will be found in *SCC*, vol. 5, pt. 2, pp. xxiv ff. His complaint against science echoes that of Roszak, but he has less appreciation of the validity of science as a form of experience in its own right. One may not like some of its attitudes and blindnesses – *mais c'est la nature de l'animal.*

[52] Roszak (1972), pp. 109 ff. [53] On all this see Bevan (1940).

[54] This is a famous Muslim expression. When the Muslims in their great expansion encountered Hindus or Buddhists they were liable to insist on conversion to Islam at the point of the sword, but Jews and Christians were tolerated as minorities, even if oppressed, because they also accepted the Old Testament as a sacred book.

in Chinese thought there never was anything outside Nature. The numinous, the holy, could be, and was, present at many points within and beyond the world of man, 'The world is a holy vessel' says the *Tao Te Ching*, 'let him that would tamper with it, beware'.[55] Thus once again Chinese culture was not under the same drives as that of Europe, and wherever you look, whether in Chinese organic philosophy or the beauty of Nature as interpreted by the Chinese artists and poets, or in the studies of Chinese historians, this 'nothing-but-ism' never came up – and most important of all it never dominated in Chinese science and technology, great and world-shaking though their discoveries and inventions were. It may well be that this absence was one of the limitations, like the presence of that tension between the separated sacred and secular in Europe, which prevented the rise of modern experimental science in China, and encouraged it in the West; but that is no ground for doubting that Chinese level-headedness may now be deeply needed and called upon to rescue the Western world from the slough of mechanical materialism and scientism which it has fallen into.

If the anti-idolatry complex of Israel and Christendom had been the only influence leading to reductionism in Europe the compulsion might not have been so strong, but it was reinforced from quite a different quarter, the atomism of the Greeks. The schools of Democritus and Epicurus held, broadly speaking, to a mechanical materialism, and it was put into prophetic imperishable poetry by the great Lucretius. One is accustomed to think of a schizophrenia of Europe, torn between the theological world-picture and that of materialist atomism, but in this particular they were closely allied; for while the former reduced the holy objects of antiquity to common wood and stone, the latter reduced not only wood and stone but also flesh and blood to the fortuitous clashing of hard impenetrable massy particles. Again in China what a difference! Buddhist philosophers must have talked much about the advanced atomic speculations of India, but never with any perceptible effect, for the Chinese remained perennially faithful to their prototypic wave theory of Yin and Yang, universal transmission in a continuous medium, and action at a distance. The Yin and Yang, and the Five Elements, never lent themselves to reductionism because they were always inextricably together in the continuum, and in the organic phantasms which came and went within it; never separated out, isolated or 'purified', even in theory. So once again there was no 'nothing-but-ism' in Chinese thought to parallel what Greeks and Hebrews joined together to create in the West.

Nothing that I am saying here, however, should be taken to indicate that there is anything different about modern science in China today from that in the rest of the world. Recently when lecturing in Brussels and Ghent I came across a curious idea that the practice of modern science in China today was as different from that in the rest of the world as modern science had been, in its turn, from ancient or medieval science. This was a notion doubtless stimulated by the success achieved in China in drawing peasant-farmers and working people of all kinds into the field

[55] Ch. 29, Waley (1934), p. 179.

of observation and experiment, but it has no force. Science is one and indivisible. The differences are essentially sociological – what you do science for, whether for the benefit of the people as a whole, or for the private profit of great industrial enterprises, or for the development of fiendish forms of modern warfare; in a word, your motive. The differences will also be great according to whom you get to do it, whether you confine it to highly trained professionals, or whether you can use a mass of people with only minimal training; and this means again how you manage the whole affair.

But basically, modern science must be identical under every meridian. There is only one logic of controlled experimentation, only one application of mathematical hypotheses, and their testing by statistical methods. There are canons here which cannot be transgressed; the basic method of discovery itself, which was discovered in Galileo's time. The 'paradigms' of course change, and have changed, as knowledge increases, for example when the Einsteinian world-system was superimposed on the Newtonian one, but that does not alter the basic character of the scientific method itself. Or again, beliefs that were held in ancient and medieval times can come obliquely true in the light of later discoveries; for example, on the atomic level, the dream of the alchemists of transmuting baser metals into gold was bound always to remain a dream, but after the discovery of radio-activity it became possible to turn one element into another, including gold, by the addition or subtraction of electrons. No, science is a unity, and it is not done differently in China from anywhere else; but what we can rightly object to is the idea that science is the only valid way of apprehending the universe. Perhaps we fell into this mistake because modern science originated among us in the West; conversely the Chinese never had the temptation under which we fell, and now is the time for them to give us help to climb back to the realm of true humanity.

(5) The Challenge of Technology and Medicine

Let us turn now to an entirely different aspect of the situation. Even in a properly balanced human society, where the natural sciences were counterbalanced all the time by what used to be called in Cambridge the 'moral sciences', and other forms of human experience such as the religious and the aesthetic, there would still be great difficulty in dealing with the all but intolerable ethical choices which applied science places before mankind and will increasingly place as time goes on. The young people of the counter-culture are revolted by the necessity of making such choices, but neither they nor we can go back to the 'bliss of ignorance' as in primitive times. Actually it never was bliss of course, because the very mission of science was to lead us out of the wilderness of ancient fears, taboos and superstitions. But the promised land will never be won by science alone. The control of applied science is probably the greatest single problem for humanity today, and one might even go so far as to wonder whether the most penetrating social critiques, such as the theory of the class struggle, and historical materialism, are not simply aspects of this basic question.

No doubt man has been facing it ever since the discovery of fire, but today it threatens his very existence. Everyone knows about nuclear power and the devastating possibilities of nuclear weapons,[56] but such apparently simple problems as the disposal of the radio-active waste from nuclear power stations are nightmares to those who worry about the social responsibilities of science.[57] Nowadays mathematical engineering is almost as dangerous, and the possibilities of 'artificial intelligence', and the vast computing machines which can and will be built, with their fabulous information storage and retrieval, are quite breathtaking. The privacy of the individual is now endangered, the rights of children to be taught by living teachers, and the safety of whole populations exposed to the danger of some electrical or mechanical fault when computers are harnessed to 'defence'.

The possibilities of biology and medicine are at least as challenging. My own professional background has made it natural for me to follow such developments. One of the largest fields in which they arise is that of generation, for this is the first time in human history that man is on the point of acquiring absolute control both of reproduction and infertility. All too soon we shall be in possession of means for controlling the sex of the human embryo. After this the sterilisation of whole groups might become a live issue. Ethical controversies have raged for years round contraception and abortions[58] but problems are also raised by the new foetal medicine, which can detect grave abnormalities long before birth,[59] and by artificial insemination, which is only by convention attached to infertile marriages.[60] Legal considerations and changes are lagging far behind the actual possibilities, such as spermatozoa banks, maintained from donors outstanding for physical or intellectual brilliance, and possibly several generations older than the receiving womb.[61]

Again, now that we know the chemical structure and coding of the semantophore molecules of deoxyribonucleic acid (DNA) which carry the instructions for making each new human organism, infinite possibilities are open for interfering with this hereditary material.[62] That would be biological engineering applied at the molecular level; one could envisage the insertion of an entirely new piece of chromosome, or the removal of another. Or one could produce hitherto unheard-of hybrids by substituting a mixed-cell nucleus for the original one of the fertilised egg itself. These may seem distant prospects, requiring enormous expenditure of money; but there has already been unexpected success in transferring genes (the hereditary units) from

[56] Cf. the recent lecture by B. T. Feld, 'Doves of the World, Unite!', Feld (1974), p. 910.

[57] Cf. Lovins (1975). [58] See Anon (1974).

[59] This is called amniocentesis, the cytological examination of cells sampled by biopsy from the embryo and its membranes.

[60] Cf. the Ciba Foundation Symposium (1973).

[61] On all these questions see Jones & Bodmer (1974), a study by a British Association Working Party.

[62] See the Trueman Wood Lecture of C. H. Waddington (1975b). My old friend and collaborator takes a refreshingly cool view of the dangers before us because of the immense expense which researches in such embryology and genetics involve, and the consequent certainty of public scrutiny. I am not so optimistic, for two reasons (a) the possible activities of totalitarian states, and (b) the doubt that public scrutiny – or debate – necessarily leads to right ethical policy. As Waddington himself says: 'one wonders whether we are intellectually, emotionally, or morally, prepared to face such choices . . .'.

one lot of bacteria to another.[63] Certain viruses can pick up genes and put them into the bacterial nuclear systems. What if an antibiotic-resistant strain of bacteria were produced, which quickly spread all over the world like wildfire, and decimated human populations? That this is a real danger has been shown by a self-denying ordinance achieved very recently in California,[64] where the scientists working in these fields agreed to establish a moratorium on such experiments, at least until more laboratories with adequate safety equipment and security became available.[65] Here there is one very tempting possibility open to mankind, namely the possible insertion into plants of genes favouring the symbiosis of nitrogen-fixing bacteria, as happens in the leguminous plants today. If that could be arranged for the staple crop cereals it would be a gift to humanity almost as great as the gift of fire. What effect would this immeasurable increase in food production have upon the human race?

Medicine is also confronting humanity with almost insoluble problems.[66] The conquest of transplantation intolerance has already led to a great proliferation of organ transplants, and no doubt the surgeons in time to come will have access to whole banks of spare parts for human beings.[67] But transplantation studies go much further, for it is now possible to make chimaeras between animal species, since certain killed viruses make the tissues stick together, and this could be used to unite human with animal tissues.[68] What is going to come of that? Ethical problems also arise in all cases where the treatment may be very expensive, needing elaborate machinery – for example the kidney machines which dialyse the blood, and can keep a person going even though his or her kidneys are only able to function very ineffectively. Who is going to choose who gets the advantage of restricted techniques in short supply?

Again, much work is being done on the fertilisation and cultivation of human eggs *in vitro* up to the blastocyst stage before their reimplantation into a uterus to go on developing until term.[69] Aldous Huxley in his famous novel *Brave New World* visualised the isolation of totipotent blastomeres so as to reproduce many identical copies of low-grade human beings, and this is not at all impossible.[70] But there are other ways of effecting such 'cloning'. For example, nuclei from adult cells can take the place of the egg's own nucleus itself, so that a whole regiment of individuals with identical genetic material could be created. The question would then arise: do all

[63] This was the great discovery of the 'transforming principle' (free genes) by Avery and others. Its importance in relation to the solution of the problem of the genetic code has been clearly brought out by Olby (1974).

[64] This was the Berg Conference at Asilomar, on which see *International News* (1975), **254**, p. 6. Self-imposed controls for research involving genetic transfer techniques had already been adumbrated in the Ashby Report (1975). This was followed by the Report of the Working Party on the Practice of Genetic Manipulation, presented to Parliament in August 1976, drafting a code of practice, under Professor Sir Robert Williams, Chairman.

[65] The opening of such a laboratory by Imperial Chemical Industries at Runcorn was announced in the London *Times*, 6 June 1975.

[66] One of the most obvious, and widely debated, is of course that of euthanasia. On all these questions, see Leach (1970).

[67] Cf. the Ciba Foundation Symposium (1966).

[68] H. G. Wells (1896). His prophetic *Island of Dr Moreau* has become only too real a possibility.

[69] Cf. Edwards (1974), p. 3, and Edwards & Sharpe (1971), p. 87.

[70] My review in *Scrutiny* (1932), has been reprinted in Watt (1975), p. 202. It was closely in tune with what is now being said here more than forty years later.

human beings have an inalienable right to individuality? Such is the fix that Faust has got himself into, and the young suspect that they know why.

(6) China's Immanent Ethics

I could go on a lot longer, but it must be evident that humanity has never hitherto had to face anything like the tremendous ethical problems posed by the physico-chemical and biological sciences.[71] Now it is not at all obvious that the traditional ethics of the Western world, even with all its tomes of moral theology and casuistry, is the best equipped to deal with these problems, and certainly not on its own. Even within the sciences it is not obvious that the traditional modes of thinking of Western philosophy are the most adequate for the extraordinary and incredible events which go on in the world of sub-atomic particles; and indeed there are those, such as Odagiri Mizuho,[72] who are showing that Buddhist philosophy may give a good deal of help to the nuclear physicist which could not come from Western ideas alone. Here we can do no more than raise this point. What needs saying is that most of those who have worried quite properly about the control of applied science in the West have so far failed to realise that there is a great culture in the East which for 2,000 years has upheld a powerful ethical system never supported by supernatural sanctions. This is where Chinese culture may have, I think, an invaluable gift to make to the world. Nearly all the great philosophers of China have agreed in seeing human nature as fundamentally good, and considering justice and righteousness as arising directly out of it by the action of what we in the West might call the 'inner light'. The Johannine light, perhaps, 'which lighteth every man that cometh into the world'.[73] Let men and women have proper training in youth, the right ideals, and a classless society which will bring out the best elements potentially within them.[74]

For the Chinese, then, ethics was accepted as internally generated, intrinsic and immanent, not imposed by any divine fiat, like the tables of the law delivered to Moses on the mountain. I should go so far as to say that never have the Chinese been more faithful to this doctrine, interpreting it in terms of selfless service to others, to people, than they are at the present day. *Wei jen min fu wu!* 為人民服務.[75] This is my second fundamental point. If the world is searching for an ethic firmly based on the nature of man, a humanist ethic which could justify resistance to every

71 Cf. the Ciba Foundation Symposium (1972).

72 See for example his paper Odagiri (1974) and Odagiri (1955–62), numerous papers. Others have been thinking along similar lines, notably Capra (1974), p. 3. We now have his book, *The Tao of Physics* (Capra (1975)).

73 John 1:9.

74 Is it not a remarkable fact that the charismatic leader of the eight hundred million 'black-haired people', Mao Tse-Tung, should be a social and ethical philosopher, not a military man. Plato's famous remark would be relevant here, though it does not seem easy to point to any European parallel. The contrary is only too obvious, from Alexander the Great to Napoleon, Hitler and Mussolini. And when philosophers or natural philosophers did appear at the top in Europe, as perhaps Alfonso X of Castile or Rudolph II in Prague, their reigns were not very successful. In the Chinese case we have millennial background of respect for scholars and thinkers. I am indebted for this point to a conversation in Montreal with Ronald Melzack and Elizabeth Fox.

75 'In everything you do, let it be done for the people.'

dehumanising invention of social control, an ethic in the light of which mankind could judge dispassionately what the best course to take will be in the face of the multitude of alarming options raised by the ever-growing powers the natural sciences give us, then let it listen to the sages of Confucianism and Mohism, the philosophers of Taoism and Legalism.[76] Obviously we must not expect from them exact advice on choices arising from techniques which they would never have been able to imagine. Obviously, also, we are in no way bound down to the formulations they gave to their ideas in ancient feudal, or medieval feudal-bureaucratic society – time marches on. But what matters is their spirit, their undying faith in the basic goodness of human nature, free from all transcendental elements and capable of leading to a more and more perfect organisation of human society.

I could put it this way in a kind of fantasy. If in some future incarnation I were ever to find myself a member of the Council of the Human Biology Authority of the World Co-operative Commonwealth, I know exactly whom I should be wanting to sit on the Board with me. I should like a few Jews, for they have a most noble ethical instinct,[77] but above all I should pray for Chinese colleagues, the descendants of the sages, with their sense of justice and righteousness (*liang hsin* 良心) and their profound understanding of what constitutes the fullest, healthiest life for man on earth. I would expect from them a firm realisation of the dangers to which science exposes us, and which we must at all costs avoid. The saying in the *Kuan Yin Tzu* book may well come true: 'Only those who have the Tao can perform these actions – and better still, not perform them, though able to perform them!'[78] Doesn't this echo amazingly the narrative of the Temptations in the Gospel?[79]

(7) China and the Matter–Spirit Dichotomy

Once again let us set off in a different direction. What else could Western society profitably learn from the Chinese tradition? What would have to come about in world society as a whole to allow of the integration, or 'oecumenogenesis', of Eastern and Western cultures? To begin with, there is one thing which seems to me to have been insufficiently discussed so far, and that is whether the extreme division of the world into spirit and matter, so characteristic of European thought, has been a good thing or not. It is certainly extremely un-Chinese, for we find in all branches of the history of science in China that there was a great reluctance to make this sharp dichotomy. China was profoundly non-Cartesian. I always remember a friend of mine at the

[76] At the present time in China Confucianism is unpopular and the good points of the Legalists are being rediscovered. This is quite justified because for far too long certain aspects of Confucius' teachings, especially the subjection of women, had been taken in a 'fundamentalist' way, without historical criticisms of his acceptance of feudal or proto-feudal society. But like all the other Chinese sages he gave his ethics a naturalist, not a supernatural basis; and he certainly propagated the doctrine that every man who could profit by education should have it, regardless of birth or wealth. Mo Ti, with his 'universal love', is now widely appreciated, and the Taoists generated a wisdom implicit in most of what all Chinese instinctively do.

[77] I have found a unique consciousness of ethical values among them, as in my friend Louis Rapkine, now long dead, whose patron saint was Benedict Spinoza.

[78] Ch. 7, p. Ia. [79] Luke 4:3–13.

B.A.'s table in Caius many years ago growing sarcastic at the expense of aberrant scientists like Oliver Lodge, and saying how 'these silly fools don't realise that you can't turn matter into spirit just by making it thin!'[80] Yet that is precisely what the Chinese were doing all through the ages. It is a curious thought that while matter in the West began as being very material, and got to be more and more ethereal as time went on and the Bohr model of the atom came to be accepted, in China on the other hand *chhi* 氣, which started as a very ethereal sort of *pneuma*, came to embrace even the most solid of matter by the time of the Neo-Confucian scholastics.

At all events, the disinclination of the Chinese to draw any sharp line of distinction between spirit and matter was at one with their deeply organic philosophy, and their psychosomatic medicine too, developed so long before Europeans came to the same outlook. Of course in modern times there have been in the West many types of philosophy, such as that of A. N. Whitehead, which have done away with the sharp distinction, and a non-obscurantist organicism characterises the best experimental biology going on today. In the old *palaestra*, or gymnasium, where vitalism battled long ago with mechanicism, a struggle continues between reductionists and anti-reductionists.[81] But it is rather shadow-boxing, because it must be perfectly obvious that while the different integrative levels in living organisms only acquire their full meaning when related with the levels above and the levels below, the properties and behaviour of all living things up to the highest levels must be implicit *in potentia* in the nature of the protons and electrons that build them up. But of course if you only knew the protons and electrons, and had no knowledge of the highest levels of integration and organisation, it would be impossible to predict what those would be.[82] This is what I mean by non-obscurantist organicism. It fits into the traditional Chinese world-picture much better than it does to that of Europe, so bedevilled as always by traditional theological philosophy.

The sharp division between spirit and matter probably has as one of its most important correlates the equally sharp distinction between the relative value and significance of mental and manual labour.[83] It is evident that Chinese society today is making a great effort to overcome this, and to convince people that the use of the body is as important and health-giving as the use of the brain.[84] Mention has already been made of the recruiting of manual workers into the ranks of scientific observation, and by the same token it is considered highly desirable today that managerial, technological and intellectual workers should do a period now and then at the bench or in the field. No one who has any acquaintance with Chinese industrial production at the present time, even if only through printed or written accounts, can doubt that a tremendous revolution has taken place in industrial

[80] Cf. *SCC*, vol. 5, pt. 2, p. 86. [81] See the interesting paper of Engelhardt (1974), p. 11.

[82] Exactly the same point of view is put by Weisskopf (1975), p. 15.

[83] For a wide-ranging study of labour and its ethics see the remarkable survey and bibliography by Weltfish (1975).

[84] China has had to mobilise a vast society and solve innumerable institutional problems, to create a viable educational system, to gain access to modern technology and to provide health services for everyone; but the success of all this depends on the acceptance of a universal value system based on solidarity, revolutionary mentality and the primacy of moral over material motives. Cf. Ganière (1974), and Oldham (1973), pp. 80 ff.

relationships.[85] Artisans, craftsmen and workers are encouraged to take initiative, bring forward inventions, tackle jobs co-operatively, and manage their own affairs within the framework of a great factory.

It is extremely interesting that the most enlightened sectors of Western industry have begun to follow similar lines.[86] For example, it is common knowledge that two at least of the great Swedish firms in the motor car industry have completely abolished the assembly-line, which in the days of Henry Ford certainly increased mass-production but reduced the individual worker to a machine carrying out one simple task at a predetermined tempo along with other machines. Such jobs were dull and tiring, and destructive of the worker's self-esteem. Widespread disaffection showed itself, and continues to show itself, in the form of high labour turnover, chronic absenteeism, poor workmanship and even sabotage. Thus the principles of job simplification, repetition and close control are now giving way in many industries, even in the capitalist West, in favour of organisation in autonomous small groups shaped to fit particular jobs. Such groups can solve problems, learn from the problem-solving, and derive satisfaction from it. They can study, reason, evaluate and strive towards goals, while doing all the time not one particular repeated thing but a considerable variety of technical operations.[87] Now insofar as autonomous group production gives the worker more opportunity to influence his own job, to take on responsibility, to solve problems, and to advance his own development, as well as co-operating much more fully and joyfully with his mates, it is following methods and ideals which are typically Chinese.[88] This does not mean that the conveyor-belt is abolished; on the contrary it is still useful, but as a tool or means of transportation, not simply as a source of unvarying stress and strain. The delegation of power and influence to the manual worker is going to become quite normal and undramatic in all forms of developed society; and it springs from the basic principle that manual and mental work should not be regarded as diametrical opposites, but ideally combined in the same person. Anyone who has ever been a working scientist understands this. As the great Ivan Petrovitch Pavlov once said, 'I have done a great deal of menial and manual work in the course of my scientific life, and I think it is the manual work which has often given me the most pleasure'.[89]

Here then is the third point I am making. Wasn't it a wise course the Chinese took when they avoided too sharp a distinction between matter and spirit? And perhaps it led to sociological wisdom too, the activity of the whole man in brain work and hand work; mind and matter no longer at loggerheads.[90]

[85] See the interesting interview with Charles Bettelheim of the Ecole Pratique des Hautes Etudes in Paris: in Bettelheim (1975), p. 9.

[86] Cf. Björk (1975), p. 17; also Weltfish (1975).

[87] Long ago Stuart Chase, in his remarkable book *Men and Machines* (Chase (1929)) enumerated the many different types of contact between them, ranging from the beneficial through the neutral to the dangerous and the lethal. Cf. Needham (1943), p. 134. Here might be mentioned the extremely valuable 'Bibliography of the Philosophy of Technology' prepared by Mitcham & Mackey (1973).

[88] See the important paper by Denis Goulet: Goulet (1975).

[89] Pavlov (1941), vol. 2, p. 53; quoted in Needham (1943), pp. 156–7.

[90] Of course all this is independent of the political question, whether in the long run the utmost satisfaction for industrial workers can ever be gained under the capitalist system, where they do not feel in any sense the owners and managers of the enterprise in which they work.

(8) China and the Co-operative Mentality

Just now the word 'co-operation' was mentioned, and again we have an extremely Chinese trait which is surely going to have a great effect upon the rest of the world. Mo Ti spoke about the 'ant attack', Ritchie Calder emphasised what could be done by 'a million men with teaspoons'. In one of our volumes we illustrated a wonderful picture of a vast mass of men co-operating in hydraulic engineering work, the opening of a great new canal;[91] and I myself during the war and since have often marvelled at the way the Chinese could manage the large numbers of people needed for labour on airfield building, communications and hydraulic works. Here again the organic philosophy of China has much to teach the West. The atomic fragmentation of the competitive, acquisitive society destroys the personality. Competitive individualism only leads to alienation – quite unlike the cohesive atmosphere of the Chinese extended family, the guild or the secret society, or those who now join together in co-operative communities and industrial units.[92] There is of course a fundamental equalitarianism here, and my fourth thought is that it will be all to the good of the world if this racial and rational brotherhood spreads throughout mankind.

In what I am saying here I am mostly commending Chinese ways of life and thought to the rest of the world, but I would be the first to admit that they do not yet solve all possible problems. Some acute Japanese friends, particularly Yamada Keiji, have pointed to a fundamental 'amateurism' which ran throughout traditional Chinese society, especially of the literati, and still today the Chinese ideal is to produce the 'all-round' man, the scientific, industrial, humanistic, aesthetic man, even though I might in some ways miss the religious element. But the problem then arises whether the Chinese will be able to conquer the 'commanding heights' of nuclear technology and similar achievements, if professionalism is too much subordinated to the all-round ideal. It is too early as yet to say. But of one thing I feel certain, namely that China will not produce those types of utterly inhuman scientists and engineers who know little, and care less, about the needs and desires of the average man and woman. The 'new man' in China will solve this problem, aided by the infinite resources for development that are in the people themselves.

(9) Logic in China and the West

Another thing I should like to say something about is the question of logic in China and the West. One of the earliest things I noticed about my Chinese friends, some forty years ago, was that so often they would not answer 'yes' or 'no' to my formulations, but something like 'well, not exactly'. Undoubtedly this was an outward and visible sign of a certain subtlety of thought which runs right through the whole of Chinese

[91] *SCC*, vol. 4, pt. 3, Fig. 876, p. 262.

[92] One must remember that, broadly speaking, the Chinese people had in their history neither a hereditary aristocracy nor an entrepreneurial bourgeoisie, only an intellectual élite often quite widely recruited. I record grateful thanks here to conversations with Stefan Dedijer and Ivan Divac.

culture. From the point of view of the 'Science and Civilisation in China' project it is obviously of the highest importance to elucidate what part logic and logical thinking played in relation to the development of the sciences in China.

First, it can be shown that formal logic was more fully and perfectly incorporated in the linguistic structure of Chinese than of any Indo-European language. Secondly, all the main methods of reasoning and forms of syllogisms can be found in Chinese philosophical and medical writings from the −4th century onwards. But thirdly, it is clear that no Aristotle, no Panini, arose in China to codify successfully the features of formal logic – Kung-Sun Lung 公孫龍 and the Mohists attempted to achieve this, but because of the lack of interest of subsequent generations their writings were only imperfectly preserved and now have to be rescued from textual corruption.[93] It may have been precisely because of the profoundly logical structure of the language that the need for codification was never felt. Fourthly, by the same token, the minds of Chinese thinkers were not mesmerised by abstract logic, so that full weight could be given to all kinds of nuances rising above the 'single-vision' of the 'either-or' dichotomy. Fifth, and lastly, comes the question, what relation did all this have to the development of science in China? The answer seems to be that it had no effect at all, whether in mathematics, astronomy, geology, physiology or medicine. Only the breakthrough to modern science did not take place, and it seems in the highest degree unlikely that that could have been due to the presence of formal logic in the West. For as we all know, the founding fathers of the scientific revolution agreed with Francis Bacon's dictum *logica est inutilis ad inventionem scientiarum*.[94]

Speaking as one who was a working scientist for many years himself, I remember always feeling how unsatisfactory the 'A or not-A' disjunction was. Of course it was obviously useful, indeed quite essential, for classification, but always as a preliminary sorting to be followed by further sortings. It was thus the basic tool of the taxonomist, no doubt. But for the chemist, the physicist, or the physiologist it seemed radically unsatisfactory because in Nature A is always changing into not-A as one looks at it, and the difficulty is to catch it on the hop.[95] In my time at Cambridge, no science undergraduate or research student ever dreamt of taking courses in formal logic, and over many years of attendance at tea-club meetings and lectures by colleagues,

[93] Later on, the Dignaga logic, also intensional, was brought in from India, but it never spread beyond the relatively narrow circles of those who occupied themselves with Buddhist philosophy.

[94] Cf. Bacon (1620) in his preface to the 'Great Instauration', quoted in *SCC*, vol. 2, p. 200. A less well-known, but equally downright, statement may be found in Webster (1654), *Examination of Academies*. He wrote: 'It is clear, that Syllogising, and Logical invention are but a resumption of that which was known before, and that which we know not, Logick cannot find out; for Demonstration, and the knowledge of it, is in the Teacher, not in the Learner; and therefore it serves not so much to find out Science, as to make ostentation of it being found out; not to invent it, but being invented to demonstrate and shew it to others. A Chymist when he shews me the preparation of the sulphur of Antimony, the salt of Tartar, the spirit of Vitriol, and the uses of them, he teacheth me that knowledge which I was ignorant of before, the like of which no Logick ever performed . . .'.

I will only conclude with that remarkable saying of the Lord Bacon: 'Logick which is abused doth conduce to establish and fix errors (which are founded in vulgar notions) rather than to the inquisition of Verity, so that it is more hurtful than profitable'.

[95] Actually the principle of the excluded middle is quite compatible philosophically with change, as was shown in a brilliant paper by Ajdukiewicz (1948) (unfortunately still not translated into an international language).

I hardly remember any occasions when people had to be criticised on account of logical fallacies. The premises and the statistical treatments were always much more important.

As for the history of Chinese scientific thought, the avoidance of rigid 'A or not-A' conceptions can be seen very well in the relations of the Yang and Yin. These two great forces in the universe were always thought of in terms of a prototypic wave-theory, the Yang reaching its maximum when the Yin was at its minimum, but neither force was ever absolutely dominant for more than a moment, for immediately its power began to fail and it was slowly but surely replaced by its partner, and so the whole thing happened over and over again. This is what Nathan Sivin has called the fundamental principle of Chinese natural philosophy, or the 'First Law of Chinese Proto-Physics'.[96] And even during those very short moments of time when Yang or Yin reach the height of their powers, still they are 'not exactly' all Yang or all Yin, because by an extraordinary feat of insight the Yang harbours a nucleus of Yin within itself, and vice versa; and within the nucleus again there is an element of Yang, and so *ad infinitum*.

What China has to teach us here is, I think, that we ought to be much less rigid in our thinking and more flexible in our argumentation. This would mean that we would be more open-minded in many things, both scientific and social. We should be more ready to entertain ideas about possibilities hitherto unheard-of, alternative technologies, experimental social groupings. In personal life we should be less conventional regarding human relationships, and more tolerant of all ways of life that do not break the law of love.[97] And of course in science we should never fear the new and utterly revolutionary. We should 'test all things' as the apostle says, 'and hold fast to that which is good'.[98]

(10) Chinese and Western Attitudes to Nature

The sixth and last major point which I would like to take up in this discussion is the question of the different attitudes to Nature held traditionally by China and Christendom. This is a subject which readily lends itself to vague generalisations, but nevertheless I think that something relatively concrete can be said.[99] Many half-truths (or even less) concerning it have been in circulation for a long time. For example, F. S. C. Northrop wanted to characterise the Chinese approach as basically aesthetic in contrast with the scientific approach of Europe.[100] Or again, Feng Yu-Lan once said that the Chinese philosophers had never sought to dominate Nature; it was themselves they sought to dominate.[101] If this sort of thing had been the whole story it would obviously have been impossible to fill many volumes with the recital of the

[96] He first enunciated this 'law of inevitable succession' at the initial International Conference on Taoist Studies at Bellagio in 1968. 'Any maximum state of a variable is inherently unstable, and evokes the rise of its opposite.'

[97] It is noticeable that in Europe pleasure has all too often been thought of simply as the absence of pain, rather than a positive thing in its own right, the Yang as opposed to the Yin, and both indispensable parts of living.

[98] I Thess. 5:21.

[99] Two papers may be recommended: Cranmer-Byng (1972) and Watanabe (1972).

[100] Northrop (1946).

[101] Feng (1922); cf. Needham (1969a), p. 115.

achievements of the Chinese in all scientific realms from mathematics to medicine, over a period of some years. But there were great differences in the attitude to Nature, and here once again there is much for the world of today to learn from the Chinese tradition.

In the first place it is evident to anyone who knows anything about Chinese civilisation that it did not have any well-developed theology of a creator deity. Unlike the thinkers of other early civilised peoples, the ancient Chinese philosophers did not give much credit to creation myths accounting for the origins of the world. There were myths of organising and arranging gods, or demi-gods, but they were not taken too seriously. Chinese thinkers did not, in the main, believe in a single God directing the cosmos, but thought rather in terms of an impersonal force (*Thien* 天), meaning 'heaven' or 'the heavens' indeed, but here better translated 'as 'the cosmic order.' Similarly, the Tao 道 (or *Thien* 天 *tao*) was the 'order of Nature'. Thus in the old Chinese world-view man was not regarded as the lord of a universe prepared for his use and enjoyment by God the Creator. From early times there was the conception of a *scala naturae* in which man was thought of as the highest of the forms of life, but nevertheless that did not give to him any authorisation to do exactly what he liked with the rest of 'creation'. The universe did not exist specifically to satisfy man. His role in the universe was 'to assist in the transforming and nourishing process of heaven and earth', and this was why it was so often said that man formed a triad with heaven and earth (*Thien*, *ti* 地, *jen* 人). It was not for man to question the way of Heaven nor to compete with it, but rather to fall in with it while satisfying his basic necessities. It was as if there were three levels each with its own organisation, as in the famous statement, 'Heaven has its seasons, man his government, and the earth its natural wealth'.

Hence the key word is always harmony; the ancient Chinese sought for order and harmony throughout natural phenomena, and took this to be the ideal in all human relationships. Early Chinese thinkers were extremely impressed by the recurrences and cyclical movements which they observed in Nature – the four seasons, the phases of the moon, the paths of the planets, the return of comets, the cycle of birth, maturity, decay and death in all things living. *Fan che Tao chih tung* 反者道之動, as the *Tao Te Ching* says, 'returning is the characteristic motion of the Tao'.[102] *Thien*, or Heaven, was more and more seen as an impersonal force generating the patterns of the world of Nature; phenomena were thought of as parts of a hierarchy of wholes forming a cosmic pattern in which everything acted on everything else, not by mechanical impulsions but by co-operation in accord with the spontaneous motivations of its own inner nature. Thus for the Chinese the natural world was not something hostile or evil, which had to be perpetually subdued by will-power and brute force, but something much more like the greatest of all living organisms, the governing principles of which had to be understood so that life could be lived in harmony with it. Call it an organic naturalism if you like; however one describes it, this has been the basic attitude of

[102] Ch. 40; cf. Waley (1934), p. 192.

Chinese culture through the ages. Man is central, but he is not the centre for which the universe was created.[103] Nevertheless he has a definite function within it, a role to fulfil, i.e. the assistance of Nature, action in conjunction with, not in disregard of, the spontaneous and interrelated processes of the natural world.

Of course the ancient Chinese hunted and fished, but by and large their civilisation was always agricultural rather than pastoral; hence perhaps a more patient, less dominating, more feminine attitude to natural resources. The 'Man of Sung', who pulled up his sprouts in order to make the crop grow faster, was a standard laughing-stock for the peasant-farmers of China through two millennia, accustomed as they were to a much more patient attitude towards Nature.[104] It is true that widespread deforestation occurred as the ages went by, but this should be put down to the pressures of social conditions, and warnings against it can be found in many texts, as in the *Meng Tzu* book (Mencius),[105] or in the *Huai Nan Tzu* protesting against the inordinate use of wood for firing metallurgical furnaces.[106] Warnings against depletion of natural resources are quite common in Chinese literature; another good example would be the action of that governor of Kuang-tung in the Later Han period, Meng Chhang 孟嘗, who made the pearlers give a rest for a few years to the pearl-oysters, and afterwards guard against over-fishing.[107] Whenever anything could be done in accordance with Nature (this was the great *wu wei* 無為 doctrine of the Taoists) it was best to do it that way. For example, if water was wanted at 50 ft. above the level of a river, it was much better to take it off by a derivate lateral canal some miles upstream and follow the contours, rather than laboriously lift it by water-raising machinery at the spot. All this was not a 'passive' attitude to Nature, as some superficial minds have supposed; it was a profoundly right instinct that to use Nature it was necessary to go along with her. The Taoists would have applauded Francis Bacon's saying, *Natura enim non imperatur nisi parendo* (Nature can only be commanded by obeying her).[108] Thus to sum it up, there was throughout Chinese history a recognition that man is part of an organism far greater than himself, and by corollary a great sensitivity to the possible depletion, and pollution, of natural resources.

How different was all this from the feudal or imperialist domination of Nature arising from the Hebrew tradition. The People of the Book were never guided towards any restraint in the utilisation of those natural resources which God had provided for

[103] As Chang Tsai 張載 said in his *Hsi Ming* (*c.* +1066), 'That which fills the universe I regard as my own body and that which directs the universe I consider as my own nature'. All the Neo-Confucian philosophers had this nature-mysticism, this one-ning of oneself with the creative life-giving force of Nature, this love intoxication with all people and all things. 'The man of *jen* 仁', wrote Chheng Ming-Tao (*d.* +1085), 'forms one body with all things without any differentiation'. See the paper by Chhen Jung-Chieh (1975), p. 107.

[104] *Mencius*, II, 1, ii, 16. Cf. *SCC*, vol. 4, pt. 2, p. 347.

[105] The classical passage on nature conservation occurs at *Mencius* I, 1, iii, 3. Kung-Sun Chhou 公孫丑 took the same line, see *Tso Chuan*, Duke Chao, 16th year, Couvreur (1914), vol. 3, p. 272.

[106] Ch. 8, p. 10a; cf. *SCC*, vol. 4. pt. 2, p. 139.

[107] *Hou Han Shu*, ch. 106, p. 13b; cf. *SCC*, vol. 4, pt. 3, pp. 670–1.

[108] Bacon (1620), aphorism 129. Cf. *SCC*, vol. 2. p. 61.

their use.[109] As long as modern science still lay in the womb of time this lordship may have done no great harm, but once science had become airborne, as it were, then whole forests could be cut down every day to provide paper for the banal (and often vicious) printed matter of the popular press; while noxious chemicals, like organic mercury compounds or radioactive poisons, could freely spread about everywhere. Lynn White in his remarkable book *Machina ex Deo*[110] has revealed the responsibility of Christendom in inducing men to have an utterly possessive and destructive attitude to the rest of Nature,[111] a record only mitigated by the relatively minor voice of St Francis.[112] Western man will have to retrace his steps.[113]

Meanwhile today there is evidence that the Chinese are conscious of the possibilities of pollution, which after all they have seen in horrifying forms very close at hand in Japan; and they are building into their industrial plants all kinds of arrangements for avoiding toxic or opprobrious effluents. Clearly this is something much easier to do in a socialist economy than when a private firm, or limited liability company, is under the constraints of competitive marketing and profit making. Modern science too is validating some of the characteristic Chinese techniques. For example, the use of human excreta as fertiliser has been a characteristic feature of Chinese agriculture for 2,000 years. It was always a good thing in that it prevented the losses of phosphorus, nitrogen and other soil nutrients which happened in the West; but it was also a bad thing because it contributed to the spread of disease. But now in the light of modern knowledge of composting technique it is quite easy to avoid the latter drawback while retaining the former advantage.[114]

One last thought which emerges from this discussion of attitudes to Nature is that the Peoples of the Book, and the West in general, have always been far too given to masculine domination. It seems imperative and urgent that the Western world

[109] In words of much interest, Marco Pallis has linked this precisely with the anti-idolatry complex spoken of already (p. 79 above). 'It may be pointed out', he said, 'that when Christianity emerged as victor from its protracted struggle with paganism, a violent reaction set in against what had come to be regarded, rightly or wrongly, as a divinisation of physical phenomena, a certain anti-nature bias was thereby imparted to Christian feeling and thinking that has persisted ever since'. And he goes on to show how the Renaissance gave a vast impetus to human interest in all the things of Nature while at the same time setting the stamp of profanation on them, removed all hesitations concerning what the medieval theologians called *turpis curiositas*, and proceeded to the wholesale ravaging of the creation on land, at sea, and in the air. Pallis (1974), pp. 77 ff.

[110] White (1968). 'Christianity is the most anthropocentric religion the world has ever seen . . . By destroying pagan animism, Christianity made it possible to exploit Nature in a mood of indifference to the feelings of natural objects' (p. 86). What a contrast to Chinese *feng shui* 風水, which went perhaps to the other extreme, forbidding mines, roads and industries in the interest of leaving the blessings conferred by Nature undisturbed.

[111] See also Lynn White's celebrated article, White (1967).

[112] Parallel currents in Israel and Islam might he found in Chassidism and Sufism. On the Chassidism of Hasidaeans see *The Oxford Dictionary of the Christian Church* (1974), p. 271, and *A Dictionary of Christian Spirituality* (1983), pp. 226–7.

[113] These criticisms do not hold good of Eastern Orthodox Christianity to anything like the same extent. The Greek and Syrian fathers had a much more sacramental appreciation of material things than the up-and-coming West. See for example Brock (1974), p. 685, and Waal (1974), p. 697. Perhaps it was a true instinct for keeping intact the distinctively religious form of experience that led the Greeks to ban mechanical clocks and musical organs from their churches.

[114] See the interesting book of Scott (1952). For current composting methods in China the files of *China Reconstructs* and *China Pictorial* may be consulted.

should learn from the Chinese the infinite value of feminine yieldingness.[115] This is the message of the 'Valley Spirit' (*ku shen* 谷神) of the *Tao Te Ching*.[116] Of course for the Chinese the greatest perfection always consisted in the most perfect balance of the Yin and Yang, the female and male forces in the universe. These great opposites were always seen as relational, not contradictory; complementary, not antagonistic. This was far different from the Persian dualism with which the Yin–Yang doctrine has often been confused. Indeed, the Yin–Yang balance might be a good pattern for that equilibrium between the forms of experience which we need so much, that harmony between compassion and knowledge-power. Here again, then, there is something vital which the rest of the world has to learn from the Chinese tradition, if it is not to be torn to pieces by the interplay of intrinsic warring psychological factors, and external aggression against Nature and between men. The *ewig Weibliche* comes to us in Chinese dress, a Margaret-Gretchen, a Hsüan Nü 玄女, who can be the salvation of the world as she was of Faust himself.

It is high time that I came at last to my general conclusion. What I have been trying to urge is that in very many respects Chinese culture, Chinese traditions, Chinese *ethos* and Chinese human beings, those living today as well as those of all the ages, have contributions of outstanding importance to make for the future guidance of the human world. Nothing that I have been saying denies the 'Everlasting Gospel' of the two great commandments; but it is time that Christians realised that some of their highest values may be coming back to them from cultures and peoples far outside historical Christendom. The question is: what is humanity going to do with the Pandora's box of science and technology? Once again I should like to say: *Ex Oriente Lux*.

[115] Even the military classics, like the *Sun Tzu Ping Fa*, counsel the 'way of weakness', and warn against driving an enemy to desperation. As the *Tao Te Ching* says: 'Weapons are ill-omened things. All beings loathe them eternally. He who has the Tao has no concern with them.' Ch. 31; cf. Waley (1934), p. 181.

[116] Ch. 6; cf. Waley (1934), p. 149.

(e) LITERARY CHINESE AS A LANGUAGE FOR SCIENCE

Kenneth Robinson and Joseph Needham

(1) Scientific Style Distinguished from Literary Style

In the course of writing the previous volumes we have become very aware of the vast store of technical terms in Chinese literature, some of whose meanings are perhaps lost for ever. Some of them may be recovered by patient study of the technical contexts in which they occur,[1] and some of them are still in use, but have kept abreast of technical evolution by changing their meanings century by century.[2] All of these must eventually be studied and documented in detail.

When writing on botany in Section 38 above,[3] we referred to the *cri de coeur* of Emil Vasilievitch Bretschneider (1833–1901) who, like other early sinological botanists, sometimes expressed a poor opinion of Chinese botanical literature. He complained of the seeming lack of punctuation, and the alleged ambiguity of classical Chinese. He was exasperated by the lack of all indexes (except the contents tables) and indexed bibliographies. He lost himself in the confusions of place name changes in different dynasties, as well as the maze of obscure names of ancient foreign countries.

Bretschneider's plight may be compared to that of such scholars as Robert Grosseteste in the 13th century, labouring to make Latin translations of ancient Greek works, long unknown in Western Europe, through the medium of Arabic. Closer and more cordial contacts with the best scholars of the Islamic world would have been as helpful to the dons of the Merton School at Oxford as would cordial contacts with the leading scholars of China have been to the good doctor in the 19th century.

Only one year before the appearance of the first volume of Bretschneider's *Botanicon Sinicum* in 1882, Georg von der Gabelentz had published his *Chinesische Grammatik*. Bretschneider's second volume came out the year after the publication of Gabelentz' second great work, *Die Sprachwissenschaft* (1891). Both of these authors, writing at the same time, gave a great forward thrust to Western understanding of Chinese scientific literature.[4] In Europe the corpus of Greek and Latin authors surviving the fires, floods and neglect of centuries, required some five hundred years of work by scholars before it had been reduced to reasonably critical editions of the

[1] For example, Keightley (1978), p. 74, has shown how the ambiguous Yin bone graph (K362), *thien* 田 (field), which can mean 'taking to the field' either for hunting or for agricultural work, can lose all its ambiguity once it is known whether the area in question was or was not a hunting area. The precise meaning can sometimes be established by building up a detailed context from many references. Li Hsüeh-Chhin (Li Xueqin) (1985), p. 422, also shows how excavation has made it possible to be certain of the meaning of such terms as *mang-thung* 盲僮, which till recently had been known only as something used in connection with dead people, but which is now known to be a wooden figurine placed in graves.

[2] 'Tracing the stylisation of a written symbol through ten or twelve centuries is one thing, and determining its changing sense from epoch to epoch is another.' Britton (1936), p. 206.

[3] *SCC*, vol. 6, pt. 1, pp. 21 ff.

[4] For an appreciation of the contribution of Gabelentz, see Harbsmeier (1981), p. 6.

various texts. Chinese literature suffered no less from the burnings of libraries and other forms of destruction than that of the classical West, and many years will be required before a comparable state of textual criticism is attained. But the corpus of classical Chinese texts (those to be dated not later than –200), is relatively small, increasing the difficulty of establishing by comparative methods doubtful points and the long lost meanings of technical terms.

Changes in technology sometimes lead to false interpretations of earlier texts. As was said above,[5] a term may continue though the thing changes:

> Thung 銅 meant copper before it meant bronze. . . . Mistakes which have been made in the past about technical terms in Chinese texts have generally been due to the fact that scholars had neither the desire nor the time nor the necessary knowledge of the natural sciences to pin down the terms in this way.

A tantalising instance of the way changes in technology may have affected social life is afforded by the section in the *Li Chi* concerning the sort of decorum that was expected of people who ride in carriages: left hand well advanced with the reins in it, and – if the driver had a lady next to him – right hand well behind the back. This passage then goes on to say: 'the ruler of a state should not ride in a one-wheeled carriage!'[6] Legge, translating, clearly thought it was an admonition against riding in wheelbarrows, for in a note he said, 'Common so long ago as now, but considered as beneath a ruler's dignity'. If the term *chhi-chhe* 奇車 really meant wheelbarrow so early (*c*. –50), it must indicate that rulers of states were thought to have been tempted even then to use this novel conveyance, but the compound literally means 'strange or rare vehicle', and perhaps in the –1st century or earlier the warning was only not to ride in carriages which were outlandishly decorated. But when the wheelbarrow was in fact invented, not later than the +3rd century, the term rare or strange vehicle would have been appropriate enough. For the present we must leave this enigma 'in the remoteness of time'.[7] Examples of a term continuing, though the thing changes, occur in all languages.

In scientific writing more than any other it is the context which gives precision to the term. This has been well brought out by Karl Popper:

> Science does not use definitions in order to determine the meaning of its terms, but only in order to introduce handy shorthand labels. And it does not depend on definitions; all definitions can be omitted without loss to the information imparted. It follows from this that in science, all terms that are really needed must be undefined terms. . . . In science, we take care that the statements we make should never depend upon the meaning of our terms. Even where our terms are defined, we never try to derive any information from the definition, or to base any argument upon it . . . we reach precision [with terms] not by reducing their

[5] *SCC*, vol. 3, p. xliii. An outstanding case of retention of characters after the meaning has changed is *huo chien* 火箭, the incendiary arrow which later was used to mean rockets. See *SCC*, vol. 5, pt. 7, pp. 11–12.

[6] *Li Chi*, 'Chhü Li' 曲禮 I, 1 (5), 43; tr. Legge (1885), vol. 1, p. 97.

[7] For data concerning the evolution of the wheelbarrow, see *SCC*, vol. 4, pt. 2, pp. 258 ff. See also *Oracle Bones, Stars and Wheelbarrows* by Frank Ross Jr. (1982).

penumbra of vagueness, but rather by keeping well within it. . . . The precision of a language depends just upon the fact that it takes care not to burden its terms with the task of being precise.[8] A term like 'sand-dune' or 'wind' is certainly very vague. (How many inches high must a little sand-hill be in order to be called a sand-dune? How quickly must the air move in order to be called a wind?). . . . We always take care to consider the range within which there may be an error; and precision does not consist in trying to reduce this range to nothing, or in pretending that there is no such range, but rather in its explicit recognition.[9]

To take a precise example, 'Newton's vagueness as to what he was referring to by the word "force" did not prevent his formulating the laws of motion, nor did it hinder his discovery of the law of gravitation'.[10]

An aphorism usually attributed to St Thomas Aquinas serves to show that Literary Chinese, if it is ambiguous, has no monopoly in such statements. The aphorism is: 'Timeo hominem unius libri', 'I fear the man of one book'.[11] We realise that the genitive case in Latin of *unius* and *libri* is going to tell us no more than *of* does in English as to what this sentence means. It could mean that this man has only read one book, has only written one book, does not possess more than one book, or puts his faith in one book only. The fear that is felt may be on behalf of the man himself. Having read so little he is quite at the mercy of his one book! Or it may be that he is to be feared in debate. A man who has read only one book, but has thoroughly mastered it, is a formidable adversary. On the other hand, if he is the author of a single book, he may be unwelcome in conversation because of his limited interests. Alternatively he may be a specialist, and though one book is enough to contain all the points he wishes to make, yet his high specialisation makes him a dangerous adversary.

Chinese and English are not alone in having words with wide ranges of meaning; it occurs no less in classical Greek, which has words with whole panoramas of different meanings. For example, *ios* means smells, odours, 'virtues', magnetic attraction, pharmacological active property, rusts, oxides, violet or purple colour, and certain refining processes.[12]

Editors were, and even today usually are more interested in literature than in technology. It is apparent that if such editors were presented with technical material, they would readily succumb to the temptation to suppress tiresome technical details. The editors of China's dynastic histories, for example, were writing for non-specialists in astronomy. As a result, 'it is clear that many of the astronomical records are mere

[8] This principle was understood by Theophrastus, in spite of the Aristotelian theories to which he subscribed, when he wrote: 'For these reasons then, as we are saying, one must not make a too precise definition; we should make our definitions typical'. *Enquiry into Plants*, I, iii, 5; tr. Hort (1916), vol. 1, p. 27.

[9] Popper (1945), vol. 2, pp. 17–18.

[10] Stebbing (1942), p. 20, referring to Newton's *Opticks*, IV, p. 261.

[11] A similar aphorism is found in Chinese: *I phien shu khan chih lao* 一篇書看至老, 'to spend one's whole life reading a single book'. Though no less open to wide interpretation, its meaning has been limited by custom to narrowness of experience.

[12] See *SCC*, vol. 5, pt. 4, p. 483, and for *iosis*, the purpling process in metallurgy, *SCC*, vol. 5, pt. 2, pp. 23, 253 ff., and many other references throughout vol. 5.

summaries of the original observations with considerable loss of detail'.[13] When confronted by apparent ambiguity in Chinese it is necessary to examine very carefully the logical framework which forms the context, and if the text is not corrupt, to decide whether an item of information has been left out which should logically have been included, because, at the time of writing, the information would not cohere without it,[14] or whether it was deliberately omitted in order to extend the range of options in interpretation, and so to enhance the suasive appeal of the writing.

To give an example of an exalted type of suasive literature in classical Chinese, we need only refer to the passage from the *Filial Piety Classic* (*Hsiao Ching*) where it is explained how a filial son should behave towards his parents. The key words are bright with potential multi-faceted meanings, some of which are picked up and enhanced by others, while some, inappropriate to the context, remain unillumined. In its total effect it resembles a sparkling chandelier. The passage reads as follows:

> The Master said, 'The service which a filial son does to his parents is as follows: In his general conduct to them, he manifests the utmost reverence; in his nourishing of them, his endeavour is to give them the utmost pleasure; when they are ill, he feels the greatest anxiety; in mourning for them (dead), he exhibits every demonstration of grief; in sacrificing to them, he displays the utmost solemnity. When a son is complete in these five things (he may be pronounced) able to serve his parents.'[15]

It is not necessary to analyse this in full, but we may make a beginning to show the interplay of meanings.

The first line of the admonition which begins, 'In his general conduct to them, he manifests the utmost reverence', reads: *chü tse chih chhi ching* 居則致其敬. The first word, *chü*, means *a settlement*, *a dwelling* or *residence*, *to reside*, *to be at home*, *to be seated*; and then, developed from the idea of sitting: *repose*, *to be comfortable*, *tranquil*, *satisfied*. From this first word one gets a delightful impression of a Chinese family house, when everyone is still and relaxed, and even the dogs don't bark – the sort of sleepy hour on a hot afternoon when a son might very well lie down and take his ease. But it is exactly at times such as this that the truly filial son is alert and vigilant towards his parents.

The remaining words sound a warning of the high standards expected of the truly filial son. What is required of him in a normally relaxed situation is indicated by the fifth word, *ching*, which means quite a spectrum of feelings from reverence to carefulness.

[13] See Clark & Stephenson (1977), p. 23, 'We have here an example of the gross disparity between the technically accurate statement found in the *chih* 志 (treatise on astronomy) and the vague notices in the *pen chi* 本紀 (Basic Annals)'.

[14] Wang Chhung 王充 puts this succinctly: 'What can reasonably be assumed is not stated'. He then goes on to give an example, translated as follows: 'If a certain type of Chinese deer is hornless and brown, one may use this as a standard of reference in describing a unicorn. A writer, therefore, who says that a unicorn is like a deer but has a horn, may be assumed to mean that it is like a deer in other ways, but differs from it in that it has a horn. One is justified, therefore, in assuming that a unicorn has the same colour as a deer, for if it were black or white, which the deer in question isn't, this would have been stated' (*Lun Heng* 論衡, tr. Forke (1907), vol. 1, p. 370).

[15] Tr. Legge (1879), p. 480. The *Hsiao Ching*, Chapter X.

The reader who wishes to get the total effect of this five-word line is now confronted by the same sort of problem as faces the reader of James Joyce's *Finnegans Wake*. In reading slowly to himself, he can hold each word in his mind, not like a butterfly on a pin, but like cut diamonds flashing as the light of his attention moves along. But how to write out a translation in normal prose? James Joyce tackled this problem with C. K. Ogden in the summer of 1929 by translating a part of 'Anna Livia Plurabelle' into a tenor line in Basic English, counterpointed by the parallel line of the original, from which secondary meanings were brought out as footnotes clustering at strategic points, giving the effect of complex chords in music.[16]

Legge's translation of the passage from the *Filial Piety Classic* gives only the tenor line. The ambiguities implicit in the original are part of the counterpoint. In the second line of the admonition, for example, one does not have to choose, as Legge does, between whether the service of a filial son causes pleasure to himself or to his parents. The pleasure is mutual. If he is truly filial, the service of his parents will give him the greatest joy, and naturally it will give pleasure also to his parents, and the greater his pleasure, the greater theirs will be. It is true, therefore, as Legge points out that the fourth word in the line, the pronoun *chhi*, 'his' or 'theirs', is ambiguous, but it is a dynamic ambiguity.

Whether or not Literary Chinese uses extra words to make a meaning clear beyond a doubt depends on the writer's aim. If his aim is suasive, it may be more telling to leave the options open. As Graham says of late Thang poetry, it 'can be damaged severely by the irrelevant precisions imposed by Indo-European person, number and tense'.[17] But a passage closely parallel to this from the *Hsiao Ching* is one from the *Li Chi* which shows how precise a Chinese author could be when a matter of real moment was at stake.[18] This passage is concerned with what is meant by 'nourishing the aged', *yang lao* 養老. The author explains that the filial son should make his parents' hearts glad, not go against their wishes, gladden their eyes and ears, make them comfortable in their sleeping places, and supply them with drink and food, thus loyally 'nourishing' them. 'Such is the filial son to life's end', *hsiao tzu chih shen chung* 孝子之身終. But does 'life's end' mean 'to the end of the parents' lives', or 'to

[16] For example:

Joyce's composite text: There's the Belle for Sexaloitez! And Concepta de Send-us-pray! Pang!
Tenor line from text: There's the bell for Sachseläute – And Concepta de Spiritu. Ding-dong.
Undertones of text: There's the lovely girl for sexual loitering – and conception, let's pray that not! Birth-pang!

Joyce's notes on text:

Sachseläute: Swiss festival. An answer to the Latin of the Angelus. Overtones of romantic encounters. But the bell summons to church. Puns on Saxons as angels.
Concepta de Spiritu: Part of the *Ave Maria*.
Send us: Sanctus.
Pang: The sound of the bell tolling; the sound of clothes hitting the rock as the washerwoman who is speaking does her work. The suggestion of birth pains beginning for the 'belle'.

Some of the difficulties of interpretation and the elucidations arrived at in this co-operative venture with James Joyce are recorded in *Psyche*, nos. 41 and 46. We record our thanks for this quotation to the Orthological Institute, London.

[17] Graham (1965), p. 22. [18] *Li Chi*, ch. 12, tr. Legge (1885), vol. 1, p. 467.

the end of the life of the filial son'? For orthodox Confucianism this was not a matter which could be left open. The idea that a son need be filial only until his parents died was quite unacceptable. Here was a case within the suasive literature where the author felt precision was called for. He dealt with it in the following words: 'As for "until the end of life", that does not mean the end of life for the father and mother, but the end of his own life', '"*chung shen yeh che, fei chung fu mu chih shen, chung chhi shen yeh*" 終身也者，非終父母之身，終其身也'.

Yet the contrapuntal method of writing was no stranger to the West even before James Joyce. One need only take a line from one of Shakespeare's sonnets[19] to see it in action:

Bare ruin'd choirs where late the sweet birds sang.

The words *bare*, *ruined*, *choirs*, *sweet*, *birds* and *sang* can all be read with at least double meanings, and *late* can be interpreted in four different senses simultaneously, giving a multiplicity of interpretations, such as:

Chancels stripped of their adornment and falling into ruin, where late in the evening the birds we love so well were singing.

or:

The ghost of choristers, whose monasteries were ruined at the time of Reformation, wearing thread-bare rags, now sing where not so long ago they sang like sweet-voiced birds.

as well as the more obvious interpretation from the context of the sonnet as a whole:

Branches of trees now stripped of leaves form wind-swept choir-stalls / or – make lamentable music, / where only recently (in the warmer months) birds used to perch and sing so sweetly.

It is fitting to end this foray into multi-meaning as illustrated by Shakespeare's evocative line with an acknowledgement to William Empson who, by drawing attention to at least seven types of ambiguity, furthered the cause both of poetry and of clear thinking.[20] We may now turn from the language of reflection to that of participation.

It will be as well to spell out at the start of our discussions what exactly is required of a scientific language, and to make some estimate of the extent to which Literary Chinese met these requirements. If the language of science is to be efficient, the user's memory should not be burdened by terms which are over-long, hard to pronounce, differentiated from other terms in small ways which can easily be missed in rapid reading or speaking, or so far removed from ordinary speech as to be quite alien to the common reader. The language must be 'stripped of all grammatical superfluities and

[19] Sonnet LXXIII. For a fuller appreciation of this line, see Empson (1961), pp. 2–3.

[20] Empson (1961).

psychological hindrances'.[21] In considering language as a tool for science, Stanley Gerr makes the point that it must satisfy the basic engineering requirement of all 'tools': efficient and economical operation. 'An efficient language, in the strictly technical sense of the word, would be one in which there was a small consumption of "linguistic energy". . . . It must not only convey the writer's meaning accurately and fully, but it must do this with a minimum expenditure of mental and physical energy on the part of both author and reader'. It must not dissipate 'energy, time, and attention on unnecessary consideration of the linguistic or symbolic forms in which the ideas are presented'. In particular 'a definite object or universe of discourse must be postulated before they [descriptive ideas] can assume concrete, specific significance'.[22]

The advantages for concise scientific writing of a language undistracted by inflections become clear when one observes the shifts that English, which still retains inflectional vestiges, compels an author to adopt, as when we read: 'This indicates that some other factor/s is participating in the eventual closure of the adrenalectomised rat'.[23] Though unpleasant to ear and eye, this is less clumsy than saying: 'Some other factor is, or factors are, participating'. Whether it is essential in such a context to indicate the possibility of there being more than one other factor in question is doubtful. But it is quite certain that the verb *be* does not need three additional changes of form – *is, am, are*. Yet English has gone further than any other of the languages of Europe in approaching the Chinese model by divesting itself of unnecessary morphological distractions, a fact which has perhaps been not unrelated to its becoming the most used language of world science.

Provided, therefore, that when necessary a language can, amongst other things, distinguish between one and more than one; group items in sets, series and hierarchies; isolate one item from a group by reason of a distinguishing characteristic; indicate the results of comparison and suggest hypothetical alternatives, there is arguably a great advantage for a scientific language if it is not compelled to adopt unnecessary distinctions. Sometimes it is so important to be able to avoid making distinctions that a collective term has to be invented or resurrected to fill this role. For example, until about 1897, English authors were compelled to write of 'brothers and sisters and half-brothers and half-sisters' with time-consuming specificity, or else to be content with an inexact general term such as 'relation'; this was no longer

[21] Gerr (1944), pp. 2–3. The astonishing economy of Literary Chinese as developed by the Mohists is demonstrated in a sentence of nine characters translated by Graham (1978), pp. 142 and 350. It illustrates how, by the specialised use of particles, one implication can be put inside another: '*Tzhu jan shih pi jan tse chü wei mi*' 此然是必然則具為麋. This sentence refers to the Père David's deer, the *mi-lu* 麋鹿 or *mi* deer, which was believed to have four eyes, and was used as a stock example by Mohist logicians. The translation reads: 'If what is so of the instance here were necessarily so of a thing that it is, all would be *mi-lu* deer'.

[22] Gerr (1944), p. 6.

[23] V. G. Daniels, 'Research Report on the Termination of Macromolecular Uptake by the Neonatal Intestine after Lactation in the Rat' (Unpub. 1972). Further examples of the difficulties caused to authors by the continuing inflections of English are afforded by Alfred Bloom (1981), when he writes of children in general: 'The English child . . . she . . .' (p. 20), or 'If John had come in earlier they would have (but didn't) arrive . . . on time' (p. 19).

necessary once the Old English term 'sibling' had been reintroduced with a new and precise meaning to cover the four specifics.

The language of science should, in the words of Thomas Sprat, aim at a 'mathematical plainness'.[24] It should be clear, simple, accurate and, if possible, unambiguous, though an unambiguous statement is not necessarily clear or simple. If, for reasons of style or concision, a statement has to be ambiguous, the ambiguity should subsequently be explicated.[25] A term used in a specialised technical sense is not to be considered ambiguous if that special sense is clear to the intended reader, and if it is used consistently in that sense. For example, the term 'gynaeceum' will not be ambiguous to the botanist, even though it means 'the women's apartment' to the Greek historian.

It may well be that authors will be handicapped if a necessary technical term is missing from their vocabulary. For example, it is difficult to talk about the laws of gravity if there is no word for gravity in the language; but the point may be made that as soon as a concept has become clear, and the need for a term is felt, it will not be long before a new term is invented or borrowed. Like pre-Newtonian writers in the West, the Chinese did not have a term for gravity. Things were heavy, and by metaphor, serious. The word *chung* 重 in Chinese carries both these meanings, as does *gravis* and *gravitas* in Latin. But as soon as Newton made it clear that we are dealing with a force, English was able to isolate the Latin word 'gravity', which till then had mainly been used in figurative senses, and to distinguish it from the general words to do with weight and heaviness, reserving gravity or more often gravitation for the new idea of this force. Similarly all that Chinese needed to do in order to cope with the new concept in the 19th century was to form a new compound, namely *chung-li* 重力, 'weight-force', the term which is still in use today. In this Chinese term the idea of force is always explicit. In English it is sometimes implied, as when one talks about the Law of Gravity, meaning the Law of Gravitational Force.

Another danger in the use of a highly developed language, especially when one is addressing a well-educated reading public, is excessive abbreviation and over-compression. This is most likely to occur when the author is seeking stylistic effects rather than conveying information. In the preceding volumes examples have been given of Chinese authors who cultivated a 'mathematical plainness'. Wang Chhung comes readily to mind, and the agricultural admonitions of the *Chhi Min Yao Shu*.[26]

[24] The founders of the Royal Society of London 'have therefore been more rigorous in putting in Execution the only Remedy, that can be found for this Extravagance; and that has been a constant Resolution, to reject all the Amplifications, Digressions, and Swellings of Style; to return back to the primitive Purity and Shortness, when Men delivered so many Things, almost in an equal Number of Words. They have exacted from all their Members, a close, naked, natural way of Speaking; positive Expressions, clear Senses; a native Easiness; bringing all Things as near the mathematical Plainness as they can; . . .' Sprat (1722), p. 113.

[25] As is done in the *Li Chi* where 'life's end' has to be explicated to make it clear whose life is intended. See above, p. 100.

[26] Although so much Literary Chinese even on technical subjects is elegant, there are exceptions. As Graham (1978), p. 135, points out: 'From the point of view of style elegance means nothing to the Mohist, syntactic clarity means everything'; and in illustration of the jarring repetition of the word *yu* in two different senses, he quotes the line: '*yu yu yü Chhin ma, yu yu yü ma yeh*' 有有於秦馬，有有於馬也, 'To have some Ch'in horses is to have some horses'.

Then there were the early Mohists with their 'notoriously graceless style . . . humourless, ponderous, repetitious', giving the impression of 'the solemn self-educated man who writes with difficulty and only for practical purposes . . .', but whose style, in the later Mohist writings, is no longer clumsy; 'on the contrary it is neat and functional, but it is equally remote from literary concerns'.[27] Chinese authors had at their command a magnificent instrument for the communication of precise scientific ideas, for their synthesis at different levels from general to particular,[28] or for suasion towards the moral ends of life. How it was used depended inevitably on the aims and abilities of the author, and how we understand the works of the past depends in large measure on our understanding of those aims, as well as on the condition of the surviving texts. Defects of style must be distinguished from radical defects such as ambiguous attribution within the sentence, which mislead the enquirer after facts. Every great age of literature towards its end develops bad stylistic habits. There are, however, few examples in Chinese literature of English-type gobbledygook, such as 'The functioning quality of this exit is now deleted', meaning 'No exit'![29]

A scientific style which is not only precise but also elegant is not easily achieved, and is unlikely to appear early in a civilisation's maturing. One reason for this is that at the outset there is no clear differentiation between science, religion and magic. The suasive use of words, highly appropriate to magic, is mingled with the language of participation in the proto-sciences. Examples of such writing would be the opening verses of the Book of Genesis, or the description of the formation of a human being by the 'condensation' of *chhi* 氣, already quoted above,[30] which reads as follows:

(When the) *chhi* of the elements (is) settled, condensation (i.e. corporeality) (is brought about); this condensation (acquires) a spirit; (after it has acquired) a spirit it comes down (i.e. is born); (after it has) come down it (becomes) fixed (i.e. complete in all its parts); (after it has) become fixed (it acquires) strength; with strength (comes) intelligence; with intelligence (comes) growth; growth (leads to) full stature; and with full stature (it becomes truly) a Man.

(Thus) Heaven supported him from above, Earth supported him from below; he who follows (the Tao of Heaven and Earth) shall live; he who violates (the Tao of Heaven and Earth) shall die.

Poetry also plays its part. Science, semi-scientific anecdotes, biographical traditions and mnemonic verses – all were grist for the poetic mill. In the early centuries when books were a rarity, putting important information into verse was a wise precaution if it was not to be lost and forgotten. Even Confucius, whose interest in science was not remarkable, nevertheless apparently felt that learning by heart the songs of the *Shih Ching* gave the learner the additional advantage of widening his acquaintance with the names of birds, beasts, plants and trees.[31]

[27] Shih Sheng-Han (1958), pp. 26–7. [28] Graham (1978), pp. 7–8.

[29] For example, 'military might' can be expressed in Chinese at three different levels, *wu-li* 武力 (military power), *chün-li* 軍力 (army power), and *ping-li* 兵力 (soldier power).

[30] Quoted by the BBC on Radio 4 on 29 September 1982. The Chinese equivalent of this vice is perhaps excessive and not always very relevant quotation.

[31] See *SCC*, vol. 2, p. 242, for this account of the inscription on a –4th-century sword handle.

In both China and the West belief in the mnemonic and educative values of verse died hard. The *Chhien Tzu Wen* 千字文, the +6th-century 'Thousand Character Classic', for example, summarised the main fields of knowledge with which a young scholar might be expected to become familiar, in verses which could be easily memorised, while as late as 1789–91 Erasmus Darwin wrote *The Botanic Garden* in two volumes of Augustan couplets, because he hoped 'to enlist the Imagination under the banner of science', and felt that verse was 'better suited to entertain and charm'.[32] Volume I, sub-titled 'The Economy of Vegetation', in fact covered the whole field of natural philosophy. His verse was, however, supplemented by extensive footnotes.[33]

In this he may be compared with Tai Khai-Chih 戴凱之 writing in the +5th century, who, in his *Chu Phu* 竹譜, provides accurate descriptions of plants, originally distinguishing no fewer than seventy different kinds of bamboo. The treatise itself was written in the form of four-character rhymed verse, but after every few lines he added a brief prose commentary. This change or alteration in style perhaps indicates an early awareness that the style of suasive writing needs to be differentiated from that of scientific writing.[34] He began boldly by pointing out the difficulties of ancient classification which divided all plants into trees (*mu* 木) and herbaceous plants (*tshao* 草), to neither of which categories could he allocate bamboo. He seems, like Erasmus Darwin, to have been radical by nature, for he attributed to scholars who do not dare to rectify erroneous ideas expressed in ancient writings an undue deference to antiquity; comparing their superstitious reverence with the dread felt by the Huns for the Chinese commander Chih Tu 郅都 who had opposed them five centuries before.

We have now given some idea of the main desiderata of scientific writing. Stanley Gerr expressed his views on the subject as follows:

The language of science is necessarily characterised by two main features: rational syntax and a 'functional' vocabulary. The first of these was understood to imply the use of a simple, logically unequivocal sentence structure in which syntactic elaboration had been reduced to the minimum required for efficient and convenient exposition; the second was the extensive use of 'functional' or 'operational' terms.

'Functional' or 'operational' terms were explained as representing the coalescence of 'nominal' (i.e. structural or descriptive) and 'verbal' (functional, operational, dynamic) significance in a single basic symbol or expression, as in – a saw 'saws', a pump 'pumps', or say $\int f(x)dx$, in which the symbol of integration $\int$ might be interpreted in a 'nominal' sense – the 'integral' of some f(x)dx – or as a sort of verbal 'imperative' – integrate f(x)dx – depending on the context.[35]

[32] *Lun Yü*, XVII, 9.

[33] Today scientific verse is almost entirely humorous, as in a verse of J. B. S. Haldane which I recall:

I cannot synthesise a bun by simply sitting in the sun. I do not answer – yes, yes, yes, when I am offered meals of S. But readers, rhizo-stomes and rats are fairly good at making fats. So let us firmly stick to this, our most efficient synthesis.

[34] *The Botanic Garden* was published in two parts, Part I containing 'The Economy of Vegetation', published in 1791, and Part II containing 'The Loves of the Plants', which was published before it, in 1789.

[35] The *Chu Phu* (Treatise on Bamboos) has been described in some detail above, vol. 6, pt. 1, pp. 378–86. As an example of his verse plus prose comment technique we may quote stanza 5.

Considering the extent of Chinese scientific literature and the high level attained in many of the late medieval Chinese sciences and technologies, it would be surprising if Literary Chinese had failed to respond to the challenges of developing science. This must now be explored in greater detail.

(2) Comparison, Scales and Measurement

Most languages have the capacity to express comparisons between two or three elements in a series, and Chinese is no exception. In inflected languages it is usual to compare items by placing them in serial order, good – better – best, in English, for example, or in Latin, *bonus – melior – optimus*, and it is a simple matter to have suffixes or changes of stem to indicate comparative and superlative forms of adjectives and adverbs. In Chinese it does not make sense to speak of suffixes or stems, but such serial gradations are still possible by adding words before the adjective or adverb in question. But the genius of the Chinese and English languages is so different that one should not look for equivalents. Much is decided by the context. In an appropriate context one may use the word *hen* 很, 'very', and may say that a thing is 'hen hao' 很好, meaning not that it is very good, but that it is acceptable and may be described as 'good'. *Hao* by itself may be used to mean 'better' if a comparison is expressed or implied. In English we might say, 'This is a good one', implying that any others are inferior, though to make quite sure it would be more usual in English to say 'this one is the better', or 'this is the best'. In Chinese an adjective may be used by itself and may seem to be comparison free, as when a person on meeting says 'ni hao pu hao' 你好不好, 'Are you well?' – to which the reply might be 'hao-a' 好啊, 'I'm fine'. But there might yet be an implied comparison if the reply was 'wo hao-le' 我好了, 'I'm fine now', or 'I've recovered'. The example given is from modern Chinese, but similar examples can also be cited from ancient Chinese to prove that it too has the capacity to express comparisons. Harbsmeier does just that in his discussion of 'relative quantifiers and comparison of degree'.[36]

Clearly the capacity to think in relative terms is necessary for scientific investigation, for such investigation may require the recording and analysis of scientific data. This often involves serial evaluation and grading. Thus we may perhaps think of comparison as a proto-scientific capacity of language.

For scientific investigation it is usually necessary to classify data into more than three categories. Even in non-scientific activities, when answers are required to questionnaires expressing choice, preference or a value judgement, as in the routine activity of grading a student or an employee, it is usually more useful to do so on a scale

The shoots are called *sun* 筍 and the sheaths *tho* 籜. In summer they are numerous and in spring few. When the roots and culms are about to rot, blossoms and seeds *fu* 箙 then appear. When the bamboos produce flowers and seeds, in that year they decay and die. The character for the seeds is pronounced *fu*.

This mixed verse and prose form has a long history in China and India. See above, vol. 6, pt. 1, p. 379, and Waley (1958), p. 159.

[36] Harbsmeier (1981), pp. 100–15.

of five rather than three, poor – fair – good – very good – excellent, for example. The scale of three is the crudest, and the scale of five perhaps the easiest to operate while yet showing some sensitivity in discrimination since it permits two extremes, a middle point and two intermediates. A scale of seven permits two intermediates between the centre and each extreme, and a scale of nine permits the placing of one intermediate between each step in the five-point scale. It is obvious that higher powers of discrimination on the part of the investigator are required as the number of points on the scale increases.[37] It is not surprising that the Chinese, with their great penchant towards the pragmatic, showed a highly developed capacity for the classification of complex phenomena early in their history.

To extend this principle further, however, soon results in there being too many items within a group or category for them to be placed in serial order by comparison of adjectives, and yet one may wish to draw attention to a particular member of the group. In a family or a small military force, individuals can be called out by name. If their names are not known, they can be allotted numbers. An office can then summon a particular man from a rank of soldiers by calling out his number. This procedure can be extended by not only numbering but also measuring. For example a set of regular pentagons all have the same shape but may vary greatly in area. If their areas are measured they can be serialised by area. Anything which can be measured can be serialised by grading – towns on a map by their latitudes, crops in the Andes by their altitudes, the ripeness of apples by their colour, and so on.

Grading items on a scale from one extreme to another as a means of serialised classification came into use in China at an early date. One can see the simplest rating on a three-point scale (high – middle – low, *shang* 上, *chung* 中, *hsia* 下) developed into a nine-point scale, possibly by the beginning and certainly by the middle of the –1st millennium.[38] This was done by making nine binomes: high–high, middle–high, and low–high, high–middle, etc.

We have in a previous volume drawn attention to the revenue list of the nine ancient provinces preserved in the *Shu Ching* 書經 (Historical Classic).[39] The provinces are listed in the '*Yü Kung* 禹貢' chapter in geographical order (see Fig. 13), proceeding round China as it was in those days in a clockwise direction, from Chi-chou 冀州 where the old Shang capital city of Po 亳 was, to Yen-chou 兗州, which in modern terms would be in East Hopei and North-West Shantung, then to Chhing-chou 青州

[37] For sophistication this could not quite compare with the traditional examination marking system of Oxford and Cambridge universities, which had permutations of alpha, beta and gamma, with plus, double plus, minus and double minus, and therefore yielded three sets of five rating positions, a scale of fifteen in all with further refinements in the placing of +, ++ and –, ––. Such a system nevertheless makes it possible to place a knife in the back with exquisite delicacy. For example, it was said of Harold Wilson when he was Britain's Prime Minister: 'What a pity. So nearly ++'.

[38] The nine-point scale is used twice in the 'Yü Kung' chapter of the *Shu Ching*, as described above. The reasons for believing that this chapter was written in the first half of the –5th century are given in *SCC*, vol. 6, pt. 1, p. 83n (e).

[39] The description of each province as recorded in the 'Yü Kung' was translated in vol. 6, pt. 1, pp. 85 ff. As an example of the rating scales we may instance 'Chhing-chou . . . Its fields are on lower uplands, and its revenues are of the upper second grade' (*Chhing-chou . . . chüeh thien wei shang hsia, chüeh fu chung shang* 青州 . . . 厥田惟上下，厥賦中上).

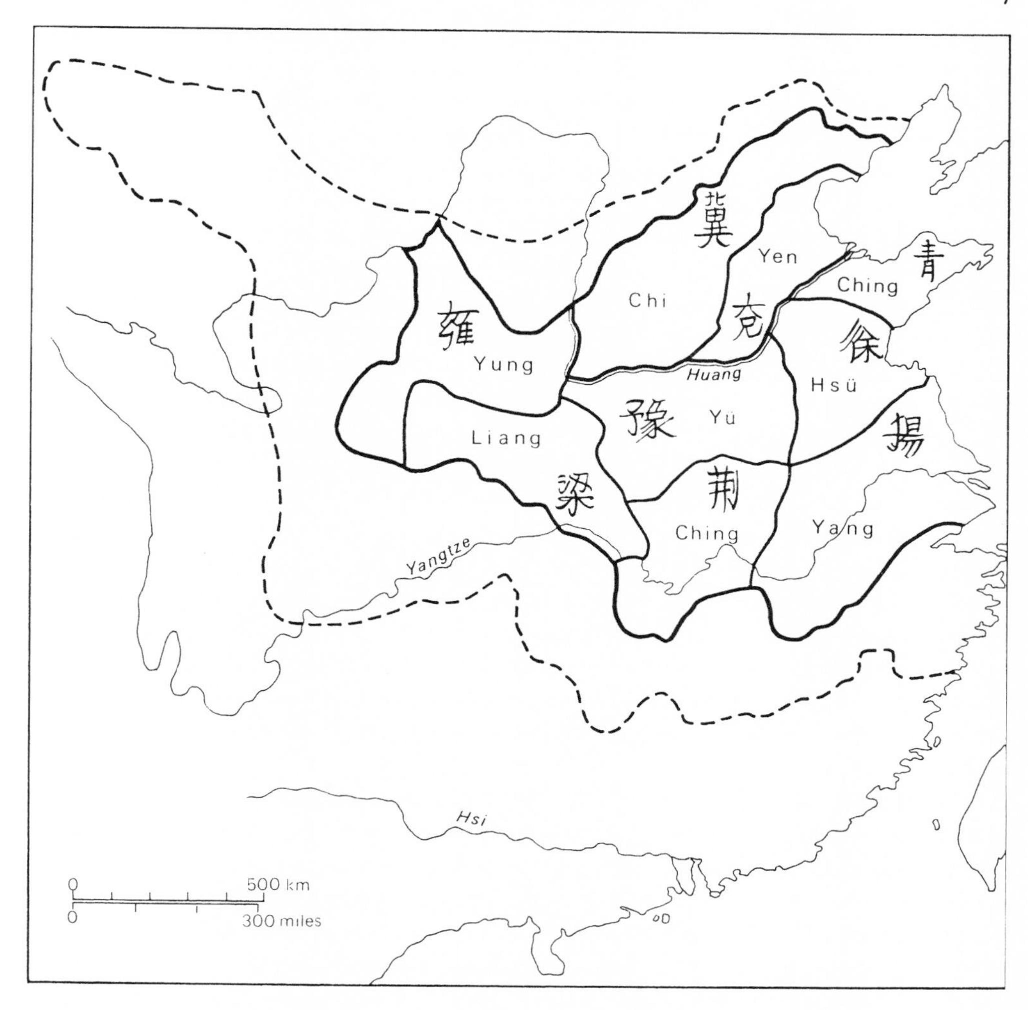

Figure 13. Map of the Nine Provinces. See *SCC*, vol. 6, pt. 1, p. 84, Fig. 23.

in North-East Shantung, and so on round.[40] However, the text also contains the ratings of each province as regards its capacity to produce revenue for the government, and we may list the provinces according to this capacity, following the serial rating order, with Chi-chou 冀州 in the first place, and its neighbour Yen-chou 兗州 in the last. The reasons why one province should be high on the rating list and another low, must have been complex because their revenues derived not only from agriculture but also from the production of textiles, the quarrying of precious stones and metals, the provision of salt and rare articles as tribute, and so on; to which must be added the factor of the difficulty of collection and transport in remote areas. Nevertheless,

[40] The provinces are listed in this order because the reader is following in the steps of Yü 禹, the culture hero, during his great work of draining the waters.

the important point is that during the –1st millennium, the rulers of China were able to assess their provinces and place them on a productivity scale in serial order.

The text contains a second rating of the provinces in serial order by altitude in three sets.[41] The lowest may be described as the 'Yangtze Valley Sequence', which consists of Yang-chou 揚州 at the eastward end of the Yangtze towards its mouth, Ching-chou 荊州 occupying a central position in its course, and Liang-chou 梁州 towards the headwaters of the river Han 漢, a tributary of the Yangtze. The middle set may be described as the 'Yellow River Sequence', with Yen-chou 兗州 in the east, where it reaches the sea, Chi-chou 冀州 further west into modern Shansi, and Yü-chou 豫州 south of this, but including the high country of western Honan. The third and highest set is not a river valley sequence, but two mountainous areas, Chhing-chou 青州 in eastern Shantung, Hsü-chou 徐州 in western Shantung, and Yung-chou 雍州 in the far west, where the land rises up towards the Tibetan massif. Naturally this province is rated the highest of the high on the altitude scale.

That items within a category should be placed in serial order is fairly inevitable once a centralised government begins to collect contributions in kind from its subjects. But the series could very well remain a traditional list learnt by heart. Once the serial order is determined by such abstractions as value, distance or altitude, the government is beginning to move in the direction of scientific thinking.

What is remarkable about the statements recorded in the 'Yü Kung' rating list is that the government of China should have started moving towards the measurement of abstractions so early. A few centuries later, however, there is evidence of far greater sophistication.

A marked difference in conceptualisation in China and the West is to be found in the manner of presenting abstractions. For example, the synthesis of opposed concepts as a mode of thought is of great antiquity in China, and is still active today in the formation of new compounds such as *shen-so* 伸縮, literally 'stretch-recoil', to mean 'elasticity'.[42] When in the past an area of thought was delimited by contrasting adjectives, such as open and closed, speakers of Indo-European languages were naturally tempted to exploit the facility of these languages in making new words by adding a suffix such as -ness to one of them, and to create a noun such as openness which would act as a label to summarise the vague area of thought delimited by the opposing adjectives. Unfortunately European languages have not taken kindly to the idea of fusing two adjectives into a combined unit to which a suffix could be added, as in 'the hot-coldness' of the country. Grammatical irregularities also add to the difficulties. They therefore chose one from the pair of adjectives (either 'high' or 'low', for example) for development as a noun to the exclusion of the other. Thus came into existence in Indo-European languages the huge vocabulary of what Newnham[43] calls 'loaded terms', such as *length* or *height*, terms 'which in themselves suggest that the item concerned *is* long or high'. In English there are

[41] The reasons for believing this scale to be one expressing the altitude of 'fields' and not their fertility are expressed with due caution in *SCC*, vol. 6, pt. 1, p. 92.

[42] Newnham (1971), p. 105.

[43] Onions (1936), p. 2145.

many irregularities in grammar and spelling and the growth of vocabulary often seems arbitrary. Whereas 'high' gives 'height', 'low' does not give 'lowth'; 'warm' gives 'warmth' and 'hot' gives 'heat' but 'cold' does not give 'coldth', and 'coldness' is not necessarily the opposite of 'heat' but may mean less warm than expected, as when one speaks of the coldness of marble. The same is true of *froid* and *froideur* in French. The trouble is that the loaded term is loaded in relation to an unnamed point of reference on a scale. The noun conceals a comparison of adjectives. For example, the height of a hill implies that it is higher than the plain or than sea level. One could only talk of 'lowth' in relation to some given standard of reference.

The coming of modern science precipitated a crisis for the users of the Indo-European languages. When heat was measured it became ridiculous to talk about ice having so many degrees of heat.

A new term was needed which would conveniently summarise all that lay on the scale between the two extremes. This could have been 'hot-coldness' but new words were invented, such as 'temperature', which had originally meant the action of tempering or moderating something. But by 1670 it had been given the new meaning of 'the quality or condition of a body which in degree varies directly with the amount of heat contained in the body'.[44] Sometimes an existing word met the requirements of modern science, as when 'distance', the space between two points, adequately summarised the relationship of 'near' and 'far'. But often the ancient word was not adequate, so that 'speed', for instance, which implies quickness, had, by the year 1550, to be supplemented with the new word 'velocity' which, though in Latin meaning 'speedy', in English came to mean 'rate of motion', covering slowness as well. And of course at a later stage we find the distinction between speed as a scalar, and velocity as a vector quantity.

The Chinese language was able to take this problem in its stride. As Gerr points out, 'An interesting and important group of terms in Japanese (and Chinese) uses the implied synthesis of "opposed concepts" to provide a general notion which includes both the extremes and a multitude of intermediate degrees'.[45]

A similar type of opposition of two ends of the scale occurs in the examples of mathematical procedures set out in the *Chiu Chang Suan Shu* 九章算書. At one end of the scale a simple example is given, at the other a complex one. This is already a concession, for a truly wise man would need only one example in order to understand the pattern; as it is expressed in the *Chou Pi Suan Ching* 周髀算經 – 'Though the principles [of mathematics] can be simply stated, their implications are far reaching. If from the knowledge of the problem one can deduce an appreciation of 10,000, a true understanding is then reached.'[46] In a society where this is the attitude to knowledge, the setting forth of a proof, particular by particular, in logical

[44] Gerr (1944), p. 37.

[45] Elasticity was itself a new concept in 1654. To coin a word for it, meaning had to be stretched from the Greek verb elaunein (ἐλαύνειν) meaning: 'To hammer out into thin plates'.

[46] *Chou Pi Suan Ching*, ch. 2, p. 16, in a dialogue between Chhen Tzu 陳子 and Jung Fang 榮方, trans., mod. auct., Chhen Cheng-Yih (1980), p. 72: '*Fu tao shu yen yüeh erh yung po che, chih lei chih ming, wen i lei erh i wan shih ta che, wei chih chi tao* 夫道術言約而用博者，智類之明，問一類而以萬事達者，謂之知道'.

steps would not only be disruptive of the Tao, but unworthy of a true scholar. There was, therefore, in China a considerable barrier to the formulation of mathematical proofs, due to the social attitude of the educated.

The *Kuan Tzu* 管子 cannot be precisely dated, for, as so often with these early works, ancient material was used by later compilers.[47] But around the –3rd to –2nd century is a reasonable working date. Chapter 58 of the *Kuan Tzu* contains a passage of the greatest interest in the development of scales for classification. It was discussed in some detail above,[48] being one of the oldest writings on geo-botany in any civilisation. But it merits further brief consideration on account of the skill with which these early writers classified the information fed in from prolonged field research.

This chapter discusses the different types of plant-clothed terrain, lists the quality of the soil and the accessibility of water, groups plant species in scales of aridity and altitude, and considers their edaphic relations. It is an analysis of habitats within the Nine Provinces (Fig. 13). These habitats are considered under five heads. The first three are types of land considered in relation to the depth of the water-table. The fourth outlines an oecological gradient on a twelve-point scale. The fifth is concerned with soil productivity (for further details see *SCC*, vol. 6, pt. 1, pp. 48–56).

Under the first head the land considered is that of a gently sloping plain surrounding a great river, categorised as potentially irrigable farmland, *tu thien* 瀆田. Five types of soil are described under this head, each placed on a scale in series according to the depth of the water-table, at intervals of 7 feet, from the 'driest' at 35 feet above the water-table to the 'wettest' at only 7 feet above. The ways in which each type of soil could be identified, by its colour and other qualities, and the plants naturally growing on it, were described in *SCC*, Volume 6, part 1, and need not be repeated.

From the gently sloping plain the survey proceeds upwards into hill country. Under the second head are considered different sorts of land in which the water-table becomes progressively further removed from the growing crops as the land rises up towards mountains.

The scale continues as before at 7 foot intervals, from a depth of 42 feet to 140 feet. Wells sunk in farms would have given the necessary water-table information. Because under this head there is so great a difference in the levels of the water-table, five subdivisions are not enough. Fifteen are named, and, in addition, warning is given of three types of area where digging is useless because no water can be found.

Right up in the really mountainous area which is studied under head three it is no longer possible to extend the water-table scale as before. Instead, three areas are described on the mountain top where water can be found by boring to a depth of 2 feet, 3 feet and 5 feet. In addition there are two areas on the 'flanks' of mountains[49] where it is possible to obtain water by boring to a depth of 14 feet and 21 feet. But

[47] On the composition and dating of the *Kuan Tzu* see Rickett (1965), pp. 1–35.

[48] See *SCC*, vol. 6, pt. 1, p. 49. This chapter is entitled '*Ti Yüan* 地員' which may perhaps be interpreted: 'On the Variety of what Earth Produces'.

[49] The terms used are *chhai* 豺, said by commentators to mean *phang* 旁, and *tshe* 側, both words meaning 'side', and evidently used here as technical terms for some aspects of a mountain's flanks.

here the sites are still serialised in descending order of depths to be dug, from 2 to 21 feet.

The usefulness of the number five in placing items on a scale is well brought out by the fact that the first three heads list five, fifteen and five items respectively. A scale consists of quite arbitrary divisions. It would, no doubt, have been possible to list seven types of soil by a different classification under head one, and to have set them out on the water-table scale at seven points dictated by the depth of the water in steps of 5 feet instead of seven. But in this instance, five types of soil seem to be the most convenient classification. Under head four, however, a different situation obtains, for here we have an oecological gradient in which typical plants are listed on a twelve-point scale of habitats ranging from lake water to dry ground. In their serial classification the early geo-botanists of China were not victims of numerology.

Under the fifth head are listed eighteen different types of soil graded for their productivity. What is particularly interesting is that these soils are serialised on, as it were, a percentage rating; that is to say, the three best types of soil are rated at ten out of ten, and the others as fractions of ten.

The three highest yielding soil types, each of which is awarded ten out of ten in the productivity rating, are described in great detail, as being the standards against which lesser soils may be matched. We shall return to this description before long. The fifteen remaining soil types range from 80 per cent to 30 per cent in productivity. As the *Kuan Tzu* writer says:

Among the Nine Provinces there are ninety different (sorts of) plants growing on their soils. Every type of soil has its regular characteristics, and every plant can be graded in an order (of luxuriance).[50]

(*Chiu chou chih thu wei chiu shih wu. Mei chou yu chhang erh wu yu tzhu* 九州之土為九十物，每州有常而物有次。).

These fifteen soil types are classified at three levels. Three are awarded 80 per cent for productivity, and are called 'upper soils', *shang thu* 上土; six are classified as 'middle soils', *chung thu* 中土, three of which are rated at 70 per cent and three at 60 per cent. The third group of 'lower soils', *hsia thu* 下土, are also six in number, but they are subdivided into three groups of two soils each, meriting 50 per cent, 40 per cent and 30 per cent respectively. Once again it may be noted that this system of classification does not look like the sort of schematisation that might have been dreamed up by a bureaucrat who had never got his hands dirty. No doubt it was put in its final form by such a person, but the data on which it is based smell of the land. For instance a bureaucrat would have been tempted to divide the fifteen types of land, which fall below the level of the standard-setting top three, into three equal groups of five each, or five equal groups of three each. But the best land is not the commonest, and it is entirely plausible that only three types should have been found to merit an 80 per cent rating. Similarly, the fact that the poorest land of all, rated at only 30 per

[50] *Kuan Tzu*, ch. 58.

cent for productivity, should form a sub-group of only two types is also persuasive. It is further to be noted that the author of this document resisted the temptation to classify all the lands together into five percentage groups: – 20 per cent, 40 per cent, 60 per cent, 80 per cent and 100 per cent – but proceeded pragmatically with his 'standard group' at 100 per cent, and then six, not five, other groups at 80 per cent, 70 per cent, 60 per cent, 50 per cent, 40 per cent and 30 per cent, which would certainly permit a finer grading than would be possible if he had classified them in only five groups.

Such, then, is the remarkable progress made in China by the –3rd to –2nd century in the art of serialised classification.

To form a scale by measuring the depth of wells to the water-table is one thing. To measure the lengths of a taut string which, when plucked at different intervals, produces a musical scale, is another. But to organise sense impressions when the data can be measured only with sophisticated modern equipment in such a way as to produce a scale is a very different matter. Yet in both China and Greece attempts appear to have been made to classify tastes, smells and colours on the analogy of a musical scale.[51] There was always the rainbow suggesting an order of colours which the Chinese followed by placing yellow in the middle, but it was not possible to apply numbers to colours except in a numerological way until Angström in his *Recherches sur le spectre solaire*, published in 1868, showed how wavelengths could be analysed by means of diffraction gratings; these were perfected by Henry Augustus Rowland in 1887, who invented a ruling engine with an exceedingly regular drive-screw which was able to rule up to 43,000 lines per inch on the metal grating.[52] Neither the ideas nor the technology for this were available anywhere until the 19th century.

Terms subtending different groups of named colours are very useful. English is at a disadvantage in not having a term in general which covers a range of greens and blues. One is forced to say duck-egg blue or peacock green or possibly turquoise, whereas scientific writing has *cyan-blue*, the green also known as 'verditer'. The Latin term *caeruleus* subtends *cyaneus*, azure blue, and *viridis*, leaf green, just as *chhing* 青 in classical Chinese usage subtends a range of greens and blues distinguished according to context. But *chhing* has acquired a broad covering power, often in order to avoid describing something as black, a word with unfortunate Yin 陰 connotations. Similarly in English the word 'pink' covers a segment of the spectrum at a certain level of dilution ranging from orange to mauve. To say that something is pink is not to speak ambiguously. It is a saving in mental energy to use a term at a more general level of classification if this meets the needs of the situation. If the speaker is required to be more precise he will descend one level and find such terms as 'salmon', 'rose' or 'cyclamen' to define more exactly the type of pink he has in mind. The same occurs in Chinese. It is often convenient for an author to use the term *chhing*, for it rarely happens that ambiguity occurs due to the sky being green or the grass blue, but if it

[51] For a detailed comment on the parallelism between colours and sounds, and the fact that the Chinese regarded them as 'tallies', see *SCC*, vol. 4, pt. 1, p. 164, n. e.

[52] *Dictionary of Scientific Biography*, vol. XI, p. 578. See Gillispie (1970).

should be necessary to be more precise, Chinese offers a wonderful range of colour terms.

It would seem that in early times in any culture, colour words are used loosely because of the complexity of the phenomenon being observed. Certainly in the classical West this was so. As Stearn says, 'Classical use of colour terminology was "too wide, too indefinite, too variable" to supply good precedent for modern scientific purposes'.[53] The terminology was used in variable fashion because of the three different factors in colour – hue, value and chroma. The word for a colour remarkable for its chroma or vividness, such as Tyrian purple, might be used because of its vividness rather than its hue, just as in Latin *purpureus* and *candidus* might be used to mean 'shining' rather than purple or white. Similarly in Chinese, the word *chhing* is sometimes used to indicate value (lightness/darkness) as when a cow described as *chhing* is black, and a horse said to be *chhing* is grey, though in Chinese the interpretation is complicated by the Yin factor mentioned above, or by the wish to indicate through the use of this word that the animal is brindled.[54]

But the rise of different technologies compels greater precision in the colour vocabulary. An artist wishing to reproduce a particular red would, even in early times, have had to distinguish between ingredients obtained from the madder plant or the kermes insect or from cinnabar, and increasing refinements in their manufacture increased the vocabulary, so that the word kermes generated two colour words *carmine* and *crimson*, for example. Rigour in the use of terms was even more important for dyers attempting to supply customers with the same colour for cloth, and the need for standardisation of colour became urgent when European armies followed the Turkish example and gave their soldiers clothing which was uniform in colour and cut. In China not only was there a high tradition in painting, dyeing and embroidery, but in ceramics the Chinese potters produced glazes with colours of a range and delicacy which astonished the world. Inevitably these colours acquired special names, but the nomenclature of Chinese glazes is fraught with difficulty, partly because time has taken its toll and the original term has become distorted, as with the *chi hung* 霽紅 or 'sky-clearing red' glaze,[55] partly because the term may describe the manner in which the glaze was applied rather than its final colour, or some other aspect of the technical process, and partly because a glaze intended for one colour might turn out as another, purple instead of red, for example, due to the vagaries of temperature or trace elements, or the absence of oxygen. But what is impressive is the fact that so many colours were named, and, as so often with Chinese technical terms, with a

[53] Stearn (1966), p. 242. But he also makes the point (p. 238) that 'the despised dyers, clothiers, artists, decorators and cavalry-men of antiquity, indeed all who in their callings then used colour terms with precision, must have had specialised vocabularies which have left little or no literary record. Colour names as used by poets tend to be metaphorically or indefinitely applied. . . . The development of a colour vocabulary depends largely upon progress in extracting and manufacturing dyestuffs and paints with consistent results.'

[54] A lacuna in the terminology of ancient Greece and Rome will also be observed in the fact that words for brown and grey were not introduced into Latin or the Romance languages until the arrival of the barbarians, who brought with them Old Teutonic *brun-o-z, whence Late Latin *brunus*, or Old High German *gris*, whence modern French *gris*. There were, however, words in Latin for dim, swarthy and tawny, such as *fuscus* and *pullus*.

[55] Hobson (1915), vol. 2, pp. 9–10, suggests possible meanings of the term.

vivid directness. Among the reds we find that of 'mule's liver', *lo-kan hung* 騾肝紅, and 'horse's lung', *ma-fei hung* 馬肺紅, as well as reds named after jewels, flowers, fruit, cosmetics, and the chemicals used in their manufacture. The use of colours stimulates classification. For example, of the various colours vaguely regarded as blue, one group in English is labelled violet, by abstraction of the colour from the general concept of the plant of that name. Of the various colours regarded as violet, some at a certain level of dilution are classified as lavender,[56] this term again being an abstraction from the general concept of the lavender plant. Similarly in Chinese the general term *chhing* subtends three rather more precise colours, indigo-blue, *lan* 藍, green, *lü* 綠, and that part of the spectrum where blue and green meet, inadequately described in English perhaps as turquoise, but in Chinese by reference to the blue-green plumage of the kingfisher, *tshui* 翠. Each of these again subtends a number of colours of still more precise application. The terms for them are labels, not definitions, just as in English 'mazarine blue' or 'shocking pink' do not describe the colours they represent, but are means of reference once the colour is known. Thus, among the greens we find some with uninformative labels like 'strong green', *ta lü* 大綠, or 'superior green', *shang lü* 上綠, but others are highly descriptive, being based on acute observation, such as the green which is the colour of bean oil, *tou-yu se* 豆油色, or the brilliant green of growing onions,[57] *tshung-lü* 蔥綠, to be distinguished from the more delicate green of onion sprouts, *chhing-tshung* 青蔥,[58] and the intense iridescent green likened to snakeskin, *she phi lü* 蛇皮綠.

Even if some sense impressions could not be graded by measurement, in the way that was possible with the levels of water-tables or by measuring the length of a string emitting notes of a particular pitch, it was still possible to promote a sense of order by grading impressions on some sort of scale.

A great step forward in scientific thinking is made when things are grouped and compared not only by likeness but by measurement. A point is reached where the comparison of adjectives is replaced by the juxtaposition of nouns, as when 'This is moister than that', is replaced by 'This contains 15 grammes of water, but that has only 10'.[59] The Chinese were, from the earliest times, concerned with the accuracy of measurement, something so distrusted by the Greek leisured class that in geometry Plato would permit only the use of the compass and ruler. This was illogical, since a straight line and a circle so drawn are only approximations, as the microscope or a good magnifying glass readily reveal. The Greek point of view was perhaps derived from some early confusion over things which can be counted and by contrast things which can only be measured, in which case the resulting numbers are approximations. This confusion between collectives, countables and measurables still persists in European languages today, as in such words as *informations* and *accommodations*, where speakers from different nations hold that information and accommodation can or cannot be

[56] For the botanical gradations of colour see Stearn (1966), pp. 240–2, and H. A. Dade's colour chart there reproduced, in which lavender is to flax blue and lilac as violet is to blue and purple.

[57] See Bushell (1906), vol. 2, p. 21. [58] Hobson (1915), vol. 1, p. 62.

[59] The importance of the supersession of Aristotle's qualitative physics by a quantitative and strictly testable physics is made by Graham (1971), p. 178.

counted. The Chinese, on the other hand, took an extremely pragmatic view of measurement, in early times as well as more recently, as Nathan Sivin has shown:

When we measure something an inch or a grain at a time there is bound to be discrepancy by the time we have counted up to a foot or an ounce. . . . Our best course is to continue using a technique so long as its inaccuracies remain imperceptible . . .[60]

In *SCC*, Volume 3, the point was made that Chinese characters which today mean 100, 1,000, 10,000 and so on, were not originally merely numbers, but indicators of place-value as well, a concept dating from Shang times.[61] It would seem that Literary Chinese was a language well equipped for describing accurate measurement, having terms needed for expressing not only very large numbers in terms of place-values but also very small ones (Fig. 14).

Over the centuries changes in the way terms were systematised occurred, and finally we find three systems developed from the simpler pre-Chhin system which had characters indicating each step from 1 to 10^8. These three systems are set out in Figure 14.

Although classical Greek used the word 'myriad' for ten thousand, Latin speakers were content with 'mille', a thousand. But by the 14th century in England, and somewhat earlier in France and Italy, the need was felt for a number-term rather bigger than the thousand, and this was provided by the Italians who added to the word *mille* an augmentative suffix, making *millione*, meaning 'a thousand thousands'.[62] This was followed in the 17th century by billion, whose ambiguity will be considered below. In Europe, then, the trend has been to increase the number of numerical terms, but in East Asia the trend has been the other way. China from an early date had not only, like Greece, a term for 10^4, a myriad (*wan* 萬); but also terms for 10^5 (*i* 億); for 10^6, a million (*chao* 兆); 10^7 (*ching* 京), and 10^8 (*kai* 垓). These terms certainly existed before the –3rd century[63] (see Fig. 15). During or after the Han some of them changed their values. It was quite unnecessary to have a special term for 10^5 (*i* 億), which could be expressed as 'ten *wan*', so *i* 億 was changed to mean '*wan* times *wan*' ten thousand times ten thousand or a hundred million, 10^8, and the word *chao* 兆, which had till then meant a million (10^6), was also unnecessary, for a million could logically be expressed as 'a hundred times ten thousand', which is how this number is expressed in China today. *Chao* therefore, during or after the Han, came to mean either 10^{12} or 10^{16} according to one or other of the systems in use. In other words, before the Han,

[60] *Tai Chen Wen Chi*, p. 99 (Hongkong, 1974), cited by Sivin (1982), p. 6.

[61] See *SCC*, vol. 3, p. 83. For a more recent survey of the early use of numbers in China see Lam Lay-Yong (1987), pp. 365 ff., where we read, 'the origin of the Chinese written number system, which is of decimal scale, could be traced to the oracle bone characters of the Shang Dynasty. . . . The counting rod system is a transcription of this decimal number system into a notational form . . .'.

[62] The word was sufficiently familiar in Venice by 1300 to be applied as a nickname to Marco Polo. See Yule & Cordier (1871), pp. 6, 54 and 67. But this early use meant only 'a large number or quantity'. Its use in relation to money was defined in 1330 by Iacopo d'Acqui as 'mille milia librarum'. See *Dizionario etimologico della lingua italiana* (1983).

[63] Chhien Pao-Tsung (1932), p. 93.

	10^{16}	10^{15}	10^{14}	10^{13}	10^{12}	10^{11}	10^{10}	10^{9}	10^{8}	10^{7}	10^{6}	10^{5}	10^{4}	10^{3}	10^{2}	10	1
PRE-CHHIN									垓	京	兆	億	萬	千	百	十	一
Lower system			載	正	澗	溝	穰	秭	垓	京	兆	億	萬	千	百	十	一
Middle system	京	千	百	十	兆	千	百	十	億	千	百	十	萬	千	百	十	一
Upper system	兆	千	百	十	萬	千	百	十	億	千	百	十	萬	千	百	十	一
Chiu Chang Suan Shu (example)					一 1	六 6	四 4	四 4	八 8	六 6	六 6	四 4	三 3	七 7	五 5	0	0
Modern Chinese	[兆]	千	百	十	[兆]	千	百	十	億	千	百	十	萬	千	百	十	一

Figure 14. Diagram showing Chinese numerical systems.

Decimal fractions / Type of measurement	Feet	Ins	Fen	Li	Hao	Ssu	Hu	Wei	Hsien	Sha	Chhen	Remaining Digits
	尺	寸	分	釐	毫	絲	忽	微	纖	沙	塵	
Linear	1	4	1	4	2	1	3	5	6	2	3	730950
Rectangular		141	42	13	56	23	73					095048801689
Cubic		1189	207	115	002	721	066					7175

Figure 15. Diagram showing methods of indicating decimal fraction systems.

there were special terms for ten to the power of two continuously up to eight, and possibly to nine.[64] A rational and economic system, however, only required special terms for 10^2, 10^4, 10^8 and 10^{16}. By eliminating special characters for 10^5 and 10^6, Literary Chinese came nearer to the ideally logical system. With the dropping of the special term for 10^7 (*ching* 京), and the use of *chao* for 10^{16} modern Chinese has now attained this ideal system, except that the word for thousand (*chhien* 千), too useful to be discarded, still remains,[65] and *chao* tends to be used on occasion for 10^{12}.

The complexities of higher numbers were not, however, entirely eliminated, for when systems had been elaborated by which the terms for higher numbers could be interpreted in three different ways it was necessary to know which system an author was using. More exactly, there were three systems, the lower, middle and upper, details of which were given in *SCC* Volume 3.[66] It is not necessary to repeat all that was said there. An example will suffice. *I* 億 in the 'lower system' meant 10^5, as it had done in early times. In the 'middle system' it meant 10^8, and in the upper system also 10^8, but *chao* 兆 in the lower system meant 10^6, in the middle system 10^{12}, and in the upper system 10^{16}. It is obviously important to know which system a mathematician is following in a given text. In our example of 1644866437500 quoted below [7.2.5, p.11] from the *Chiu Chang Suan Shu*,[67] Problem 24, and used in Figure 14, it is quite clear that the author is following the 'upper system' because his first figure, 1, is followed by twelve ciphers or places. In other words it is ten to the power of twelve. As can be seen from Figure 14, in the 'middle system' this would be expressed as *chao*, and only in the 'upper system' is it expressed as *wan-i* 萬億, i.e. $10{,}000 \times 100{,}000{,}000$.

This is the way the term is used in *Chiu Chang Suan Shu*, Problem 24. If it is true, as Shen Kua 沈括, the +11th-century astronomer, asserts,[68] that the lower system is the earlier, the pattern in the Chinese numeral system would seem to be similar to that which developed in India, where the Indian genius for nomenclature and classification showed itself in a numeral system in which, as in the 'lower system' in China, a new term was invented for each succeeding power of ten. But whereas the Chinese system does not appear to have exceeded eleven terms above 1,000,[69] which

[64] On this see *SCC*, vol. 3, p. 87.

[65] For the present Chinese numerative system see Figure 14. We are indebted to Li Wen-Lin 李文林 of the Institute of Mathematics, Academia Sinica, for help in this classification.

[66] See *SCC*, vol. 3, p. 87.

[67] The example in the *Chiu Chang Suan Shu*, ch. 4, Problem 24 reads: *i liu ssu ssu pa liu liu ssu san chhi wu* 一六四四八六六四三七五, i.e. 1,644,866,437,500.

[68] Shen Kua (*Meng Chhi Pi Than*, ch. 18, para. 7). See *SCC*, vol. 3, p. 87, n. b, where continuing confusion in modern scientific usage is discussed.

[69] The eleven terms are:

wan 萬,	*i* 億,	*chao* 兆,
ching 京,	*kai* 垓,	*tzu* 秭,
jang 壤,	*kou* 溝,	*chien* 澗,
cheng 正,	*tsai* 正.	

is two more than are found in the *Yajurveda Samhitâ*,[70] when the Indians elaborated a centesimal scale they invented twenty-three additional terms.[71] One cannot but admire the determination of Indian mathematicians who continued to find names for ever higher denominations far beyond the point where other civilisations had called a halt to the process. Nevertheless a big vocabulary of special terms for large numbers taxes the memory, and in India as elsewhere, simpler systems have now taken their place, though even today it is normal for large numbers to be expressed in lakhs (100,000) and crores (10,000,000).

The question of how large numbers should be expressed is no longer a real problem in writing,[72] since international conventions now largely obtain. All use the so-called Arabic numerals, but Western authors still group the ciphers in threes indicated either by a comma or by a space between the groups of figures. This is not usual in India where commas, if used, indicate lakhs and crores, and are placed to indicate 10^5 or 10^7, nor in China where the numbers are conceived in terms of the power of 4, 8 and 16, and where the logical point of view is taken that if commas or spaces are not needed for ciphers to the right of the decimal point, they should not be necessary to the left either. In fact, there are occasions when it would be convenient to group figures in a long sequence of decimal places in fives, in order that the reader may form a quick appreciation of how many decimal places are involved. The same would seem to hold for figures both to the left and the right of the decimal point. A glance would show within what power of ten the total number fell. The number of cubic feet in the sphere mentioned in *Chiu Chang Suan Shu*, Problem 24, for example (see Fig. 14), if grouped as 164 48664 37500 would immediately be seen to have thirteen ciphers, or in other words to be within the range of 10^{12}.

It is interesting that not only does Literary Chinese have terms like thousand and myriad for large numbers, but, at least from the +3rd century, it has also had terms for the positions of decimal fractions (see Fig. 15). These are nine: *fen* 分, *li* 釐, *hao* 毫, *ssu* 絲, *hu* 忽, *wei* 微, *hsien* 纖, *sha* 沙, and *chhen* 塵. But whereas a term like thousand or myriad indicates an actual number as well as a decimal place-value, these nine terms came to indicate merely the order of a number in a sequence of numbers. *Fen* is usually a tenth, and ten *fen* make a whole. But a *fen* is $\frac{1}{100}$ of a tael, a Chinese 'ounce' of silver (*liang* 兩). A *hao* is $\frac{1}{1,000}$ of a *mou* 畝 or Chinese 'acre', but $\frac{1}{10,000}$ of a tael. *Fen* and *hao* are simply guides to the position in the numerical sequence. A foot is divided into 10 inches, and a tael of silver was divided into

[70] *Yajurveda Samhitâ*, xvii, 2.

[71] See Datta & Singh (1935), pp. 10–12, for details of other systems in India.

[72] The same cannot be said of spoken numbers. Chinese numerals are admirably consistent, but Indo-European languages are plagued by anomalies, such as the trick of inverting the word order, as in 'three-and-twenty'; reversion to outmoded counting systems as in 'four score and ten', 'quatre-vingt-dix'; change of calculating method from addition to multiplication as in 'fourteen' and 'forty', with additional spelling anomalies in English; and failure to agree at the international level on the meaning of the terms *billion*, *trillion*, etc. This resulted not from ill-will but because Western civilisation is compartmented by nation-states. 'The American system of numeration for denominations above one million was modelled on the French system but more recently the French system has been changed to correspond to the German and British system' (*Webster's Dictionary*, 1976 edn, p. 1549 under Number Table). This confusion in terms is also discussed in *SCC*, vol. 3, p. 87, n. b.

10 'cash' *chhien* 錢. Only after these subdivisions of the basic unit do the nine sequential terms begin. Etymologically they may have begun as minute measures of length, the *hu* being described as the diameter of a freshly reeled silk fibre newly produced by the silkworm.[73] In fact they must have been invented or used as a later scholastic construct to express results of calculations. Liu Hui 劉徽 had no names for decimal places beyond *hu* 忽[74] but when names ran out he continued by expressing the fraction as part of a tenth, a hundredth, a thousandth, etc. of the last term. An unusual feature is that each of these terms may indicate not one digit in the sequence but two or three. Thus we find the 16th-century mathematician and musicologist Chu Tsai-Yü 朱載堉 expressing large decimal fractions in three different ways, according to whether they represented a linear, rectangular or cubic measure. For example, one side of a rectangle, or a square or cube root, might be expressed as shown in the 'linear' row in Figure 15, but the area of a rectangle might be expressed in square feet with double digits and the volume of a cube in cubic feet in triple digits. The examples in Figure 15 are taken from Chu Tsai-Yü's calculations to establish measurements for equal temperament and in the original text have from 10 to 23 places of decimals.[75] Chu Tsai-Yü expresses his calculations as follows: The side of a triangle measuring 14.142135623730950 . . . inches is written in Chinese characters as: 1 foot, 4 inches, 1 *fen*, 4 *li*, 2 *hao*, 1 *ssu*, 3 *hu*, 5 *wei*, 6 *hsien*, 2 *sha*, 3 *chhen*, followed by unclassified digits 7300950, etc. But when this side is multiplied by 10 inches and the resulting area is expressed in square inches, he writes it as: 141 inches, 42 *fen*, 13 *li*, 56 *hao*, 23 *ssu*, 73 *hu*, etc., and again the figures 1189.20711500272106671 75, representing a cubic measurement, are expressed in cubic inches as: 1,189 inches, 207 *fen*, 115 *li*, and so on. The decimal fraction terms are used to express not place but the sequential order of digits.

It is clear from these examples that Literary Chinese was as well able to meet the demands of expressing complicated numbers as any other pre-modern language. But counting and measuring are not the greatest tests of a language, as we shall see in what follows.

(3) Problems of Classification and Retrieval of Information

We have spoken of the importance of numbers in promoting a sense of order in a seemingly chaotic universe, and of how items which have been compared may be set out in a serial order of increasing usefulness and accuracy as the arts of measurement are refined. The next problem to be considered is that of the retrieval of information. Retrieval becomes a problem when information is available in such bulk that the reader can only find what he is looking for when it is arranged according to a systematic classification.

[73] See *SCC*, vol. 3, p. 85, on the expression of decimal fractions and all types of weights and measures. This follows a reference to Sun Tzu's having noted (by *c.* +300) that the diameter of freshly spun silk threads formed the standard for the *hu* measure.

[74] Donald Wagner *personal communication*, 9 June 1983.

[75] *Lü Hsüeh Hsin Shuo*, ch. 1, fol. 10a. See Robinson (1980), pp. 112–16.

Let us consider first the characteristics of the Chinese language as a source of labels for the items that are to be included in any system of classification designed for the easy retrieval of information.

In pursuing the question of systems of classification a considerable emphasis will be given to botany. This is because the relationship of the plants with which it is concerned are of such complexity as to present the highest challenge to systematic classifiers. The need for getting some order into the confusing world of plants was not lost on the thinkers of the Lyceum in ancient Athens, nor on their contemporaries in ancient China. Herbalists and doctors whose income depends on patients anxious for an instant remedy, are not in a position to delay over-long their decision on what would be the best medicine. They need an efficient finding mechanism.

(i) *Levels of specification*

The Chinese language is well endowed with words which do not compel a speaker or writer to be prematurely specific. This is less so with Indo-European languages. English has been compelled to invent such recent neologisms as 'chair*person*' to avoid the male specificity of *man*, an embarrassment which does not occur in such Chinese compounds as *kung jen* 工人, workman, where *jen* 人, like *person*, is non-specific. English is still unable to avoid the specificity of *he* and *she* which is possible in spoken Chinese with the single word *tha* 他. If specificity is required in Chinese it is always possible to say 'that male person' or 'that female person'. Similar obligatory distinctions, as between male and female friends, are forced on speakers of inflected European languages even more often than in English, as when the German speaker must say either *Freund* or *Freundin*. The law and government in the West often make it necessary to find a general term or a collective which can be relied on to cover all instances. The word *vehicle*, for example, is needed in English to cover many different types of carriage and conveyance. The Chinese equivalent, *chhe* 車, is a word of the people. This word may be analysed now as an example.

Originally depicted in writing as a chariot or wheeled vehicle, it was generalised and broadened to cover not only a vehicle but any sort of wheeled contraption.[76] Below this highest level of generalisation the word was distinguished into compounds of two types – (a) vehicles, and (b) machines with wheels, as opposed to machines which did not have wheels, such as looms. At the first level of specification occur such terms as (a) *ping-chhe* 兵車, 'war-vehicle' or chariot, and (b) *hua-chhe* 滑車, 'slip-contraption' or pulley, the simple machine with wheels that reduces friction and makes the work smoother, or *feng-chhe* 風車, the wind-machine with internal wheels, one of whose specific applications is winnowing.

At the next level of specification the word *chhe* in the sense of a vehicle may be further defined according to how it is powered, for example, 'horse-vehicle' (i.e. some type of cart or carriage), 'hand-vehicle' (i.e. a wheelbarrow or push-cart),

[76] On wheeled contraptions see *SCC*, vol. 5, pt. 5, p. 225.

'fire-vehicle' (a locomotive), or 'gas-vehicle' (a car or automobile of some sort using internal combustion for its power). Each of these may then be further defined, for example, with reference to their use or purpose. Thus the 'gas-vehicle' which is used for the transport of goods becomes a 'goods vehicle', *huo-chhe* 貨車, in which it is no longer necessary to repeat the word 'gas', and the term can conveniently be abbreviated. Similarly a 'gas-vehicle' which protects or saves lives, *wei-sheng-chhe* 衞生車, and one which carries the wounded, *fu-shang-chhe* 負傷車, are both in English regarded as ambulances, though in Chinese the differentiation of term distinguishes between ambulances for civilian and for military use.[77]

What becomes clear from this example is that Chinese terms can be simplified by cancellation of unnecessary elements almost as if there were a mathematical rule to that effect. The rule would be that if in its context one component of a term can be dropped without loss of clarity, then that component will be cancelled for the sake of brevity. In a really well-established context it is even possible to go back to the highest level of generalisation, equivalent in the above example to a wounded soldier on the battlefield saying, 'Where is the contraption?' This rarely happens in English or other Indo-European languages for two reasons. First, the terms are seldom organised, as they are in Chinese, in hierarchic levels reminiscent of botany. Instead of the four levels of *chhe* in the above example one finds in English at each level completely individual words which betray no connection or organisational relation to each other, i.e. contraption, vehicle, automobile, ambulance. Secondly, to form compound terms in which the components can be moved around like mathematical symbols, it is essential that they should be short, preferably monosyllabic. In English a technical term such as 'carry-wounded-vehicle', seven syllables where Chinese uses three, would be too clumsy to gain acceptance.

For the purpose of science a language needs to be well provided not only with general and collective nouns, but with nouns at different levels of specificity, appropriate to the level of detail at which a matter will be treated.[78] If things are to be properly classified, each descent to the more specific creates a term which becomes a collective for the category of specifics at the next descent. For example, in botany an ovule in the specific condition of belonging to a phanerogam and having been fertilised is termed a *seed*. A particular type of seed together with its envelope is described as a

[77] The facility with which Chinese technical terms can make distinctions often lacking in Indo-European languages should be noted. The word 'armature', for example, is distinguished in Japanese according to whether it is engaged in producing or consuming an electric current, *hatsuden-shi* (*fa tien tzu* 發電子), being the armature of a generator, and *dendo-shi* (*tien tung tzu* 電動子), the armature of an electric motor. A technical term is not, of course, self-defining. It is a label. The Latin components of the English word *locomotive* give a rough indication that the machine in question is not static but moves from place to place. But they do not define it. Also words may remain when the things they label have changed beyond recognition, as is shown by the English word *mass* which originally meant a cake of barley meal and ended up meaning a quantity of matter. Or again, accidents can happen, as when the newly invented wheelbarrow borrowed the name *fang-chhe* 紡車 from the silk-winder, since both were mono-wheels. Subsequent generations could not distinguish one from the other as they appeared in ancient texts, just as future generations might be puzzled to read in English, 'He took his wheel for a spin'. See *SCC*, vol. 4, pt. 2, p. 267n.

[78] For further thoughts on Chinese technical terminology see *SCC*, vol. 3, p. 574, 'We know the technical terms for all the parts of the very complicated bronze crossbow triggers of the Han . . .'.

Table 1. *Soil identification characteristics, from* Kuan Tzu, *Ch. 58, 'Ti Yüan'.*

Soil types / Check-list	*Hsi Thu* 息土	*Chhih Lu* 赤壚	*Huang Thang* 黃唐	*Chhih Chih* 斥埴	*Hei Chih* 黑埴
Water-table (approx.) depth in feet	35	28	21	14	7
Units for soil 'resonance'	64	72	81	96	108
scale note or position	chio	shang	kung	yü	chih
animal making similar sound	pheasant	sheep	cow	horse	pig
colour of water	grey-green	white	yellow	red	black
taste of water	–	sweet	stinking	salty	bitter
colour of soil	–	red	yellow	–	black
characteristic of inhabitants	robust	long-lived	unsettled	unsettled	–
crops grown	all cereals	all cereals	glutinous millets	soy beans wheat	rice wheat
distinctive plants and trees (some examples only)	thatch grass jujube tree	floss grass wild pear	floss grass white mulberry	reeds willows	[not certainly identified]

fruit. An indehiscent fruit is a *nut.* Found on bushes belonging to the genus *Corylus* is the type of nut known as a *hazel.*

For quick description, and the ability to direct another person rapidly to the objective, without having to search for the precise word, it is important to have at a certain level general words which do not compel a speaker to make choices before choice is necessary. If one has a general colour word like *black*, this is enough to begin with. It can be qualified later as regards value and chroma. If one has a general weather word like *wind*, it can be qualified later to indicate a high wind, gale, blizzard, typhoon, etc. Items in series of five are easy to memorise. They can be used as check lists. To see the check lists in action in Chinese we may turn to the chapter of the *Kuan Tzu* already referred to,[79] named '*Ti Yüan* 地員', which, as we

[79] See above p. 110 and for botanical details *SCC*, vol. 6, pt. 1, pp. 48 ff.

said in our discussion on ecology and phyto-geography in Volume 6,[80] could well be translated as 'On the Variety of what Earth Produces'. The word *yüan* in fact has, as one of its meanings, 'number', explained in the *Shuo Wen* as 'Thing numbers'.[81] If the chapter title here is 'Earth Numbers', this might be developed a little, as: 'On the use of numbers for grading crops from different soils at varying depths of water-table'. This chapter begins with a study of five main types of soil. Its findings may be reduced to a table, as shown in Table 1. It is certainly not a work of rural uplift, written to benefit the peasantry. On the one hand, a peasant whose family has been cultivating the same fields for generations does not need to be told that the soil is sticky clay but will produce a decent crop of beans. This he knows already. On the other hand, even if he could read, it would do him no good at all to discover that the musical pitch note of his soil is expressed in terms of a length of ninety-six units. Unlike the *Georgics* of Virgil which may have been intended to help persuade those yeomen farmers who had not yet been put out of business by the slave-manned latifundia of Italy that they were on to a good thing, this 'Ti Yüan' chapter was perhaps designed for officials whose task was to assess the productivity of the land with a view to taxing it, or possibly, in times of benevolent rule, to avoid overtaxing peasants whose crops could not possibly meet their obligations.[82]

If a tax official is to make a fair assessment, he must be quite sure what the land is capable of. If not much of a practical farmer himself he would need some clear guidelines to go by, in helping him to assess the land. The 'Ti Yüan' gives him ten such indications. Some of these are very sensible. For example, there is in some cases the colour of the earth itself as shown by the name. There is red '*lu* 壚' soil, yellow '*thang* 唐' soil, and black '*chih* 埴' soil.[83] Another excellent way of recognising soil types is by noting what plants grow on it. But if a man is not a botanist he will be confused by too much detail or profusion. The 'Ti Yüan' therefore selects just one or two 'grassy plants' (*tshao* 草), and one or two trees for each type of soil, plants which would in those days have been easily identifiable, such as thatch grass or the jujube tree, which grow more readily on one type of soil than another. The 'Ti Yüan' also indicates what are the staple crops which grow well and can be expected to bring in the revenue, and sometimes indicates other sources of profit. There is also a brief description of the people living on the different soils, summing up their chief characteristic, such as 'sturdy', 'long-lived', and in two cases 'wandering', meaning perhaps that the people living on the poorer soils were obliged to follow shifting cultivation, getting only a few crops of millet from the same place before moving on. There is no comment on the people living in the swampy land where rice was grown, but the text has suffered in several places. Six rows in the check list (Table 1), however, still remain to be considered, and these are of great interest, for they are

[80] *SCC*, vol. 6, pt. 1, p. 49. [81] The phrase is *wu shu* 物數.

[82] See *SCC*, vol. 6, pt. 1, p. 93, which in this connection cites the *Chou Li*, ch. 4, p. 34b (ch. 16); tr. auct., adjuv. Biot (1851), vol. 1, pp. 368 ff.

[83] For further details of the five soils, *hsi thu* 息土, *chhih lu* 赤壚, *huang thang* 黃唐, *chhih chih* 斥埴 and *hei chih* 黑埴, see *SCC*, vol. 6, pt. 1, pp. 49 ff.

all to do with water. Any form of precipitation in North China is of importance. If there is no snow and the rains fail, the level of the water-table is of vital importance, and so also is the purity or degree of salination or alkaline content of the water.

When the number of items of information, or of species and varieties, is very large, they can be systematised in one of two ways. If the main aim is to create a finding-system, items can be pigeon-holed with the help of a system of co-ordinates. If the main aim is to place an item in a network, in such a way that its connections and relationships with other items in a system, and with other systems, become clear, then it must be classified in a hierarchic or dendritic system, its position being decided as accurately as possible by a close study of its every detail. There are, however, drawbacks to both of these devices.

If a pigeon-holing system is used, one or both of the parameters may consist of artificially selected criteria, and a particular item may not properly fit into any of the available pigeon-holes. Alternatively it may fit so many as to render the system useless as a finding device. With the dendritic system, on the other hand, the main drawback is that any system of classification arrived at must be regarded as provisional. Indeed, all science is provisional, and nowhere is this more apparent than in zoology where classificatory systems are continually being eroded or transformed.[84] Moreover, when a system has been so perfected as to appear relatively permanent, the complexity resulting from amassed detail is likely to hinder universal intelligibility, that vital ingredient in social stability, of which Marcel Granet said, when discussing the Chinese conception of Order and Peace: 'In all the Schools we find the idea . . . that there is no difference between the principle of universal good understanding and that of universal intelligibility. . . . All authority rests on Reason.'[85]

The supreme example of the pigeon-holing system is the *I Ching*[86] which we have described as a Universal Concept Repository and as the 'Administrative Approach' to natural phenomena. We have already drawn attention to its deadening effect on scientific enquiry, quoting the passage where *blood* is pigeon-holed under the twenty-ninth hexagram.[87] This hexagram is one of the sixty-four criteria which form one of the axes of the device, and represents the abstract concept of *flowing motion*, with which are arbitrarily correlated other things such as being blood-red in colour. If an item is pigeon-holed under No. 29, it must then inevitably be blood-red, and there is therefore no need for further enquiry into blood's redness or its causes. At the same time it must be admitted that the damage is done not by the criteria as such, though they may not always have been well chosen, but by the qualities, properties and phenomena with which they were correlated not on scientific grounds but on those of sympathetic magic, similarity or even sheer whimsy.

[84] 'Finally, it should be emphasised that classification schemes are simply cataloguing systems, devised by scientists to help them in their work. A classification therefore does not constitute proof of relationships between organisms, nor does it necessarily indicate a pathway of evolution, although classifications provide a conceptual framework for the discussion of these matters.' Friday and Ingram (1985), 'A Classification of Living Organisms', p. 397.

[85] Granet (1934), p. 591. [86] *SCC*, vol. 2, p. 304. [87] *SCC*, vol. 2, p. 334.

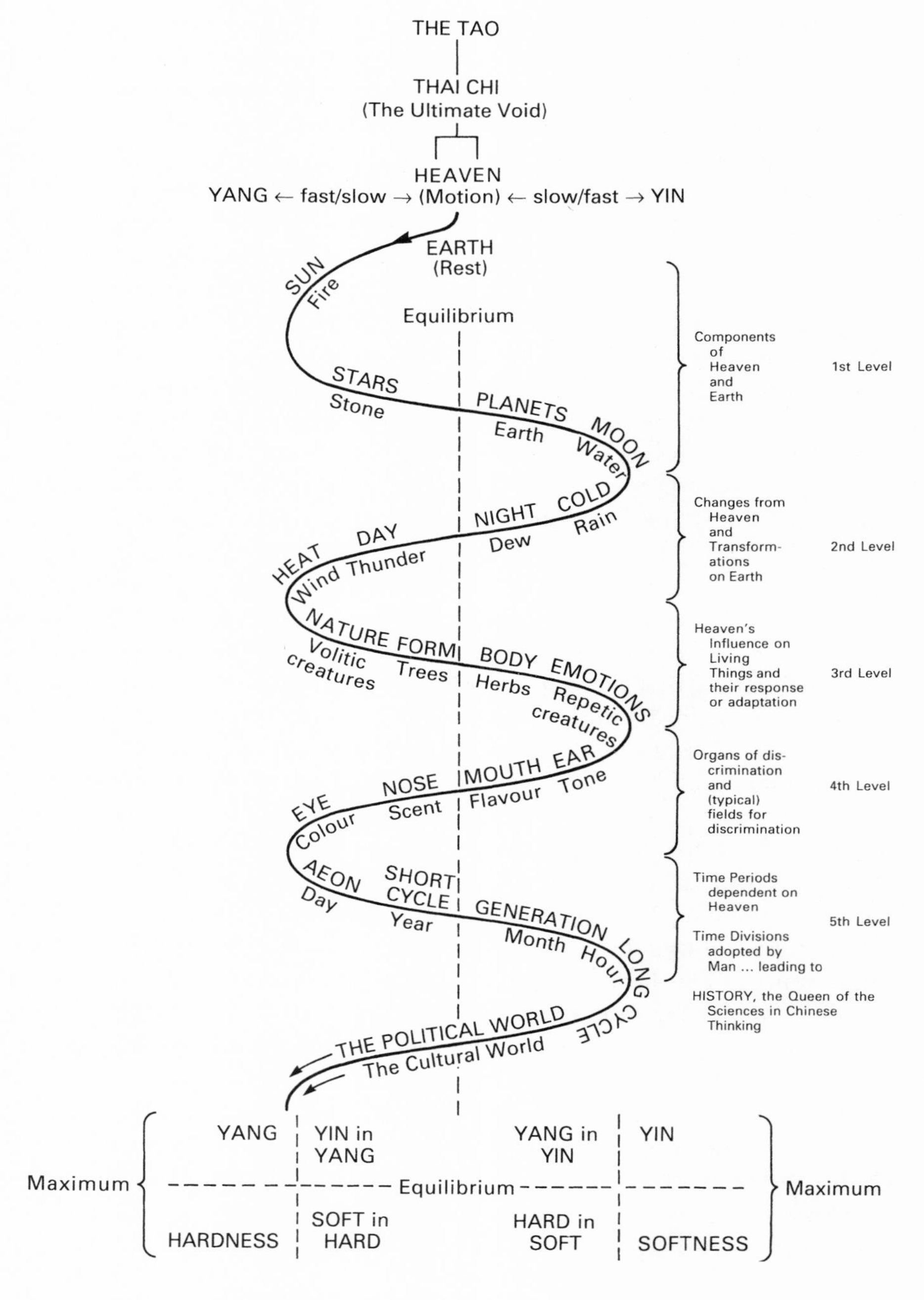

Figure 16. Diagram showing the Tao and the parameter of change as described by Shao Yung. From Needham & Robinson (1991), *Comparative Criticism*, **13**, p. 14.

There is much scope for classifying information by co-ordinates, and if the aim of the *I Ching* had been simply to introduce order into the bewildering world of natural phenomena through classification, it would have served a useful purpose for a time, as did the artificial classificatory system of Linnaeus. Though the criteria chosen would not seem to be ideal for classifying things, i.e. objects, artefacts and products, it must be remembered that its original purpose was by divination to introduce order and decision into the world of apparently random human events. Once the *I Ching* was used for more than classification, and attempted to indicate connections between events, past and future, it ceased to be of possible benefit to science.

The actual mechanism of the *I Ching* system of classification was extremely simple. One may visualise it as an oblong nest of boxes. Along the top, which forms one of the axes, are the names of the hexagrams, the sixty-four *kua* 卦. Vertically down one side are the categories into which the 'ten thousand things' may be grouped. When one of the ten thousand things comes up for classification it must first be allotted as accurately as possible to its group, and then, as accurately as the information available allows, to one of the sixty-four *kua*, and so to its pigeon-hole. Once in its pigeon-hole it was allowed – and this was the fatal error – to acquire the characteristics of its *kua*. So *blood*, on being inserted in the pigeon-hole under *kua* No. 29, acquired the characteristics of redness, of flowing motion, of the dangers associated with torrents of water, of the edges of ravines and of danger in general and the reaction to it, regardless of the fact that it was because it had clearly got some of these characteristics, and could easily be associated with others, that it had been allotted to pigeon-hole No. 29 in the first place. The complexity of the *I Ching* system arose from the difficulty in deciding into which category and pigeon-hole a thing should first be posted. Here the divinatory process played its part. Once categorisation had been achieved there was little incentive to ask difficult questions which might lead to its being taken out for reclassification.

(ii) *Shao Yung's contribution to classification*

It is clear that the diviners who first created the *I Ching* did not intend it primarily as a scheme of universal classification. In the work of Shao Yung 邵雍 (1011–1077), however, we find a deliberate attempt to create such a scheme.

To appreciate Shao Yung's work, it must be borne in mind that he broke away from the theory of the Five 'Elements' (*wu hsing* 五行), and developed a new one based on fours, or rather, on mathematics of a simple sort, that of the *I Ching*, in which one is divided into two, two into four, four into eight, and so on, up to sixty-four, the total number of hexagrams (*kua* 卦). It must also be remembered that in his view of nature he was not dealing with inert particles which can be isolated, but with a vibrating interacting network in continual movement, from Yang to Yin and back again to Yang, changing as regularly as the full moon ebbs to new moon and back again to full. It is better, therefore, not to tabulate Shao Yung's ideas in static lists. If they are to be represented diagrammatically, the diagram must embrace the idea of motion and rest.[88] This is to be represented in Figure 16 by the sinuous dark line emanating from the Ultimate Void, the *Thai Chi* 太極, passing by way of Heaven and Earth through all the manifestations of the Tao in Time. What he gives us, however, is not a cosmogonic history, nor a proto-evolutionary theory, but a number of charts embodying a system of classification of far greater sensitivity than that of the old pigeon-holing system of the *I Ching*.[89]

The terms which appear on this diagram are those which Shao Yung selected in order that the 'the myriad things' might be classified under new principles. His hope was evidently that if the right criteria are selected as co-ordinates, and they are permutated to exhaustion, an appropriate pigeon-hole will be found for every item. But how to select them, and how to permutate them? It is obvious that the world does not consist of simple mixtures – fire with earth, with stone, or with water, for example. In any mixture the proportions are important. In allowing for this the Yin–Yang theory would be a help, for the proportions of the Yin in the Yang, or of the Yang in the Yin were held to be continually varying. If creatures or plants were influenced by the Yang or the Yin when they came into being, one would expect to find them ranged in a series of infinite subtlety.

First, therefore, following the indications of the *I Ching*, from the unknowable Tao he derived two primary manifestations, motion (*tung* 動) and rest (*ching* 靜). Heaven was born of movement, and Earth of repose or rest.[90] This is a considerable advance on the more primitive statement put forward in the *Tao Te Ching* some fourteen centuries before: 'It was from the Nameless that Heaven and Earth sprang'.[91] In the heavens, or rather, in Heaven, motion and rest manifest themselves as the Yang and the Yin influence. Both Yang and Yin according to Shao Yung are motion, and at the apogee the motion becomes extreme. When, however, they are united, harmonised or in balance, motion gives way to rest, and the union of these two forces enriches the world. The motion of the Yin–Yang forces is therefore visualised as in clear contrast to the motion of a pendulum where motion is swiftest at the nadir of its swing, and the briefest instant of repose is found only at the two extremities. On Earth motion and rest are manifested not by the Yang and the Yin influence, but by the *Kang* 剛 and the *Jou* 柔, the Hard and the Soft, which we may term Durity and Lenity. These

[88] See *SCC*, vol. 4, pt. 1, pp. 8–10, for the relationship of cyclical and wave conceptions, and Chinese views on periodic phenomena in Nature, due to resistances which act on matter in a state of change.

[89] Wyatt (1996), p. 243, whose work appeared one year after Joseph Needham's death, denies that Shao Yung could have been developing an 'all-encompassing explanatory theory'. [Ed.]

[90] Shao Yung's work consisted partly of narrative works and partly of unexplained charts and diagrams, which are however preserved with explanations in the work of Tshai Chhen 蔡沉 (1167–1230) entitled *Ching Shih Chih Yao* 經世指要, 'Important Principles in the Cosmological Chronology'. Shao Yung does, however, explain his theories in that part of the *Huang Chi Ching Shih Shu* (ch. 5) entitled *Kuan Wu* 觀物, 'Observation of Things'. Parts of this are translated by Feng Yu-Lan (1953), vol. II, pp. 455 ff., and the above paragraphs are derived from the *Kuan Wu* (nei phien), ch. 1, fol. 2a ff. in *Huang Chi Ching Shih Shu*, ch. 3, which is found in abridged form in *Hsing Li Ta Chhüan Shu* 性理大全書 (ch. 9). It is interesting to contrast the views of Shao Yung with those of Copernicus (1473–1543) some four centuries later. Copernicus held that 'the condition of *being at rest* is considered as nobler and more divine than that of *change* and *inconsistency*; the latter therefore is more suited to the earth than to the universe'. *De revolutionibus orbium coelestium*, 1, 1, Cap. VIII.

[91] *Tao Te Ching*, ch. 1, tr. Waley (1934), p. 141.

two entities were newly introduced to Chinese natural philosophy by Shao Yung.[92] The Yin and the Yang influence all that is on Earth. All that is on Earth responds to this influence by manifesting durity or lenity in a greater or lesser degree.

It can scarcely be disputed that heavenly forces such as sunlight do have an influence on creatures which are on the surface of the Earth, such as crocodiles and jellyfish, which manifest varying degrees of durity and lenity. The question may then arise why the jellyfish, when washed up into the full glare of the sun, should dehydrate, while the crocodile, emerging from the cold water, enjoys the heat. Before any theory of evolution had arisen, there were only two possible answers – either that someone in His wisdom had made the creatures that way, or if, as in China, there was no known Creator, then because the materials composing each body were different or because their proportions varied. But why should they vary? Thinkers in China could scarcely avoid trying to discover what it was that apparently influenced the proportionate mixtures in different bodies. But in attempting to answer this question, it was necessary to enquire how exactly such influencing was done. In modern parlance, where Heaven and Earth meet, what happens at the interface?

In the West this problem of interface was dealt with for a time by the linguistic device of personification. If the Creator wishes something to be done, he sends a message, which can be read in the livers of animals, for example, or the flight of birds, or be delivered by an angelos or diviner, or by the prophet who reads God's message and announces it. Later, when Heaven had acquired an adequate bureaucracy, the mortal message-reader was superseded by a heavenly messenger or angel. In China, on the other hand, as there was no Creator there could be no messenger. Activity at the interface was therefore explained in terms of *Wu Hsing* 五行 or Five Avenues or passage-ways up and down between Heaven and Earth, the five ways in which Heaven influences things on Earth, by descending and soaking (water), by ascending and burning (fire), by submitting to cutting and accepting form (wood), by moulding or remoulding (metal), or by growth and the production of vegetation (earth).[93] For Shao Yung this ancient explanation was no longer adequate.

His new system may now be considered, guided by Figure 16. When Heaven's influence is Yang in its most extreme form, it is manifested by the Sun. In diluted form it is manifested in the Stars. There can be no objection to classifying both stars and sun as Yang, for stars are in fact suns, and Shao Yung in the +11th century was not to know that some stars are even larger and more intensely hot than our sun. When Heaven's influence is Yin in its most extreme form, it is manifested in the

[92] This point was made in a brief estimate of Shao Yung's work in *SCC*, vol. 2, p. 455.

[93] Nevertheless this early theory compares rather favourably with Kepler's 'explanation' of heavenly influences, that 'the sun represents the Father; the sphere of the fixed stars, the Son; the invisible forces which, emanating from the Father, act through interstellar space, represent the Holy Ghost'.

Koestler (1960), p. 60, conveniently summarises this as: 'the Sun carries the image [cf. *hsiang* 象 manifestation] of God the Father and Creator . . .'. *Wu Hsing* is translated in many different ways by different authors into English, for example, *modes*, *phases*, etc. Here we venture to speak of 'avenues', for the word *hsing* suggests movement along a path, but usually in these volumes we have retained the traditional though misleading word *element*, but in inverted commas. See *SCC*, vol. 2, p. 243, and Needham & Lu (1975), p. 498, where we suggest 'elemental methistemes' for use in some contexts.

Moon. In diluted form it is manifested in the Planets. There can also be no objection to classifying both moon and planets as Yin, for they are all either cold or cooling worlds. As little was known about the nature of the stars and planets at the time, however, Shao Yung was, perhaps, lucky in his classification here.

Sun, Stars, Planets and Moon are termed the Four Manifestations, *ssu hsiang* 四象. They do not, of course, 'exhaust' (*chin* 盡) Heaven's possibilities, for no mention is made of such phenomena as comets or meteorites, but they provide co-ordinates for classification. A comet would certainly be Yang, exhaling fire, and could be placed midway between the Sun and the Stars. A meteorite might be a body in transition from Yang to Yin, beginning hot and ending cold.

Whereas the dominant principle of Heaven is motion expressing itself as the Yin and the Yang, the dominant principle of Earth is rest, expressing itself as durity and lenity,[94] which is perhaps to be thought of as concretion and dispersal, *chü* 聚 and *san* 散, since that which is concreted becomes hard, and that which is dispersed like earthy particles forming mud, becomes soft. When each of the four manifestations of Heaven, the Sun, Moon, Planets and Stars, influence Earth, there is a reaction. The Sun, for example, when concentrated into the utmost durity, produces fire. This may be seen to occur when the sun's rays are focused by a burning glass. Fire, in Shao Yung's view, before telescopes had revealed the existence of flames on the surface of the sun, was a physical embodiment of the Yang. It may seem strange that stone, which is often so hard, should have appeared to Shao Yung to have less durity than fire, but when one recalls that fire in the heart of volcanoes renders stone so soft as to be liquid, his reasoning is clear.

The four terms at Earth level, which relate themselves to the Four Manifestations of Heaven, are a good example of how, in Shao Yung's thinking, natural phenomena should be classified not in rigid boxes but on a scale shading from one extreme to another. When sufficient Yin influence has entered fire to enable liquefaction to begin, one enters the area of stone. Stone covers a wide range of minerals from the hardest granite to the softest chalk, and merges imperceptibly through aggregates and clays to earths, muds, silts and waters. The feeling of the inter-relatedness of all phenomena comes over strongly in Shao Yung's schematisation. For example, the moon which influences the tides, is placed in direct correspondence with water. Water when heated becomes vapour and expands. But when the Yin influence returns it becomes cold, condenses, and, in the world at large, falls to earth as rain.[95] Moon – water – cold – rain is thus the beginning of a vertical classification under the concept of maximum Yin influence, maximum softness. There is, however, also a horizontal classification of what may be regarded as the key components of Heaven and Earth. Sun, Stars, Earth (or soil[96]), and Water he calls the Four Embodiments (*ssu thi* 四體). 'When hard and soft blend (or permutate) (*chiao* 交), they give bodily form to all that is useful on earth.'[97]

[94] *Kuan Wu* (nei phien), 1, 1a ff. [95] For the water-cycle in Chinese thought see *SCC*, vol. 3, pp. 467 ff.

[96] Earth, if contrasted to Heaven, is *ti* 地 in Chinese, but in contrast to stone or water is *thu* 土.

[97] *Kuan Wu* (nei phien), 1, 2a.

These Four Embodiments are reminiscent of some of the ancient 'Five Elements'. Metal and wood are missing, but stone has been added. We are not, however, dealing with the five ancient avenues at the interface, but something nearer to what the elements meant in the West. Fire, stone, earth and water were seen as materials capable of aggregation and dispersal, contraction and expansion, and from their blendings and permutations other 'bodies' could come. Shao Yung explains that metal is not a primary body. It comes from stone, '*chin chhu yü shih* 金出於石',[98] and wood is not primary, for it comes from earth, '*mu chhu yü thu* 木出於土', as in the sense that vegetation grows out of the soil.

He offers us, therefore, four elemental materials or 'bodies', fire, stone, earth and water, and suggests that the formation of the world came out of the blending of contraries, the Yang with the Yin, water reacting to the planets.[99] The original body, *pen thi* 本體, that is to say, the first materialisation of all this cosmic energy, was hydrotic fire, *shui huo* 水火, and geotic fire, *thu huo* 土火. Hydrotic fire we may visualise as igneous streams running as freely as water. Geotic fire, being less 'soft', we may imagine as grinding walls of lava. Yet though Shao Yung has introduced us to elemental forces which have much in common with the ideas of such Ionian philosophers as Thales and Heraclitus, he can still not abandon the ancient five avenues or modes of action at the interface.[100] They are, as they had always been in China, fire, water, earth, wood and metal. They are at this level considered to be secondary bodies, whereas fire, stone, earth and water at the higher level are primary bodies, from which wood and metal were generated. The result is that Shao Yung avoids alienating conservative or traditionalist thinkers of his day, but ends up with a stock of six rather than five elemental materials from which the Myriad Things may be generated by permutations, of which three, fire, water and earth, occur in two modes, the 'before Heaven' (*hsien thien* 先天) mode as primeval materials, and the 'after Heaven' (*hou thien* 後天) mode, in which they are 'avenues at the interface' and as Shao Yung says,[101] 'for use' (*chih yung* 致用). The traditional five – fire, water, earth, metal and wood – together with Shao Yung's addition of stone, make six materials to aid in classification.

It would be fascinating to describe Shao Yung's cosmological ideas in some detail. Here we can only suggest the main thrust of this strange work. Shao Yung apparently sought to show that the whole world is a unity reducible to a few main principles which are manifestations of one sublime principle – the Tao. To 'exhaust the Tao' by giving details of every conceivable permutation and physical operation would be an impossible task. Shao Yung therefore produced a bare skeleton of permutations with very few examples to flesh it out. Yet his is more than a work of mere numerology. There is a strange groping attempt to quantify – not, as was done in Europe from the 17th century onwards, by measurement, but by starting from some *a priori* preconception. Yet it was a form of quantification nevertheless.

[98] *Kuan Wu* (nei phien), 1, 4a. [99] *Ibid.*
[100] 'How could one consider that I omitted the "Five Avenues" and did not use them?' *Kuan Wu* (nei phien), 1, 4a.
[101] *Ibid.*

Table II. *Shao Yung's Quantification of the Myriad Things*

I	volitic flighter	(e.g., an albatross)	1,000,000	units
I	repetic flighter	(e.g., an ostrich)	100,000	,,
I	dendritic flighter	(e.g., a tree-creeper)	10,000	,,
I	herbaric flighter	(e.g., a bumble-bee)	1,000	,,
I	volitic goer	(e.g., a flying fish)	100,000	,,
I	repetic goer	(e.g., an elephant)	10,000	,,
I	dendritic goer	(e.g., a sloth)	1,000	,,
I	herbaric goer	(e.g., a harvest mouse)	100	,,
I	volitic tree	(e.g., a sycamore seed)	10,000	,,
I	repetic tree	(e.g., a mangrove)	1,000	,,
I	dendritic tree	(e.g., a giant redwood)	100	,,
I	herbaric tree	(e.g., a tree-fern)	10	,,
I	volitic herb	(e.g., a dandelion seed)	1,000	,,
I	repetic herb	(e.g., a *Mimosa pudica*)	100	,,
I	dendritic herb	(e.g., a bamboo)	10	,,
I	herbaric herb	(e.g., moss)	I	,,

Starting from the numerological association of numbers with the 'elements' and so on, he is able to quantify all living things according to a scale of values. (See Table II.) The permutations of things volitic, repetic, dendritic and herbaric will be described below. Thus creatures which are most at home in flight, the 'volitic flighters' (*fei-fei* 飛飛), are rated at I in I, or 100 per cent. This may conveniently be expressed as 1,000,000 units of worth or merit of some sort, perhaps because that which is most influenced by the Yang is most appreciated, or because they must struggle against the force of gravity to remain up, whereas Yin creatures are sustained on land or in the water. Earth-bound flighters are rated at only 100,000 units, as are the wingless creatures which have nevertheless, after a fashion, taken to the air. So it goes down the scale.

Table II is derived from Shao Yung's *Huang Chi Ching Shih Shu* 皇極經世書 preserved in the *Hsing Li Ta Chhüan Shu* 性理大全書, ch. 10, fol. 30 ff. The examples we offer in Table II are, however, *conjectural*, since Shao Yung offers few himself – horses and dragons as volitic goers, and chickens and ducks as repetic flighters.

From this sort of evaluation it is only a short step to the evaluation of people, where naturally scholars are rated highest, while those who indulge in commerce come lowest in the human scale. Ultimately everything can be classified and all the Myriad Things can be quantified.[102] Our aim is now to show how he designed his system by which the Myriad Things might be described and classified. In the topmost

[102] For details of this statistical approach see *Hsing Li Ta Chhüan Shu*, ch. 10, fol. 30 ff. [Don J. Wyatt's *The Recluse of Loyang*, 1996, was not available to Needham, but may now be consulted. See pp. 97 ff. for Shao Yung's use of Number. Ed.]

level of the system illustrated in Figure 16 he named eight essential components, sun, fire, stars, stone, planets, earth, moon, water. The aim of the second level is to show how the heavenly powers change and blend, and how Earth reacts and adapts to these changes. The sun rises and sets, the moon goes through its phases, the stars come out at night and disappear by day, the planets wander through the sky on their courses. The changes in the sun are expressed as Heat which varies from winter to summer; in the moon as Cold, which is greater in winter than in summer, and most noticeably at night; in the stars and the planets which between them indicate the alteration of night and day, stars, being Yang, indicating Day, and planets, being Yin, indicating Night. At the second level there are therefore four more influences to operate on the ten thousand things. Is it any wonder, they might well have remarked, that moon-flowers open at night, whereas the morning-glory is open by day? Such an attitude is understandable, but it does not lead to modern science.

The four terms representing Heaven's changes also form a scale from hot to cold. Between day and night comes the period of near equilibrium where light and darkness, heat and cold are in balance, the twilight times. Corresponding to them comes another curious scale, that of the Earth's responses. In hot countries the coldness of rain is particularly striking. If it falls from a height of 3,000 feet, it will usually be some 10°F (6°C) colder than the air at ground level. Rain is therefore classified under water at the Yin end of the scale. Dew comes next. It does not appear until the temperature has dropped sufficiently to promote precipitation, and this is unlikely to occur until night comes. Dew, being cool, is therefore classified with night under the term earth, for it is on the earth's surface that it is detected. Following the diagram's stony path, things watery are now left behind, and things fiery return.

It may not be immediately obvious why wind should be classified under heat and fire. Yet on reflection even the flame of a lamp produces a small wind above it. A large fire sucks in a strong current of air, and a hot desert produces dust-laden whirlwinds. All winds are of course produced by the action of the sun in heating parts of the atmosphere irregularly, though Shao Yung was not, perhaps, aware of this. Thunder is, meanwhile, classified like dew at one of the intermediate positions, for as dew is precipitated when there is a drop in the temperature of moist air, so thunderstorms occur where air with high moisture content is swept upward with 'a strong vertical lapse rate of temperature. . . . Thunderstorms over land are . . . most frequent in the afternoon and in the warmer seasons of the year.'[103] It is appropriate, then, that thunder should be classified in an intermediate position between hot winds and cold rain. When rain falls, the 'shattering of large drops produces ionization and separation of electrical charges between drops and air, so that it is thought to be the process (or one of the processes) whereby the electrical charges in thunderclouds are built up'.[104] It would, of course, be absurd to attribute to Shao Yung an understanding of phenomena which was impossible in his day. But he may claim to have used such knowledge as was available to him to such good effect in the rational grouping of

[103] Lamb (1972), vol. 1, p. 380. [104] *Ibid.*, p. 353.

phenomena that much of it still makes sense today, and it enabled scientific thinking in China to advance a further stage in the classification of phenomena. Naturally parts of his schematisation no longer carry conviction. It may seem odd today that while rain was very reasonably classified under water, thunder should be classified under stone. But when one recalls Shakespeare's Cassius declaring that he has bar'd his bosom to the thunder stone,[105] one is swiftly back in the centuries when men were not clear whether 'thunderbolts' caused thunder, were caused by thunder, or merely happened on occasion to accompany thunder or to make a noise in falling like thunder.[106]

The third level indicates how Heaven influences living things by endowing them with Natures (*hsing* 性), Form (*hsing* 形), Bodies (*thi* 體) and 'Sensitivities' (*chhing* 情), and how living things respond to these influences and adapt to their environment. It is at this level that biological classification begins, and Shao Yung's approach to the problem is of extraordinary interest. The battle over the fixity of species never had to be fought in China. The vegetable world was distinguished from the animal world not by a sharp division, not by an Aristotelian dichotomy, but once again along a scale in which one type merges into another almost imperceptibly. In such a system carnivorous plants and rooted animals are immediately at home.

The vegetable world is, as it were, 'defined' in the traditional way by the 'synthesis of opposed concepts', by using as a compound the two terms *mu* 木, trees, and *tshao* 草, herbs and grasses, which, ever since the days of the *Erh Ya* 爾雅 (–3rd century) had subsumed the world of plants. When Shao Yung speaks of Trees-and-Grasses one must visualise a scale in which the tallest trees, shown in Figure 16 on the Yang side, are reaching up for the light, the sort of trees which by their very size and height invite the 'thunder stone', and so by infinite gradations to the Yin side of the diagram where the lowliest types of herbs and lichens creep on the earth's face, luxuriating in the coolness and the deep shade.

Even more arresting is Shao Yung's manner of dealing with the animal kingdom. In the *Erh Ya* itself there were already several categories into which animals were divided, and subsequent centuries had increased their number. But Shao Yung at this point has only two. One would expect to find at least four – beasts, birds, fish and insects – but Shao Yung uses only two categories, for which it is difficult to find any term in English. One might call them 'flighters' and 'goers', but unfortunately it is possible to go through the air as well as through water and over or under ground. The Chinese terms are *fei* 飛 and *tsou* 走. *Fei* covers all types of locomotion through the air, *tsou*, which we translate 'go' for lack of a better term, covers all types of locomotion on or under the surface of land or water. It is very logical not to make any distinction between swimmers and walkers, for swimming is for many animals merely walking in the water. A horse, for example, continues to walk till it reaches the other side of a river, and a snake crosses water in the same way as it crosses land, while an eel crosses land just as it swims through water. Moreover, certain fish which have fins

[105] William Shakespeare, *Julius Caesar*, I, 3.

[106] Shakespeare's contemporary Li Shih-Chen was also still not clear about the relationship between stone and thunder, for he wrote, 'When thunder or thunderbolts turn to stones, there is a transformation of the formless into that which has form . . .'. *Pen Tshao Kang Mu*, ch. 8, preface. See *SCC*, vol. 3, p. 637.

also 'walk' through grass to reach desired ponds. This is an excellent example of Chinese being able to cope with high levels of generalisation where sometimes Indo-European languages are over-specific. The salient fact, however, is that no creature can fly without wings or special surfaces to carry it on the air. A crab can go on its legs under the water as well as on dry land, but it cannot fly. A fish can 'fly' only when it has a planing surface or membranous fins sufficiently large and efficient to sustain it in the air. Animals which go about on legs can be regarded as volitic when they have membranes between their limbs to support them in flight. Shao Yung avoids the tedious linguistic questions which bedevil Greek biology, such as 'When is a limb not a limb?' or 'Is a claw the analogue of a hand?' – questions which could not be answered until phylogenetic classification had made considerable progress. He asks a different question, 'Does it fly?' or rather, 'Does it have flight?', for certain things such as seeds can be in flight though they themselves cannot fly. And we see very clearly the inter-connections of Chinese scientific thinking, for not only does the vegetable kingdom form a developmental scale, but the animal kingdom does also, and there is no sharp dividing line between them. Some members of the vegetable kingdom may begin with flight in seeds and spores, and end up by being rooted, just as some insects may begin as 'goers' when feeding on a leaf, and end up in flight. One may begin with the most pedestrian of animals, heavy ones such as elephants on land, and hippopotami which rejoice in water. By reason of weight they will always be goers and never flighters. Proceeding towards the Yang one meets goers which are light of body and fleet of foot, such as gazelles, but though they travel fast, they no more fly than such birds as ostriches. The horse, however, is rather surprisingly said to be a volitic goer, for it 'flies' across the ground; the man on its back is perhaps near to flying. The dragon is also volitic, for, as is well known, the Chinese dragon soars into the clouds![107] Still on the Yin side of the equilibrium line but rather nearer the Yang, one would encounter, we imagine, goers capable of real flight, such as flying squirrels, associated with trees, to which they have reacted by obtaining height for the downward swoop. Fish with limited powers of flight, such as the giant ray, will also find their place at some point on the scale. Then, trespassers across the line of equilibrium, one meets going-flighters; Shao Yung instances chickens and ducks, which are Yang modified by Yin. To these we may add other creatures like ostriches and emus, which rely on leg-power for their locomotion, though their wings can still be flapped. There are some creatures like the *lilytrotter* which deserve to be placed on the very point of equilibrium, for they are equally well adapted to locomotion in the water, on its surface, on land, or in the air. Birds and insects continue on the other side of the equilibrium line towards the Yang end of the scale, spending more and more time in the air, till, with the albatross one encounters a creature which is an almost 100 per cent flighter.

Shao Yung obtains terms for the main divisions of the scale by combining them. Thus we get 'herbing-goers' (*tsou chih tshao* 走之草), 'treeing-goers' (*tsou chih mu* 走之木), 'going-goers' (*tsou chih tsou* 走之走) and 'flighting-goers' (*tsou chih fei* 走之飛),

[107] Shao Yung, *Kuan Wu* (wai phien, B), fol. 4a, in *Hsing Li Ta Chhüan Shu*, ch. 12.

and so on, permutating all the other classes.[108] Such permutations may be set out in a table (see Table II). But since they stand for categories of Plant-Animals not conceptualised in the West, one is handicapped in inventing terms for them, and literal translations such as Treeing-Goers or Flighting-Flighters are almost meaningless. But it is convenient to use, as adjectival forms for each, the words herbaric, dendritic, repetic, and volitic as in *SCC*, Volume. 6, part 1.[109]

As Shao Yung gave so few examples to illustrate his schema, it has been found helpful to cut across the genera and species of modern classification. For example, under 'dendritic-goers' one might classify certain types of monkey and sloth which spend their lives in trees, and also those coelenterate polyps which build the coral branches they have spent their lives in. (Table II.) Some of the categories resulting from the permutation of the terms may have no claimant in the world we know. But their spaces are quickly filled by the writers of science-fiction. Trees, for example, are capable of undulatory movement, but not – once rooted – of locomotion, other than by slow distribution of their kind over many generations. But fiction writers have not been slow to visualise the menace of a form of plant life capable of swift movement and action.[110] Again, the only completely 'volitic-flighters', that is to say, creatures which are airborne all their lives, are perhaps certain types of bacteria and viruses floating through the air on particles of dust. The albatross has been included in Table II as a volitic flighter, though this 'prince des nuées' when not on the wing, must, nevertheless, hobble like a cripple.[111]

Because we are now accustomed to a post-Theory of Evolution mode of thinking, the speculations of Shao Yung have a quaintly medieval air. But in the +11th century no one anywhere had a properly global view of the world, and even of that part of the world where he lived he had only a small fraction of the knowledge which we now possess. To see Shao Yung's achievement in perspective, we need only take a world which is equally unknown to us, and consider how we would set about organising our ideas about it. Clearly what is wanted is a grid which sets things in broad categories, as has been suggested by Carl Sagan in his imaginary description of 'life' on Jupiter sustained by 'sinkers', 'floaters' and 'hunters' in the upper atmosphere of that giant gas planet.[112]

The sixteen classes formed by permutating the four third-level categories which represent Earth's adaptation to the influences of Heaven obviously require further subdivision. It is part of Shao Yung's system that lower-level entities can be qualified or modified by higher-level entities. For example, clouds are a phenomenon which may be classified as of the second level at a certain point between Thunder and Rain. But there are many different sorts of clouds. Shao Yung makes sub-classes by modifying them with the first-level concepts of Fire, Stone, Earth and Water,

[108] *Kuan Wu* (nei phien), 1, in *Hsing Li Ta Chhüan Shu*, ch. 9, pp. 8b–9a.

[109] *SCC*, vol. 6, pt. 1, p. 307.

[110] As, for example, in *The Day of the Triffids* by John Wyndham (1951).

[111] 'Le poète est semblable au prince des nuées / Qui hante la tempete et se rit de l'archer. / Exilé sur le sol au milieu des huées / Ses ailes de géant l'empechent de marcher.' Charles Baudelaire (1821–67), 'L'Albatros'.

[112] Sagan (1980), pp. 40–1.

distinguishable also by their colour, i.e. igneitic clouds which are red; lapidic clouds which are white; geotic clouds which are yellow, and hydrotic clouds which are black. It is true that thunder clouds threatening rain are dark blue-black, and that sunset clouds in the aftermath of volcanic eruptions are often brilliantly red, but the lapidic and geotic clouds seem less happy, and could not hope for a very long scientific future.[113]

It is, however, very clear that Shao Yung's compound terms require further definition in order that they may be logically or systematically subdivided. His trouble is linguistic. When two nouns form a compound, the first qualifying the second, the real nature of their relationship is concealed. This frequently occurs in English also. For example, the compound terms *dog-ape*, *dog-cart*, *dog-fox*, *dog-kennel* and *dog-skin* conceal completely different relationships between the noun *dog* and the second noun which follows, namely:
an ape with a head resembling that of a dog
a cart normally pulled by a dog
a fox with the sex not of a bitch but of a dog
a house made specially for a dog
a leather made from the skin of a dog.

So with Shao Yung's compounds we cannot always be certain what relationship is intended. When, for example, the word *fei* 飛 forms the first element in a compound, it may mean that the qualified noun:

(a) spends much time in flight;
(b) has limited power of flight,
(c) has seeds which fly.

But sometimes each qualifying noun may offer a choice of interpretations as when the word *tree* qualifying a 'flighter' may mean not only that the 'flighter' spends much time up a tree, as does the hornbill, the female being completely walled up inside a hollow tree when nesting, but it may derive certain characteristics from a tree, as with the tree-creeper, which is perfectly camouflaged against its trunk, closely resembling the bark. Resemblance itself may occur in many ways. It may be resemblance in form, in texture, in colour or camouflage, and so on.

It would seem that Shao Yung believed that things in a group share the characteristics of that group simply by being in that group. This is the reasoning of the *I Ching*, and as Shao Yung had been immersed in the study of the *I Ching* from an early age, it is not unlikely that he had failed to clarify the functions of the members of the groups. For example, he says that beasts born in the grassland have hair like stalks of grass (*tshao fu chih shou* 草伏之獸). This would be a good description of the shaggy hair of the bison, and for people accustomed to the fineness of silk, even wool must have seemed as coarse as grass. Nevertheless it does not appear that he believed that this coarseness was due to the animals' eating grass, but simply that being in that

[113] *Kuan Wu* (wai phien shang) in *Hsing Li Ta Chhüan Shu*, ch. 11, fol. 51b.

category caused it to be thus, *lei shih chih jan yeh* 類使之然也.[114] On the other hand he may have wished to distinguish the influence of environment from heredity. The species (or as we should say now, the gene pool) causes it to be thus.

Shao Yung seems to have been aware that there was a need for an improved system of classification. This might mean subverting the ancient system by the creation of new categories of thought; to make these new categories it was necessary to understand, or at least to have a theory about how the items which composed them had come into existence – an early groping towards a phylogenetic system. Once the principles on which the creation of a class is based are understood, it is possible to subdivide each class according to those principles, and so to elaborate a system. But inevitably there are at first errors and misconceptions.

For example, if it is found that in the category of repetic creatures or 'goers' there are a large number with hoofs and horns, i.e. creatures with hard places at their extremities, there must be some reason for this hardness. In Shao Yung's day there was no Theory of Natural Selection; the hard extremities would therefore have to be explained as far as possible by the theory then current. This took the form that as repetic creatures were at the Yin and Lenity end of the scale, the hardness must be due to something from the opposite end. If that were so, one would expect to find the converse in operation at the other end of the scale. Among 'flighters', some might be almost solidly under the Yang[115] (dragon-flies, for whom no sun is too hot, would make an example), but one might also expect to find others with an admixture of Yin and Lenity; sure enough, birds, though Yang, have in their feathers much softness, and soft protuberances in the form of wings, very different from their hard protruding beaks. This would undoubtedly be due to a Yin admixture in the Yang.

Shao Yung's theory had the great advantage that it broke open the pigeon-holes of earlier systems and made it possible for things to be classified on scales of minute gradation, both horizontally and vertically. 'Among the grasses', he says, 'there are trees, and among the trees there are grasses'.[116] Indeed, now that the organic world has been divided into sixteen sub-types, each of these can be permutated by various influences to yield 1,920 further sub-types. With such a system, the classification of bamboos and tree-ferns, which might well be described as fern-trees, presents no linguistic dilemma – When is a tree not a tree? It enabled natural philosophers, undistracted by the side issues of definition according to the rules of scholastic logic, to get on with the task of ever more accurate scientific description, which will be a topic for our later consideration.

Considerably more important than Shao Yung's general theory, which, like most theories, could only hope for a limited working life, was the stimulus he gave China's intellectual world for centuries to come, a point worth noting by those who believe in the stagnant 'Cycles of Cathay'. When little was known about the animal and

[114] *Hsing Li Ta Chhüan Shu*, ch. 11, fol. 58b.

[115] It is in the nature of the Yin–Yang theory that neither can be total. At the very moment when the Yang reaches its maximum extent, the Yin is already present in minute degree and steadily increasing, and vice versa.

[116] *Kuan Wu* (wai phien, A) in *Hsing Li Ta Chhüan Shu*, ch. 11, fol. 36a.

vegetable kingdoms, about such remarkable changes in appearance of certain life forms as pupation and the creature which emerges from the pupa, about the appearance of animals at different seasons with their changes of coat, about whether a species breeds predictably or erratically (as was believed in the West of geese developing from barnacles, for which there are several Chinese parallels[117]), about whether animals or plants are spontaneously generated, hatched by warmth and moisture from Mother Earth, or reproduced in various unlikely ways,[118] when so much was unknown there was little point in trying to make out 'family trees' for species, little motivation for developing phylogenetic or taxonomic systems. It was hard enough even to know whether outlandish creatures should be classified as beast, bird or fish. At such a stage in the development of scientific knowledge it is important to place items systematically in categories, and when the category is clear, and the criteria required for it are understood, then to give the category a name. Such is the pressure of language, however, and the need for common, sensible, recognisable terms, that the opposite method was usually followed.

Shao Yung's classificatory system anticipated the need for categorising items according to clear criteria and using broad provisional labels. Once the items have been grouped, the criteria can be refined and subdivisions made with fresh labels. In his system it would be possible to classify the whale provisionally as a 'goer' since it does not fly. More precisely it would be cross-classified as a 'repetic goer'. Just as clouds could be subdivided according to the higher level of earthly embodiments influencing them, so presumably could animals and plants. 'Repetic goers' could be subdivided into the classes of those which are happiest in a watery domain, the 'hydrotic repetic goers', such as whales, and those which are at ease in fiery places such as hot deserts, which include various types of lizard, and, no doubt, medieval salamanders secured by the most ardent collectors, these being igneitic repetic goers. This, however, is to anticipate the work of later natural philosophers.

Shao Yung had permutated Flighters, Trees, Herbs and Goers to make sixteen classes (Table II). These four, when permutated, exhausted the response (*ying* 應) to the natural influence of things.[119] Sub-groups could, however, be formed according to the proportions of Yin and Yang influence in the mixture, i.e. according to the exact position on the scale, and by the extent to which the group was influenced by such factors as heat or dew. An example would be the 'hoof and horn' sub-group, or the 'feather and wing' sub-group. He was also aware that the time-dimension could be used as a factor in classification, citing the dimensions of a day, a month, a season and a year, and then in tens of years up to 100, 1,000 and 10,000, which takes us presumably through the life-span of giant redwood trees to the formative

[117] For example, Wang Chhung mentions that snakes and reptiles were believed to turn into fish or turtles during heavy downpours of rain, and that frogs become quails and sparrows turn into clams. *Lun Heng*, 'Unfounded Assertions', tr. Forke (1907), vol. 1, pp. 325–6.

[118] Such beliefs can still be encountered in country places in Western Europe, as for example, that maggots are produced by rotting meat. (Personal experience in Oxfordshire in 1956.)

[119] '*Tsou fei tshao mu chiao erh tung chih chih ying chin chih i* 走飛曹木交而動植之應盡埴矣.' *Kuan Wu* (nei phien, 13) in *Hsing Li Ta Chhüan Shu*, ch. 9, fol. 9a.

Figure 17. Portrait of Shao Yung. From Lunkbaek (1986), *Dialogue Between a Fisherman and a Woodcutter*, p. 4.

lives of stalactites.[120] Such, very briefly, was the contribution of Shao Yung to the art of classification (Fig. 17).

(iii) *The development of hierarchical classification*

When, five centuries later, Liu Wen-Thai 劉文泰 began work on the imperially commissioned *Pen Tshao Phin Hui Ching Yao* 本草品彙精要 (Essentials of the Pharmacopoeia Ranked According to Nature and Efficacity), which was presented in 1505, he stretched Shao Yung's system far beyond such limits.[121] He took the Four Embodiments – Fire, Stone, Earth, Water – added Metal, and then permutated these five, getting twenty-five classes consisting of such compounds as lapidic water or geotic

[120] *Kuan Wu* (wai phien, 13) in *Hsing Li Ta Chhüan Shu*, ch. 12, fol. 40b. The time dimension was of special importance in alchemy, where, for instance, 'rock salt is said to change after 150 years into magnetite, then after 200 years into iron, which, if not dug up and smelted, will turn into copper . . . that into "white metal", and that finally into gold'. See *SCC*, vol. 3, p. 639, and vol. 5, pt. 4, pp. 223–4.

[121] See *SCC*, vol. 6, pt. 1, pp. 264 ff. for a survey of Chinese pharmacopoeias.

metal. With such permutations he was able to begin the classification of minerals and chemical substances following an at least rudimentary analysis. That is to say, he was beginning to think in terms of chemical analysis, an enormous advance on the ancient classification by assumed moral qualities and virtues.[122] By further dividing all things into two groups, 'spontaneous' (*thien jan* 天然), that is to say 'natural', and 'man-made' (*jen wei* 人為), he was then able to double the number of possible classes. For the classification of plants he went still further, for not only could they be put in a class, but they could also be described by terms of such a sort as would make them easily recognisable, terms such as straggly (*san* 散), creeping (*man* 蔓), or epiphytic (*chi* 寄), 'in order to facilitate collecting for use'.[123]

But, instead of following up Shao Yung's lead in which animals were classified according to their means of locomotion, and in which animals and plants were sub-classified according to their habitat and the nature of their excrescences (as living in the grass-land, or bearing horns or tubers), he reverted, for animals, to the traditional four categories – beasts, birds, insects and fish – which inevitably raises problems when it is necessary to classify snakes, worms, or other creatures outside these four categories. Liu Wen-Thai's book did, however, introduce very interesting criteria for their sub-classification, such as the method of reproduction, or the creatures' teguments. What interests us here is the way in which the appropriate information for each entry is marshalled.

After a few necessary preliminaries, the material is divided into twenty-four heads (see Fig. 18). These fall into three main groups. First, the raw material is considered. Then it is treated as a refined product. Finally cautions are given concerning its use. Under the first six heads which are to do with the raw material, four are concerned with the identification of the plant or animal. This includes earlier descriptions, and clarification of linguistic and terminological problems. The description of the item is followed by information concerning where it is found, and at what times of the year. These are followed by two heads concerning methods of preservations, which describe not only how this is done, but which parts should be preserved. The item is then described in the form of a refined product. This is covered by thirteen heads, of which the first six describe its appearance, colour, sapidity, pharmaceutical properties, effectiveness and odour. The next four are to do with its therapeutic use, and any adjuvant factors. The following head is concerned with methods of processing, and the last two heads in this group cover the entry's other therapeutic virtues and its effects in combination with other substances. The last five heads which we have grouped together as cautions, deal with warnings about the use of the medicine, about imitations, substitutes, fraudulent samples, and with contraindications and antidotes for excess in administration.

To appreciate the high level to which the organisation of such information had now risen, we may compare Liu Wen-Thai's system with that adopted in *The Herbal*

[122] See for example *Kuan Tzu* on 'the nine virtues of jade' discussed above in *SCC*, vol. 2, p. 43.
[123] See *SCC*, vol. 6, pt. 1, p. 305.

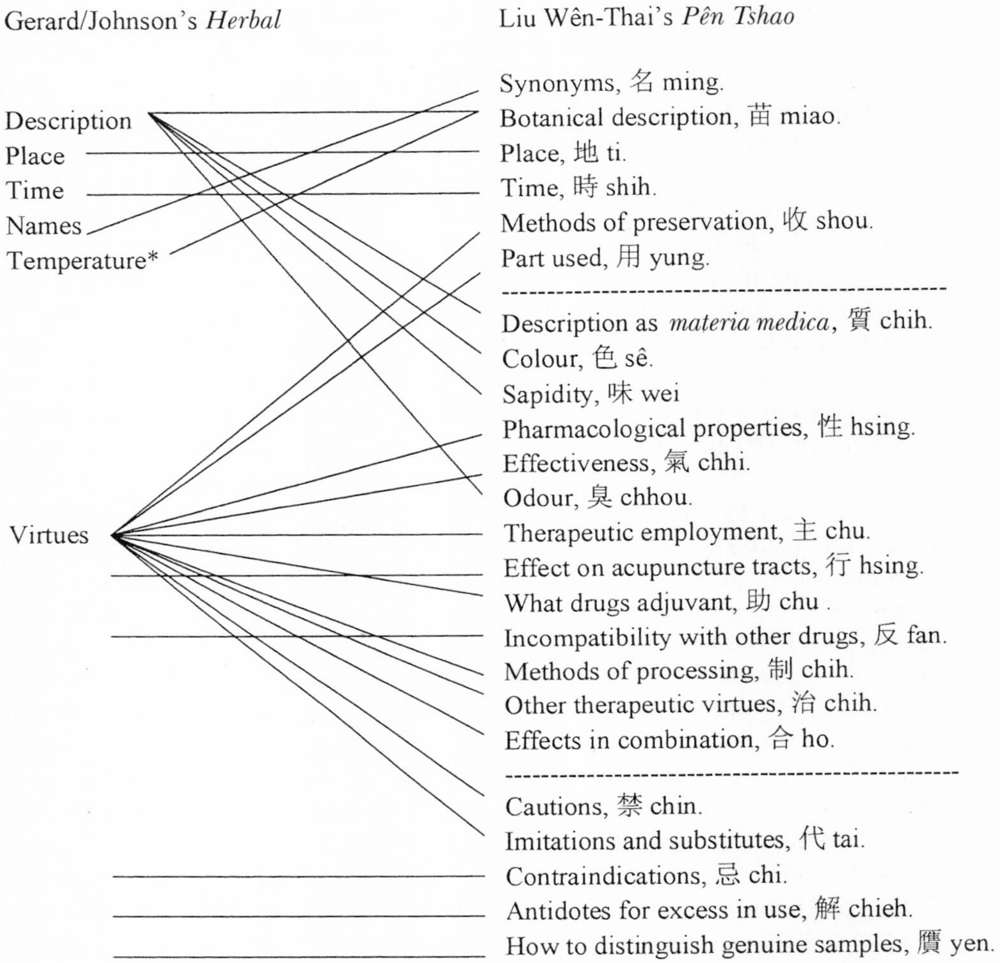

Gerard/Johnson's *Herbal*

Description
Place
Time
Names
Temperature*

Virtues

Liu Wên-Thai's *Pên Tshao*

Synonyms, 名 ming.
Botanical description, 苗 miao.
Place, 地 ti.
Time, 時 shih.
Methods of preservation, 收 shou.
Part used, 用 yung.

Description as *materia medica*, 質 chih.
Colour, 色 sê.
Sapidity, 味 wei
Pharmacological properties, 性 hsing.
Effectiveness, 氣 chhi.
Odour, 臭 chhou.
Therapeutic employment, 主 chu.
Effect on acupuncture tracts, 行 hsing.
What drugs adjuvant, 助 chu .
Incompatibility with other drugs, 反 fan.
Methods of processing, 制 chih.
Other therapeutic virtues, 治 chih.
Effects in combination, 合 ho.

Cautions, 禁 chin.
Imitations and substitutes, 代 tai.
Contraindications, 忌 chi.
Antidotes for excess in use, 解 chieh.
How to distinguish genuine samples, 贋 yen.

*By temperature is meant the power of tempering heat in the body.

Figure 18. Diagram contrasting Gerard/Johnson's *Herbal* and Liu Wen-Thai's *Pen Tshao*.

or General History of Plants, 'the best known and most often quoted herbal in the English language',[124] originally written by John Gerard (1545–1612), published in 1577, but revised with great care in 1633 by Thomas Johnson. Though appearing somewhat more than a century later than the Chinese work, and containing many more entries – 2,850 as against 1,815 – the information supplied concerning each entry is organised quite crudely, as the lists of heads under which the information is classified, shown in Figure 18, make clear. The Gerard/Johnson book classified all the data under just six heads, of which the last, the 'virtues' of a plant, is a general

[124] *Dictionary of Scientific Biography*, vol. v, pp. 361–2, sub nom. Gerard. Also Stearn (1966).

undifferentiated rag-bag collection which, in the Chinese work, is separated out into no fewer than eleven different headings. It will be noted that the English work has only one type of description, that of the appearance of plants, whereas the Chinese work separates this from the description of the *materia medica* after processing. Occasionally Gerard includes some description of the appearance of the *materia medica* under his over-general heading 'virtues'. For example, he describes how infusions may be made from the flowers of roses for 'purging gently the belly'. He concludes: 'Unto these syrrups you may adde a few drops of Vitriol, which giveth it a most beautiful colour, and also helpeth the force in cooling hot and burning fevers and agues'.[125] That may well be so, and it is an excellent example of his adding to his over-worked 'virtues' section, what drugs were considered adjuvant; but what beautiful colour, we may ask, did the syrup actually become after the oil of vitriol was added?

The strength of Gerard's book is in the description of the botanical specimens. Without the illustrations, however, it might be difficult to identify many of them. This could not really be done until a whole new technical vocabulary had been invented. But within the limitations of the vocabulary of his age he manages rather well.[126] There was then, for example, no word for *petal*. He deals with this by saying of elecampane (p. 793) that in the flowers, which are great, broad and round, 'the long small leaves that compass round about are yellow'. Today we should say that such members of the *compositae* have *capitula* composed of a number of *florets* densely packed together, the heads being each surrounded by an *involucre* of *bracts*, the *calyx* being superior, the *corolla gamopetallous*, the five *stamens syngenesious*, and the *style* single. But in writing for recognition rather than classification his descriptions were adequate when supported by illustrations.

What was less satisfactory was that so many important aspects of plants were not dealt with systematically at all, but only included among the 'virtues' in a random fashion. For example, he happens to mention among the 'virtues' of elecampane that the root is with good success mixed with counter-poisons. It resisteth poison. Unfortunately he does not say how much of the root should be administered, nor what poisons it resists. But we search in vain in his book for contraindications, incompatibilities and so on, and of course for any reference to acupuncture tracts. Nevertheless the unsystematic way in which such linkages were dealt with, and the tendency to classify anything unusual under 'virtues' suggests that in the pre-Renaissance West there had been no strong tradition of training in the use of check-lists.

A mere two generations later Li Shih-Chen 李時珍 (1518–93), building on Liu Wen-Thai's foundation, brought classification in botany to a Magnolian or Tournefortian level.[127] Perhaps unconsciously he drew on Shao Yung who had said that our world began with fire, water, stone and earth, but had dismissed metal and wood to the secondary level, and whereas Liu Wen-Thai had added metal again, Li Shih-Chen said that:

125 Gerard (1633), p. 1264.
126 For problems and progress with the phytographic language of China see *SCC*, vol. 6, pt. 1, pp. 126 ff.
127 See *SCC*, vol. 6, pt. 1, pp. 12–13.

Water and Fire existed before the myriad things, and Earth is the mother of all the myriad things. Next come metals and minerals, arising naturally out of earth, and then in order herbs, cereals, [edible] vegetable plants, fruit-bearing trees, and all the woody trees. . . . Then the tale continues with insects, fish, shellfish, birds and beasts, with mankind bringing up the rear. Such is [the ladder of beings], from the lowliest to the highest.[128]

So it was that step by step thinkers made their way from the early 'Ladder of Souls' towards the Theory of Evolution.[129]

Three stages in the development of classification have now been traced. First, grouping things in sets with a common characteristic. Secondly, placing those which have a common characteristic in serial order, leading to the concept of scales. Thirdly, using a series of terms two-dimensionally to form a grid or nest of squares which can be used for pigeon-holing information, as when high–middle–low, read in two directions, makes it possible to classify objects in nine different sets from the highest of the high to the lowest of the low.

Chinese writers do not appear to have experienced any difficulty in making their literary language carry out these functions. The order of plants listed in sequence on the 'oecological gradient' described in the *Kuan Tzu* book under the fourth head, referred to above,[130] is quite simply and clearly indicated by the phrase *hsia yü* 下於, 'low in relation to', for example, plant A is low in relation to (lower than) plant B, plant B is lower than plant C, etc.[131] The idea of series is conveyed by a technical term *tshui* 衰 in the phrase, '*fan pi tshao-wu yu shih-erh tshui*' 凡彼草物十二衰, 'all these plants have twelve precedences'. The word *tshui* originally meant a mourning garment. When the metaphor was fresh it must have evoked a vivid picture of plants as differentiated as people wearing mourning garments, which were strictly graduated in their style according to precedence. Again the idea of serial order or gradation is conveyed by the word *tzhu* 次, as in the phrase '*wu yu tzhu* 物有次', 'plants have their order', translated above in its context as 'Every type of soil has its characteristics, and every plant can be graded in an order (of luxuriance)' (*mei chou yu chhang erh wu yu tzhu* 每州有常而物有次).[132]

A fourth form of classification was, however, evolved, whose purpose was to obtain a clear idea of how items in a group are related, or how, through them, something is transmitted. This type of classification is usually described as 'hierarchical', since items are arranged in levels just as, in the +5th century, Dionysius the Pseudo-Areopagite, also known as Pseudo-Dionysius, arranged the angels into three classes, higher, middle and lower, each class subtending three choirs, from the Seraphim,

[128] Quoted in *SCC*, vol. 6, pt. 1, p. 315, from *PTKM*, ch. Shou, Fen li, p. 37 b; on the conception of the *scala naturae* in Chinese thought, cf. *SCC*, vol. 2, pp. 21 ff.

[129] Some of Li Shih-Chen's statements, mediated through Jesuit writings, formed part of Charles Darwin's Chinese sources, for example, *The Origin of Species* (1st edn), p. 34, mentions that the principle of selection was known to the Chinese from antiquity. For the various doctrines of the 'ladder of souls', see *SCC*, vol. 2, p. 22.

[130] See *SCC*, vol. 6, pt. 1, p. 49 ff.

[131] See Kuo Mo-Jo *et al.*, in *Kuan Tzu Chi Chiao* 管子集校, vol. 2, p. 922, where the gradient from lake water to dry ground, and the twelve plants representing the twelve habitats, are illustrated, following the commentary of Hsia Wei-Ying 夏緯瑛.

[132] See *SCC*, vol. 6, pt. 1, p. 55.

nearest to the All Highest, to the ordinary angels, nearest to man, so hierarchical classification developed.

As was seen above, grading into three categories – higher, middle and lower – was practised in China at an early date. But such grading was not necessarily hierarchical. It is only hierarchical if there is a connection between the higher and lower levels. If no mutual relationship is expressed, or if nothing is transmitted from one level to another, then the items are simply forming a series divided for convenience into groups, rather as an iron rod may be cut into convenient lengths and formed into a chain. A concatenation does not, however, constitute a hierarchical classification.

The [*Shen Nung*] *Pen Tshao Ching* 神農本草經 (Classical Pharmacopoeia of the Heavenly Husbandman), already referred to[133] as a work essentially of the –2nd or –1st century, provides an example of this. The 365 entries which at that time it is supposed to have contained, were placed in three groups and entitled *shang phin* 上品, *chung phin* 中品 and *hsia phin* 下品, or 'upper grade', 'middle grade' and 'lower grade', the upper grade being named *chün* 君, 'prince', the middle grade' *chhen* 臣, 'minister', and the lower grade '*tso shih* 左使', 'adjutant'. With such terminology one would seem justified in assuming that the various medical substances were being graded hierarchically, with the 'sovereign' remedies at the top, and the common remedies at the bottom, but in fact these ancient terms were being used as complex adjectives to indicate that some substances (the princes) were entirely benign, and could be used at all times; others (the ministers) could be used, but with circumspection, while the third category (the adjutants) such as lead tetroxide, lead carbonate or arsenolite, were highly dangerous and could only be used with extreme care.[134] Here, then, there is still simple grouping with some serialisation.

The prototype of hierarchical or dendritic classification is the family tree. Today it implies the transmission of genes, and from this type of classification it is possible to see whether the genes of a certain ancestor have been transmitted to another member of the family. In earlier times what was thought to be transmitted was 'blood', and the problems arising from related blood were problems of consanguinity. Along with blood was transmitted authority, at the least the authority of a father, at the highest level the authority of an emperor.

It is interesting that the modern terms used in zoological dendritic classification in both English and Chinese are extremely ancient but now used in a new and precise manner to indicate the different evolutionary levels. The main terms in the hierarchical sequence of modern science are: kingdom – phylum – class – order – family – genus – species. Of these seven terms, three in English derive by analogy from the realm of power and authority. They are: *kingdom*, denoting the widest network, as in 'the animal kingdom'; *class*, originally one of the six groups into which the Roman people were divided for the purposes of taxation; and *order*, one of the three groups into which the Roman people were divided 'horizontally', with the order of senators at the top, of knights in the middle, and of plebeians at the bottom. The three terms

[133] See *SCC*, vol. 6, pt. 1, p. 242. [134] See *SCC*, vol. 3, p. 643.

which derived from the family network are *phylum*, deriving from the Greek word for 'tribe'; *family*, a Latin word which has not altered very much in meaning; and *genus*, an assemblage of persons, animals or plants which, in Roman thinking, resembled each other by reason of consanguinity or shared characteristics. The seventh term, *species*, may be traced back not by analogy from blood or authority, but to that in a thing's appearance, or its 'essential qualities' which, in formal logic, enabled it to be distinguished from a genus and an individual.

Chinese terms used in modern dendritic classification also have a venerable history, turning up again and again as the practice of classification developed and became more refined. The ancient terms now used in China in this modern sense are taken not from the family, important though it was in China, nor yet from the structures within which authority was transmitted, but from the world of thought and learning. Kingdom is *chieh* 界, world, as in 'world of ideas'. The phylum or tribe of Western classification is in Chinese *men* 門, door, particularly meaning the door of the sage, his school of thought. Class, *kang* 綱, and order, *mu* 目, go together in encyclopaedias to indicate headings and sub-headings. The equivalent of 'family', *kho* 科, is one of the most interesting terms of all, for, as we have written elsewhere:

> It happens to be the most fundamental word for classification of any kind in the Chinese language. . . . The basic meaning of *kho* before modern times was that of the divisions of successful candidates in the imperial examinations for entering the bureaucracy. It thus meant class or degree. The oldest meaning of the word, however, was that of any hollow cavity, as in an old tree, so that the speculation is permissible that the original semantic content was the distribution of different objects derived from plants into a series of 'pigeon-holes' so as to classify and separate them.[135]

Genus and species are indicated by *shu* 屬 and *chung* 種 respectively, the former being very similar to the Latin *genus* in its meaning of 'an assemblage of persons or things', though in Chinese they are assembled by the force of political or logical dependence, whereas in the Latin term the assemblage is due to kinship or 'blood'. *Chung* 種, meaning class or kind or sort, was used by Li Shih-Chen in the 16th century in his *Pen Tshao Kang Mu* (Great Pharmacopoeia) as synonymous with his particular individual entries, classified under the various *lei* 類 or genera, many of which were in fact true genera in the modern sense.

When language and thought have advanced to the point at which it seems necessary for a particular thing to belong to a genus within which it is in some way differentiated from other items, but when it is not clear what that genus should be, an important decision has to be made. One is that existing names in a language provide the genera, and the species or sub-groups within the genus must then be classified in such a way that their differentiae support the method of classification, though this is often impossible. The other is that the various items must be classified according to the distinguishing differentia, where this is possible, and a name for

[135] Needham (1968), p. 128. See *SCC*, vol. 7, pt. 1, 'The Concept of a Class', pp. 218 ff. on the importance of the word *lei* 類, 'category'.

the genus must be found or invented. Theophrastus, 'father of botany' (*c.* −372 to −298), began his classification of plants by grouping them under existing names used generically, rather than by grouping them according to characteristics, and inventing a term to summarise what was subsumed.[136] It is to his credit, however, that when classification on traditional lines was clearly breaking down, he was prepared to adopt current colloquialisms such as dendrolachana (δενδρολαχανα) or 'tree-herb', as a term to subsume plants like the cabbage, which were found to have something of the nature of a tree.

The idea of descent through the family was no less important in China than in the West in the development of hierarchic classification.[137] The keeping of genealogies had long been practised in the West. It would appear that the Romans may have started schematic genealogies in which the lines in a diagram indicate the relationship between ancestors, for it was their custom to keep wax models of the heads of ancestors, and at one time, according to Pliny, 'extremely correct likenesses of persons'.[138] Pliny goes on to say that 'the pedigrees too were traced in a spread of lines running near the several painted portraits'.[139] In addition 'the archive rooms [in private houses] were kept filled with books of records and with written memorials of official careers'.

The importance of genealogies for scientific thinking lay, however, in the use made of them by metaphor and analogy. The key idea was that things breed. Just as a man breeds sons, so does his money breed interest. The Greek for 'interest' was tokos (Tókos), meaning literally 'child'. Moreover, as money can have a child in the form of interest, so in time can that interest have a child, tokoi tokon (Tókoi Tókwr), the children of children, or, as we should say nowadays – compound interest.

The same metaphor was also much used in China. For example, one musical note gives birth to another to produce the scale in the cycle of fifths.[140] The same word *sheng* 生, give birth to, or generate, is also of great importance in alchemy, and mineralogy, and in fact the whole conception of breeding metals and their different generations is well brought out in the passages from +5th-century Taoist authors which we quoted at some length in the Section on Mineralogy, part of which is as follows:

We read that according to the *Ho Ting* books, when cinnabar is acted upon by 'the *chhi* 氣 of the green Yang' it forms ores, which after 200 years become green ('the girl is then pregnant'), and gives birth after 300 years to lead which in a further 200 years changes to silver; lastly,

[136] His use of the terms 'tree', 'shrub', 'under-shrub' and 'herb' are examples.

[137] See *SCC*, vol. 6, pt. 1, p. 212, where reference is made to the −2nd-century *Shih Pen* 始本 (Book of Origins), which enlarges on the origin of the names of the clans and gives elaborate genealogies of the ruling houses.

[138] Pliny, *Natural History*, IX, xxxv, 2 (4), tr. Rackham (1952), p. 263.

[139] 'Stemmata vero lineis discurrebant ad imagines pictos.' *Ibid.*, 2 (6), tr. Rackham (1952), p. 265.

[140] That is to say, if a musician wishes to play on a zither the deepest or *huang-chung* 黄鍾 note, he must strike the longest string. If he then stops that string at two-thirds its length and strikes it again, the string which had formerly given birth to the *huang-chung* note, now delivers itself of the *lin-chung* 林鍾 note, which is a perfect fifth higher in pitch (*huang-chung sheng lin-chung* 黄鍾生林鍾). This formula may be found in the −3rd-century *Lü Shih Chhun Chhiu* 呂氏春秋 (Master Lü's Spring and Autumn Annals), *SPPY* edn vol. 1, ch. 6, fol. 3a.

after 200 years, the '*chhi* of the Great Unity' is obtained, and gold appears. Gold, they say, is the son of cinnabar . . .[141]

Similar ideas were current in 13th-century Europe:

Gold is produced in the earth with the aid of strong solar heat, by a brilliant mercury united to a clear and red sulphur. . . . White mercury, fixed by the virtue of incombustible white sulphur, engenders in mines a matter which fusion changes to silver. . . . Tin is generated by a clear mercury and a white and clear sulphur . . .[142]

An important step forward was taken when it was realised that the transmission of power and authority can be represented by the same diagram or vocabulary as is used for the transmission of 'blood'. New thinking often precipitates a crisis, but the crisis can perhaps be averted by the judicious use of fictions. Such fictions can be tangible, as when Egyptian queens acting as monarchs wore false ceremonial beards on ritual occasions,[143] or verbal, as when the Roman emperor Antoninus Pius adopted Marcus Aurelius as his fictive son and successor. Magical *potestas* can also be transmitted through such artefacts as sceptres, crosiers and gavels. It is, therefore, no great leap from the lines, or perhaps linen threads (*linea*), which connected the ancestral portraits described by Pliny, to the diagrammatic lines connecting the names of ancestors in a family tree, to the schematic lines indicating the transmission of the authority of St. Peter from pope to pope, or of the military authority of the commander of a division to his brigadiers, and through them to colonels, majors, captains, lieutenants and warrant officers.

Once officers, officials and functionaries are thought of as being in special relations to each other by virtue of their hierarchical levels, further developments become possible, though this may take time.[144] For example, in an early stage of tax collection, when contributions are paid in kind, one official may be given the task of arranging for the goods to be kept in store rooms. As the catchment area increases in size, and the number and range of contributions also increases, delegation of duties becomes essential. A hierarchy of senior and junior officials then comes into being, and *ipso facto* the work they do follows hierarchically, as occurred in late Chou times in China, when the taxes on fish and salt, and on the produce of mountain, marsh and market were classified as particular taxes under the general tax *tsu-shui* 租稅.

It is not essential for hierarchical types of classification to be expressed diagrammatically. The hierarchical arrangement may be implicit in the technical terms

[141] See *SCC*, vol. 3, p. 639, quoting Li Shih-Chen's *Pen Tshao Kang Mu* (1596), *PTKM*, ch. 8, p. 11a. Li Shih-Chen is quoting from books of perhaps the 12th century, but this is not certain.

[142] *SCC*, vol. 3, p. 639, quoting *Speculum Maius* (see Sarton (1931), vol. 2, p. 929).

[143] Free-born Egyptians were clean-shaven, but wore false beards on occasions. 'The allegorical connection between the sphinx and the monarch is pointed out by its having the kingly beard, as well as the crown, and other symbols of royalty' (Wilkinson (1854), vol. 2, p. 329). The upturned divine beard permitted to dead Pharaohs is illustrated in the British Museum (1972) *Treasures of Tutankhamun*, item 9, 'Canopic Coffin'.

[144] Creel (1937a), p. 103, writes, 'Not even the titles for feudal lords of various ranks had been arranged into a graded hierarchy in the early Chou period. In the beginning these titles were used rather indiscriminately. The system of five ranks emerged and became clarified only gradually.' This is developed in more detail in Creel (1937b), p. 328.

themselves, as when one speaks of a man, his father, and his grandfather. But Chinese characters are particularly well suited for indicating the different levels in this type of classification, if it is so desired. For example, it is possible to make clear to which generation individuals belong by giving one character of their personal names a common radical in each generation, as when the emperors of the Ming dynasty, and other members of the Chu 朱 family, from Emperor Hui Ti 惠帝 (personal name Yün-Wen 允炆) onwards, all had in common, as a radical in one of the characters of their personal names, the character denoting one of the Five Elements; the different members of each generation all used the character denoting the same element.[145] Thus generation by generation their personal names included the symbol for wood, fire, earth, metal or water, in that order,[146] as in Ti 棣 who succeeded his nephew Yün-Wen 允炆, followed by Kao-chih 高熾, also of Yün-Wen's generation, and he by Chan-Chi 瞻基, Chhi-Chen 祁鎮, Chhi-Yü 祁鈺, Chien-Shen 見深, and so on.

This principle of instant recognition of hierarchical order might well be exploited for scientific purposes whenever there is a need to indicate hierarchical levels. For instance seven components of the seven characters which denote the seven main levels in the biological hierarchy could be abstracted from the full character and used as indicators of hierarchical level, and added to any plant or animal name, thus making clear at a glance the position of a term on the hierarchical ladder, just as the addition of the characters for gas, metal, liquid, and so on, makes the physical nature of an element immediately apparent in modern Chinese chemistry.[147]

This is already done to a certain extent in international botany, for the generic name of every plant is the first of the two words comprising its binomial. The genus, then, contributes the basic name to the related nomenclature, and ascending from genus through family and order to the sub-classes, the terms are made by the addition of Latin suffixes to the generic term's root. For example, the genus to which a great many species of roses belong, namely the *Rosa*, adds *-aceae* to the stem to create the term *Rosaceae* meaning 'rose family'.[148] In the next higher rank of 'order', the family of *Rosaceae* adds the suffix *-ales* to its stem *Rosa-*, making *Rosales*. This suffix affords instant understanding that terms such as *Malvales* or *Cycladales* which have this ending belong to the rank of 'order'. Below the 'family' there may be a division into 'sub-families' or 'tribes'. This is indicated by the suffix *-eae*, as in *Roseae*. If *-eae* for the sub-family, *-aceae* for the family and *-ales* for the order were the only suffixes in use, there would be no problem, but unfortunately there are variants and additional suffixes for more refined distinctions. Sub-orders are distinguished by *-ineae*, as in *Malvineae* from the stem *Malva-*. Some families are known by the variant ending *-osae*,

[145] Wood, *mu* 木; fire, *huo* 火; earth, *thu* 土; metal, *chin* 金; water, *shui* 水.

[146] This is, not unreasonably, the 'mutual production order'.

[147] See Gerr (1942), p. 55, where he cites examples of gases indicated by the radical No. 84. The zoological levels could be indicated by radicals 102, 169, 120, 109, 68, 44 and 115, which are present in the characters for these terms.

[148] If there are many genera within a family, the family name will derive from the most important of the generic names. So *Rosaceae*, the rose family, derives from the genus *Rosa*, other members of the family having such different names as *Rubus*, *Prunus* and *Speraea*.

as in *leguminosae*, developed from the word *legumen* by way of its root *legumin-* and sub-families may also be distinguished by the suffix *-oideae* as in *Boroginoideae*. These suffixes may now be summarised in alphabetical order as: *-aceae*, *-ales*, *-eae*, *-ineae*, *-oideae* and *-osae*. The drawback is that these Latin suffixes are so well provided with the vowels [a] and [e] that they may not be readily distinguishable to ear or eye, especially for those of the world's inhabitants who are unaccustomed to inflected languages, and may be uncertain how much of the word is its root. Classification of plants is, however, still at the provisional stage, since 'the present state of man's knowledge is too scant to enable one to construct a phylogenetic classification and the so-called phylogenetic systems represent approaches towards an objective . . .'.[149] It may well be that in the centuries to come alternative methods of distinguishing the levels of classification such as are suggested by the use of logographs will be found to have advantages.

Organising botanic knowledge in hierarchic levels from general to particular, or as the Mohists put it, 'from the wide to the narrow',[150] nevertheless entailed a linguistic problem. Human beings are conservative, and resist changes in their thinking habits. When classification in botany came to be organised on the revolutionary lines which culminated with the binomial nomenclature of Linnaeus in 1753–8, it had to be accompanied by a linguistic revolution.

The result was a new semantic field with a structure totally different from the old. Botanists decided that the Latin nomenclature begun in 1753 (and still used today in the international rules) was indispensable, because the change in semantic field had not been accompanied by a total reconstruction of vocabulary. The Latin words used by Linnaeus and his successors in a new and precise sense had previously been used – but in a different sense – within the older traditional semantic field; and it was therefore necessary to abolish all pre-1753 meanings.[151]

This, a drastic intellectual cleansing of Augean stables, was somehow achieved. It could scarcely have been done with Chinese, for a Chinese character seldom dies, and Literary Chinese in the 18th century was far more alive than Latin, even though Latin had only ceased to be the administrative language of France a hundred years before Descartes wrote his *Discours de la Méthode*.[152] Had it not been for the rise of the vernacular languages of Europe during the Renaissance, and the relative indifference of the people who spoke them to what the botanists were doing in Latin, modern botany might well have got bogged down in endless disputes about terminology. As it was, botanists were able to take many Latin words from low levels of specificity, and exalt them to the level of genus or family where they subsumed countless species of plants whose names had not existed or had been ambiguous in the ancient Greco-Roman world, and were now redirected, reconstituted, or inverted from Greek and Latin roots. Botany was thus equipped with a new, properly organised and coded language. But how was this new language to be made available for use by the common people?

[149] Lawrence (1951), p. 13. [150] See *SCC*, vol. 2, p. 174, on Mohist 'Classification'.
[151] Haudricourt (1973), pp. 267–8.
[152] The Edict of Villers-Cotterets was in 1539. The *Discours de la Méthode* was published in 1637.

France attempted the logical answer by translating the neo-Latin terms back into French.[153] For example, the nightshade (in French *morelle*) became classified as *solanum*. To this genus belongs the potato. What had formerly been called an earth-apple or pomme-de-terre therefore became, in botanic Latin, *Solanum tuberosum*. This was then translated back into French as 'morelle tubereuse'. However, people in shops and restaurants do not order 'morelles tubereuses', or in English – 'tuberous nightshades'. They prefer 'pommes frites' or perhaps 'boiled potatoes'.

The botanic code must be kept as it stands. It cannot be translated at an early date into any living language without destroying the linguistic hierarchies, for people do not think of egg-fruit and potatoes as species of the genus '(non-deadly) nightshade'. Chinese, therefore, like English and the other pragmatic languages, continues to use the familiar terms for plants, whether recent or handed down from antiquity, and when the hierarchic level of the precise meaning is in doubt, refers to the Linnaean or some similar code as a cross-check. It is not that Chinese is compelled to do this because it uses logographic script, though the contrast of scripts in a Chinese publication makes the use of the cross-check more apparent. It is used for the same purpose in every language of the world that includes modern botany in its literature. It is possible that little by little, as resistant words lose their power, they can be edged into the right slots, and the world's vernacular languages can be systematised as Lamarck hoped they would. But it will certainly take centuries.

An illustration of the difficulties facing those who wished to do for Chinese what Linnaeus did with Latin can be seen in the following example. The well-known Chinese flower term *tzu-wan* 紫菀 (purple-luxuriant flower), which is the *Aster tataricus*, was chosen to be eponymous to the Astereae tribe of the *Compositae* family.[154] Thus, *tsu* 族 being 'tribe', the Astereae are called *tzu-wan-tsu*, and *shu* 屬 being genus, the aster genus is *tzu-wan-shu*. There are many species within the aster genus, but *Aster tataricus* which gave its name to the genus in Chinese remains plain *tzu-wan*. According to the rules of international botanical nomenclature, this genus term should be followed by a second term if the name of a species is to be formed, as *Aster* is followed by *tataricus*. It is the second term which distinguishes the species. But in Chinese this is not done. *Tzu-wan* stands alone as the species name. This is understandable, for to add a qualifying term to an old traditional name would do violence to the living language. It would be like requiring an Englishman to call a common daisy a *daisy vulgaris*. Furthermore, the similar species known as *Aster fastigiatus*, in which the flower is white, cannot really be called *tzu-wan*, for as *tzu* is purple, it cannot easily be applied to such a flower (though English does have white violets) and therefore has the abbreviated name of *nü-wan* 女菀, 'womanly-luxuriant [flower]'. But there are other species of aster whose traditional names have been kept, though they have even less connection with the generic term than did *nü-wan*. For example, there is *ma-lan*, 'horse-orchid', which is *Aster indicus*, also known as *chi-erh-chhang* 雞兒腸 'chicken intestine'; there is *Aster scaber*, which is called 'east-wind-vegetable', *tung-feng-tshai* 東風菜; there is *Aster hispidus*, *kou-wa-hua* 狗娃花, which is literally 'dog-pretty-baby-flower', and many more. What

[153] Haudricourt (1973), p. 270, quoting Lamarck's preface to his *Flore française*.
[154] Steward (1958), pp. 395 ff.

they lack is a common term or character to indicate a common genus. Linnaeus was able to take terms from the Latin language and change their meaning to meet the needs of modern botany in a way which would be very difficult and perhaps impossible in any living language. But this does not mean that the Linnaean system cannot be expressed in logographs. On the contrary new logographic terms could be invented just as new Latin terms were invented, and just as they have been in Chinese for other fields such as musical notation. The generic *Aster*, for instance, derives from the Greek word for star, and gives a rudimentary impression of a flower with petals like the rays of a star. Similarly the Chinese ideograph for star, *hsing* 星, which already combines with other elements to make a range of composite characters, could be combined with Radical No. 140 which indicates things of vegetable origin to make a new character indicating the genus *aster*, which would then be suitably qualified with second characters to distinguish each species. Such characters would be no less international than the Latin names, and would avoid the horrific difficulties of pronunciation which attend such terms as *Warszewiczella* or *Brassaocattlaeia*. Chinese characters are already given local pronunciation in many countries of the Eastern world. The same process may well continue.

The establishing of hierarchic classification is an essential first step in the systematic retrieval of information. The most efficient machine for finding information quickly was, until recently, the human brain. But for the herbalist the number of plants that could be used, and the ways in which they could be combined in recipes, and the particular ailments and diseases to which they could be applied form a total of entries too great for efficient retrieval from the human memory. Nor was it enough to classify individual items in series under such general headings as Trees/Herbs in China, or Trees/Shrubs/Undershrubs/Herbs in Greece. A branching or dendritic system of classification was clearly needed in order that the enquirer could follow a route from one point of bifurcation involving a choice to the next. In Greece this was aided by the dichotomies of Aristotelian logic, and is still with us, in, for example, the notion of vertebrates and invertebrates, used by Lamarck and Cuvier to replace Aristotle's earlier dichotomy of Enaima, animals having blood, and Anaima, animals not having blood.

The first need in the grouping of plants was to promote ease of finding, without having to run through large numbers of items grouped in series. The first step to this end was to increase the number of general headings, as from trees and herbs to trees, fruit trees, kitchen vegetables, cereals and herbs, thus reducing the number of entries listed under the head words. Very soon it is found that certain plants such as melons cannot be regarded as classifiable either under fruit trees or kitchen vegetables, therefore more precise definitions are required for the terms in the main headings, and new terms must be added.

In such a climate of opinion it is not surprising that the *Erh Ya* ('The Literary Expositor', better translated perhaps as 'The Semantic Approximator'), China's earliest dictionary (consolidated approximately −3rd century), should have listed its headings in an order of evolving complexity, from herbs to trees, and so to insects, fish, birds and beasts. This was a work of severely practical intent whose aim was to help scholars track down the meaning of obsolete terms and words.[155] Naturally the many hundreds of terms for things and relationships whose meaning had become obscured, or whose very existence had been almost forgotten during the previous tumultuous centuries, needed to be grouped according to some logical system. Nineteen sections were found to be enough for this purpose. The first eight reflect the unknown author's concern with the preservation of the ancient lore to do with the ritual of the Way. They deal in turn with *ku* 詁, ancient sayings, *yen* 言, words, and *hsün* 訓, explanations of what should be given in moral instruction. Naturally this is followed by *chhin* 親, human relationships, *kung* 宫, palaces and important dwellings, and *chhi* 器, the utensils used in traditional rituals, and by *yüeh* 樂, music, a most important part of those rituals, which might well be translated as 'glee'.[156]

The second half of the book deals with what is found on the earth, *ti* 地, and the earth itself is the heading for section nine. Section ten deals with *chhiu* 丘, grave mounds, mounds and hillocks, followed naturally by *shan* 山, hills and mountains. Since water runs off hills and forms rivers, section twelve deals with all things watery, *shui* 水. Sections thirteen and fourteen together comprise the world of botany. Section fourteen is concerned with trees, *mu* 木, and the one before it with *tshao* 草, which may be translated as herbs and grasses. But herbs and grasses may be held to subsume all the non-tree part of the botanical universe. In entering the realm of biology we find ourselves even today in a world of indefinite frontiers. Sections thirteen and fourteen are but two of the seven sections which are devoted to natural history or nature. The last five are, in serial order, *chhung* 蟲, insects or very small creatures; *yü* 魚, fish; *niao* 鳥, birds; then *shou* 獸, wild beasts; and finally *chhu* 畜, domesticated animals.

Much can be learnt from this listing. In the first place, the sequence is so clear and logical that it can be memorised from a single reading.[157] But of greater interest for classification are the terms used as headings for the sections to summarise their contents. In some cases the terms are already sufficiently general to subsume all that is required of them, but in others they are in the position of English before it had discovered, for example, 'precipitation' to subsume rain, fog, frost, and similar terms. An example of the former is the section whose title is *shui* 水, water. This section sets out to save from oblivion terms to do with springs, wells, streams and rivers, and also with such things as waves and marshes, and even such less obvious terms as that for the margin where grass and water meet.

An example of where nomenclature had not yet evolved with adequate covering power is afforded by the two terms just mentioned, *tshao* 草 and *mu* 木, literally grass and trees. Together they in fact mean *vegetation*, as is clear from the parallel sentences which occur in the section on mountains.[158]

[155] On the *Erh Ya* see also *SCC*, vol. 6, pt. 1, pp. 186 ff.

[156] In earlier times this English word carried, as does the Chinese, the multiple connotations of music, joy, entertainment, beauty and colour, and a sense of exaltation, as Cooper (1973), pp. 56–7, points out.

[157] This is not the opinion of Bauer (1966), p. 667, who regards the arrangement as clumsy, with many overlappings.

[158] *Erh Ya*, Sect. 11, 'Mountains'.

[Mountains which have] much vegetation [are known as] '*hu* 岵'. [Mountains] devoid of vegetation [are known as] '*kai* 峐'. *To tshao mu hu* 多草木岵. *Wu tshao mu kai* 無草木峐.

It is obvious that when the writer speaks of mountains with or without trees and grasses, he does not mean that they might nevertheless be covered with shrubs or bushes. This is an early example of the tendency to convey in Chinese by the 'synthesis of opposed concepts' what is covered in Indo-European languages by an abstraction, as was discussed above[159] in connection with the word *height*. Here the scale is formed by the tallest vegetable growth, the tree, at one end, and low-growing plants at the other. In between may be ranged all the creeping, climbing, rampant, shrubby, reedy and bushy plants which in Indo-European languages are subsumed by an abstraction. It was only a matter of time before the two words fused into a technical term doing the work of the abstraction *vegetation* in English. The two ends of the 'animal' scale are represented by the words *chhung* 蟲, insects and similar very small creatures, and *shou* 獸, beasts. But here the synthesis of opposed concepts was not used to make a term for 'animal'. It was only after the 17th century with the changing of perspective and classification that it became necessary for the beasts of the field and for the fowls and other denizens of the air to be subsumed by the abstraction *animal*, *tung-wu* 動物 and for the 'grass-tree' synthesis to be partly replaced by the modern term *chih-wu* 植物, *plant*.

The formulation of general terms such as *precipitation* is essential for scientific thinking, and as science develops, and things are observed to be in new or different relationships, new terms are continually being invented to express these relationships. For example, the concept of animal, distinguished from that which is vegetable or mineral, scarcely existed in English before the year 1602. The word does not occur in the Authorised Version of the Bible, though there is frequent mention of beasts of the field, fowls of the air, fish and things creeping innumerable. Latin on the other hand used the word *animal* in the widest sense, as we use it now, including both man[160] and insects,[161] and indeed the whole of creation thought of as animated matter,[162] but for *vegetable* classical Latin had only the phrase 'a thing which comes to being from the earth',[163] and for *mineral* had only *metalla*, meaning specifically mined products. The term *mineral* in any language is quite clearly a convenience label. In Western languages there have been two approaches towards getting the reader on to this network. One was by using the ancient Gaullish term *mine*, meaning a shaft sunk in the ground, and calling the things which were dug out of it 'mineral'. The other was by excluding those things which it was not, 'any natural substance that is neither animal nor vegetable', as it was regarded in England in 1602.[164] It was only at the

[159] See above, p. 108.

[160] 'This foreseeing, shrewd, many-sided . . . animal, which we call man', 'Animal hoc providum sagax multiplex . . . quem vocamus hominem.' Cicero, *De Legibus*, 1, 7, 22.

[161] 'Formicae, animal minimum', 'The ants, least of animals.' Pliny, *Natural History*, VII, xv, 13 (65).

[162] 'Hunc mundum animal esse, idque intelligens . . . , [That] this world is animal, and that it has understanding . . .'. Cicero, *Timaeus*, 3, 4.

[163] 'Res quae gignitur e terra'. Cicero, *De Finibus*, 4, 5, 13.

[164] Murray *et al.* (1933), vol. 6, p. 466.

beginning of the 17th century that the classification of things as animal, vegetable and mineral required new terms. The same need arose in China rather later, on contact with the West, with the result that the terms, instead of evolving haphazard, could be devised on a logical system, so that living things, *sheng-wu* 生物, are subdivided into moving things (animals), *tung-wu* 動物, and growing things (the vegetable kingdom), *chih-wu* 植物, as opposed to mined things (minerals), *khuang-wu* 礦物.

Once the *Erh Ya* had established a framework, the progress of formal classification, especially in botany, proceeded apace. To illustrate it, one may refer to the early Han work, whose author is uncertain, the *Shen Nung Pen Tshao Ching* already described in some detail above.[165] From the fortunate survival of a manuscript of the preface to a later work by Thao Hung-Ching 陶弘景,[166] the *Pen Tshao Ching Chi Chu* 本草經集注 commenting on and amplifying the *Shen Nung Pen Tshao Ching*, we know that, like the *Erh Ya*, though for quite a different reason,[167] it contained three *chüan* 卷 or chapters. These dealt with 365 substances – animal, vegetable and mineral – which could be used for sustaining health and curing illnesses. Thao Hung-Ching lists these substances and others which he added to a total of 730, under what he called 'the categories of things' (*wu lei* 物類), as follows: gems and stones (*yü shih* 玉石), grass and trees, i.e. vegetation (*tshao mu* 草木), these four covering 356 items altogether; and then fruits (*kuo* 果), vegetables (*tshai* 菜), rice (*mi* 米), and lastly 'food' (*shih* 食), perhaps meaning other food grains. As all but the preface of this text is lost we cannot be sure.

Already some of the problems of classification have begun to appear. Classification implies a network, and a network implies an objective. If the objective of a network is eating, both rice and fruit will be classified as food. It will be noted that the items listed above, which can be eaten as food, are grouped together, and separated from the minerals and the vegetation. If the objective of a network is healing, then fruit, and the bark of trees, such as cinnamon or quinine, will not be regarded as food, though they can be used in cooking, but as medicaments. If the objective of a network is linkage, in order to explain behaviour, as when it is noticed that some trees will accept grafts without difficulty, but others not at all, or in society how some nobles pay deference to others of higher rank, a quite different form of classification will be required, involving different levels, family trees and hierarchies. Early Chinese pharmacopoeias combine the networks of the pharmacist and the dietician.

As categories were increased and specialisation developed, particularly when materials never seen in China before began to pour in from the southern lands, and later from beyond the southern seas, just as they did into Europe from the Americas and the Indies, specimens began to be classified under wrong categories, terms were

[165] See *SCC*, vol. 6, pt. 1, pp. 235 ff.

[166] This was described in *SCC*, vol. 6, pt. 1, pp. 244 ff., and n. f, where a portion of the manuscript recovered from either Turfan or Tunhuang was quoted in translation. The title of the lost work was *Shen Nung Pen Tshao Ching* (Classical Pharmacopoeia of the Heavenly Husbandman).

[167] The classification of items according to pharmacological grading in three chapters cutting across natural categories is discussed in *SCC*, vol. 6, pt. 1, pp. 242–3.

duplicated, and confusion was rife.[168] The heroic work of reducing to order the mineralogy, metallurgy, mycology, botany, zoology, physiology and other sciences as they existed in China in the 16th century was undertaken by Li Shih-Chen 李時珍, occupying him from his thirtieth year till his seventieth in the most strenuous labour and travel. His great encyclopaedia, the *Pen Tshao Kang Mu* 本草綱目 was published in 1596, three years after his death. We have already quoted from it in a previous volume, and will do so again,[169] for his words make it clear how far China had come on the road towards scientific classification.

In the writings of old, gems, minerals, waters and earths were all inextricably confused. Insects were not distinguished from fish, nor fish from shellfish. Indeed some insects were placed in the section on trees, and some trees were placed in that on herbs. But now every group has its own Section. (The sequence is as follows:) at the head come waters and fires, then come earths, for water and fire existed before the myriad things, and earth is the mother of all the myriad (inanimate and animate) things. Next come metals and minerals, arising naturally out of earth, and then in order herbs, cereals, (edible) vegetable plants, fruit-bearing trees, and all the woody trees. These are arranged according to their sizes in an ascending order, starting with the smallest and ending with the largest. A Section on objects that can be worn by human beings follows (this is logical since most of them come from the plant world.) Then the tale continues with insects, fish, shellfish, birds and beasts, and mankind bringing up the rear. Such is (the ladder of beings), from the lowliest to the highest.[170]

In this broadly conceived pharmacological encyclopaedia the trend in classification is clear. The general terms are very general, the specific terms increasingly specific, and within each 'envelope' the items are serialised in a logical order. Moreover, as with the *Erh Ya*, there is a logical sequence running through all the topics treated. In the *Erh Ya* it was the thread of interest in ancient ritual and tradition. Man did not feature as a general category at all. Man was implicit throughout. The book began with his words, went on to his relationships, his buildings, utensils and music, and so out to the wide world and all its creatures. It is interesting that even in the *Erh Ya* animals were listed in an ascending order, from the lowliest to the most highly developed. In the *Pen Tshao Kang Mu* a sense of man's evolution is developing.

In the *Pen Tshao Kang Mu* there are in fact fewer general categories than in the *Erh Ya*, sixteen as compared with the earlier nineteen. But these sixteen terms have either far greater covering power, or more precise frontiers. The *Erh Ya*, for example, placed domestic animals and wild animals in separate categories. The *Pen Tshao Kang Mu* covers all beasts with the now more generalised term *shou* 獸 and whereas the *Erh Ya* had covered things that swim with the general word *yü* 魚, regardless of whether they had scales or carapaces or lacked them, the *Pen Tshao Kang Mu* made this necessary distinction. It is also a significant indication of the rise of scientific thinking that Li Shih-Chen made free use of the old theory of the Five Elements (*wu hsing* 五行), adopting it when it was useful, discarding it when it was not. Thus the first three categories in his system of classification were water (*shui* 水), fire (*huo* 火) and earth (*thu* 土). A devotee of the Five Element Theory would no doubt have followed these

[168] *SCC*, vol. 6, pt. 1, p. 311. [169] *SCC*, vol. 6, pt. 1, p. 315.
[170] On the 'ladder of souls', see *SCC*, vol. 2, p. 22, Table 10.

with metal and wood. But Li Shih-Chen used the first three because they fitted in with his evolutionary theories, whereas *mu* 木, (wood) he used in the precise sense of woody trees as opposed to fruit-bearing trees, and *chin* 金 (metal) he combined with *shih* 石 (stone) to make a compound meaning minerals, capable of subsuming far more under its general classification than *metal* alone.

In the world of plants he had five categories – herbs, cereals, edible vegetables, fruit trees and woody trees. The first of these, herbs (*tshao* 草), was subdivided into ten classes. The characteristic of these classes is not the plant's sex organs, or even its structure, but ease of finding.[171]

When ease of finding was the main criterion of arrangement, much would depend on the aim of the seeker. To divide trees into two categories, fruit trees and other trees, may have helped a physician or apothecary quickly to find the recipe he needed, but was not conducive to the working out of phylogenetic connections. There were, however, scholars whose interests were wider ranging, who sought to establish the reasons for things being as they are, and for their inter-relations, as, for example, Khou Tsung-Shih 寇宗奭 whose book entitled *Pen Tshao Yen I* 本草衍義 (Dilations upon Pharmaceutical Natural History), printed in +1119, carries a wealth of interesting observations and which points out, for instance, that 'stone snakes' (ammonites) could not properly be regarded or classified as snakes in the strict sense of the term.[172]

In this same century we are given a clear appreciation of the classification of living things at different levels by Chu Hsi 朱熹 (+1130 to +1200) who is reported as follows:

> Someone said: 'Birds and beasts, as well as men, all have perception and vitality, though with different degrees of penetration. Is there perception and vitality also in the vegetable kingdom?'
>
> (The philosopher) answered: 'There is indeed. . . . But the vitality of the animals is not on the same plane as man's vitality, nor is that of plants on the same level as that of animals.'[173]

Classification at different levels had already had a long history expressed in the form of the doctrines of the 'ladder of souls'. In *SCC* Volume 2 we tabulated the doctrines of Aristotle (–4th century), Hsün Chhing 荀卿 (–3rd century), Liu Chou 劉晝 (+6th century) and Wang Khuei 王逵 (+14th century).[174] Chu Hsi introduced new ideas which reached a point of considerable sophistication.

Organic things he classified in three levels, low (plants); middle (animals); and high (man), these being placed above inorganic things, with which are to be included things dried and withered, from which life has died. These four vertical levels may also be classified horizontally according to six degrees of functional development. All things, according to Chu Hsi's philosophy, have pattern (*li* 理), natural endowment (*hsing* 性) and specificity (*hsin* 心). Ascending to the level of plants, life, or the vital impulse

[171] This analysis is derived from the passage cited above, p. 156.

[172] See above, *SCC*, vol. 6, pt. 1, p. 287. For further details on the *Pen Tshao Yen I*, see *SCC*, vol. 6, pt. 1, pp. 283 ff.

[173] For the vivid sense of unity of men with all other living things see *SCC*, vol. 6, pt. 1, pp. 442–3, referring to *Honan Chheng Shih I Shu*, 河南程氏遺書 ch. 18, pp. 8b, 9a. See also above, *SCC*, vol. 2, pp. 568 ff.

[174] *SCC*, vol. 2, p. 22, Table 10.

(*sheng-i* 生意), and some form of perception, such as may be observed in plants which react to contact (*chih-chüeh* 知覺), are now present. At the higher level of animals may be found 'moral concepts' (*i-li* 義理), rudimentary and limited, however, or as we should perhaps say now, instinctive, but, by a subdivision to the highest level of all, that of man, they are 'comprehensive, embracing everything in some degree, but diffused, and therefore more easily obscured'.[175]

Hierarchic classification leads to the thesaurus-type encyclopaedia in which information is grouped according to an inverted dendritic pattern. This is the pattern followed in the great encylopaedias of China, and in the series of which this volume forms a part. To retrieve information from such books it is necessary to consult the dendritic pattern and decide under which head the required information is most likely to be found. Nevertheless retrieval of information from books composed on the dendritic pattern is inefficient without the supplementary aid of an index, on which Christoph Harbsmeier has written in *SCC*, Volume 7, part 1. First, however, it may be noted that the logographic script of China itself favours a classificatory mode of thought. In Volume 1 (p. 30) we wrote concerning *determinative-phonetic* characters: 'A determinative is an element (a radical) added to a phonetic to indicate the category within which the meaning of the word is to be sought', and to illustrate this the determinative *shui* 水, in its most general sense of 'liquid', was shown combining with different phonetics to indicate words meaning froth or foam, waves or billows, lacquer, branching streams and the sea. There are today some 450 characters which have this determinative, and most can be classified as being related in some way to that which is watery, though naturally some have been distorted by time, and in some the connection is fanciful or far-fetched. But the classificatory tendency of the Chinese script is still at work, and with beneficent effect, in modern scientific terminology, where the chemist immediately knows that a substance is normally a gas because it is classified under the gas determinative *chhi* 氣 or a liquid or a metal because it has the determinatives *shui* 水 or *chin* 金 signifying this condition or structure.[176]

(4) A Case Study: Description of the Plough

It is time we turned to actual examples of how Literary Chinese functioned as a medium for scientific and technical discourse. Our first case study chosen for analysis will be a straightforward +9th-century description of a Chinese plough, the most important of agricultural implements. It would, however, be instructive to compare this Chinese account of it with descriptions in other languages, and preferably to compare classical Chinese with classical Latin. Unfortunately no very detailed description of a plough survives in Latin. A shorter passage must therefore suffice. But by contrast there is an interesting description of a plough written in 16th-century English, at a time when that language was just beginning to develop

[175] *Chu Tzu Chhüan Shu*, ch. 42, pp. 29a ff. See *SCC*, vol. 2, p. 568.

[176] See *SCC*, vol. 1, pp. 30–1, and vol. 2, pp. 220 ff. The point is also made by Gerr (1942), p. 55, who refers also to Spooner (1942).

its technical and scientific applications. In each of these accounts no use was made of illustrations. The authors had to rely on language alone to paint the picture and express the ideas.[177]

The Latin example is provided by Pliny the Elder, writing in the +1st century, with the full resources of the libraries of the ancient world still unaffected by barbarian invasion and economic collapse, with the scientific schools of Alexandria still productive, and with communications as good as they ever would be till modern times. Pliny was less interested in agricultural machinery than in different types of crop and seed, but he does describe in his *Natural History* the different types of ploughshare known to him.[178] He was certainly well aware of the inefficiency of the ploughs used in many parts of Italy, where 'eight oxen strain panting at one ploughshare'.[179] He writes:

Ploughshares are of several kinds. The coulter (*culter*) is the name for the part fixed in front of the share-beam (*dentale*), cutting the earth before it is broken up and marking out the tracks for the future furrows with incisions which the share (*vomer*), sloping backward is to bite out in the process of ploughing. Another kind is the ordinary share consisting of a lever, i.e. a bar ending in a pointed beak (*rostrati vectis*), and a third kind used in easy soil does not present an edge along the whole of the share-beam but only has a small spike at the extremity (*exigua cuspide in rostro*). In a fourth kind of plough this spike is broader and sharper, ending off in a point (*mucro*), and using the same blade (*gladius*), both to cleave the soil and with the sharp edge to cut the roots of the weeds. An invention was made not long ago in the Grisons [Switzerland, then Raetia Galliae] fitting a plough of this sort with two small wheels – the name in the vernacular for this kind of plough is *plaumorati* (sic! for *plaumoratum*); the share has the shape of a spade (*effigiem palae*).[180]

[177] Before the invention of printing there was less recourse to illustrations to make clear the intricacies of machinery or the structure of organisms. One must, however, mention the three ancient Greek artillery manuals preserved in the *Codex Parisinus inter supplementa Graeca 607* (see Marsden (1971), p. 11), in which there are diagrams in colour. These manuscripts were bound and placed in the library at Buda *c.* 1450. Their early history is not known. We have mentioned in vol. 6, pt. 1, p. 3, how the medieval text of Apuleius Platonicus' *De Herbarum Virtutibus* (*c.* +5th century), when printed in Rome soon after 1480, provided the first printed illustrations of plants. (Sarton (1927), p. 296.) Cratevas, the physician to Mithridates, King of Pontus, had, however, in the −1st century published an illustrated herbal in at least five books (Sarton (1927), p. 213). From these perhaps derive the wonderful illustrations of plants in the Juliana Anicia Codex of Vienna of +512. One reason for this paucity of illustrations was perhaps that whereas many were able to copy texts, few had the talent to draw. After the invention of printing, one good woodcut of a flower or mechanical contrivance could be repeated again and again. The study of botany in particular benefited. No modern textbook would attempt to describe the working of a machine without an illustration. See for example the description of the functioning of a plough in Anon. (1972), vol. 1, p. 430. The invention of printing was vital for the development of the scientific revolution in two ways: (i) the dissemination of information, and (ii) improvement in the quality of the information disseminated.

[178] Graeco-Roman literature offers no good description of the plough as a whole. Virgil (−1st century) in *Georgics* I, 169–72 and 174, gives five lines of poetic description, and Varro (−1st century) in *Lingua Latina*, 5.134, gives the technical terms for eight parts of the plough, but adds details of construction only if of etymological interest. It must, however, be remembered that of the 490 books written by Varro, 'the most learned of the Romans', only two survive.

[179] 'Cum multifariam in Italia octoni boves ad singulos vomeres anhelent.' Pliny, *Natural History*, XVIII, 47, 170, tr. Rackham (1950), vol. 5, p. 297.

[180] 'Vomerum plura genera: culter vocatur infixus prae dentali priusquam proscindatur terram secans futurisque sulcis vestigia praescribens incisuris quae resupinus in arando mordeat vomer. alterum genus est volgare rostrati vectis. tertium in solo facili non toto porrectum dentali sed exigua cuspide in rostro. latior haec quarto generi et acutior in mucronem fastigata eodemque gladio scindens solum et acie laterum radices herbarum secans. non pridem inventum in Raetia Galliae ut duas adderent tali rotulas, quod genus vocant plaumorati; cuspis effigiem palae habet.' Pliny, *Natural History*, XVIII, 48, 171, tr. Rackham (1950), vol. 5, p. 297.

Being less interested in machinery for its own sake than in its use, Pliny only describes those features of the plough which affect the method of working. A heavy plough needs eight oxen to pull it. A ploughman must stoop if the furrow is to be straight. A certain type of coulter can be used on a certain type of soil. Even in the passage just quoted there is no indication of how it was used, and how much work it would do in a day. The nearest he comes to describing the appearance of any part of a plough is when he says that this share has the shape of a spade.

Within the limited area of his description Latin provides Pliny with a fair vocabulary of technical terms indicating the simpler shapes. He has *vectis* for the solid bar used as a coulter, *mucro* for a sharp point such as one finds on a sword, *rostrum* for that which curves to a point in the manner of a bird's beak, and the useful word *fastigare* meaning 'to taper to a point'. *Dentale* on the other hand can be ambiguous, for in Latin it means both the share-beam and the share, though which is intended is likely to be made clear from the context. In attempting to describe the shape of a plough-share he is less fortunate. Spade-shaped only helps if one already knows the shape of the spade in question. *Resupinus*, meaning literally 'lying on its back and then bent backwards', is some way towards suggesting how a share with mouldboard is shaped, but in the 18th and early 19th centuries all such words had to be tightened up to meet the needs of botanical description. We find, for example, such terms as *inversus*, *revolutus*, *involutus*, *convolutus*, *reclinatus*, *resupinatus*, *inclinatus*, *inflexus*, and so forth, all with very precise meanings.[181] When it comes to directions, classical Latin is even less happy. *Rectus* is a straight line, whether horizontal or vertical. If a furrow cuts across it, that would be *traversus*, and this word is used to indicate cross-ploughing, as is also *obliquus*. But as *obliquus* means 'to one side and not straight on', it can also mean 'on its side', 'awry' or 'sloping'. The sentence '*omne arvum rectis sulcis, mox et obliquis subiqi debet*'[182] could then mean not only that every field must be cross-ploughed, but also that in the later ploughing there should be an adjustment of the share and mouldboard so that the earth is turned at a greater angle to the furrow, if on Roman ploughs that was technically possible. Again, in talking about ploughing on hillsides, the phrase '*traverso monte*' should mean 'at right angles to the slope', but as this has to be done with the share 'sometimes pointing uphill and sometimes downhill', and as no one ploughs straight up or down a steep hill, in practice this can only mean 'diagonally to the slope'.[183]

The most important part of the ploughing operation is when the share either cuts through the roots of the weeds, or slices the earth below the weeds and turns the whole sod over, burying the leaves and exposing the roots. This operation is described by all three of the authors here to be considered. Pliny says that the fourth type of plough-share cuts through the roots of the weeds, but he does not say that it turns

[181] Stearn (1966), pp. 345–6.

[182] Pliny, *Natural History*, XVIII, 48, 178, tr. Rackham (1950), vol. 5, p. 301, 'Every field must be worked with straight furrows and then with slanting furrows as well'.

[183] '. . . in collibus traverso tantum monte aratur, sed modo in superiora modo in inferiora rostrante vomere'. *Ibid.*

them upside-down, nor whether the earth is pushed to one side of the share only, as in modern ploughs, or to both sides, as in primitive ploughs, though it may be inferred that he means the latter, since he writes that the *sides* (i.e. the two sides) of the share cut the weeds.

When technical terms are needed, no civilisation seems unable to produce them, providing a concept is clear. The Romans, for example, needing a term for ploughshare, produced *vomer* from the idea of the share vomiting the earth sideways as it goes along – imagery not inappropriate to the inventors of the vomitorium! Chinese also is rich in the metaphorical application of words to create technical terms, as was illustrated in *SCC* Volume 5, part 7, in such names as 'great wind-chasing gun'. The importance of words which precisely establish position, direction, insertion, appearance and relation of parts, as well as temporal relations and sequences, and of course precise measurements, are all better appreciated after the study of a text such as those describing the plough, though the plough itself is far easier to describe than some of the plant forms which challenged 18th-century botanists to improve their technical terminology, till they had arrived at a total of over 474 adjectives essential for this purpose.[184]

Precision in terminology results from conceptual differentiation of the components of a complex entity.[185] For Pliny the word *calyx* stood for such an entity, but it had not yet in his day been differentiated. It stood not only for the coverings of fruit and of heaps of charcoal during its preparation, and for the shells of eggs and of molluscs, but also for all those parts of a plant which have since been differentiated and are now known as the *involucre*, the *cupule*, the *perigon*, the *calyx* itself, the *capsule* and the *pericarp*.[186] In the words of H. W. Rickett, 'A rational terminology mirrors that upon which it is based, an understanding of the things concerned'.[187] 'No science can advance without forming a specialised vocabulary economical and precise in designating things and concepts; just as the lack of a suitable word hampers discussion, so the provision of one often leads to better understanding of the object or concept concerned.'[188] When an idea is new, lack of terms may put a strain on the writer who is trying to express the new idea. But if motivated, he will find a way. Pliny was not motivated to give accurate descriptions of ploughs, or to seek the causes of their inefficiency, because in his day there were slaves in abundance to do the work.

The author of the next passage, however, was highly motivated. This was John Fitzherbert, an English yeoman farmer, who in 1523 published *The Boke of Husbandrye*.[189] His reason for writing it is made clear in the opening section on

[184] Stearn (1966), pp. 314–49.

[185] For example, Georgius Agricola (1490–1555) used the term *aquavalens* to mean hydrochloric, nitric and sulphuric acid, because in the 16th century the complex of acids had not yet been conceptually differentiated. Only after this had been done could more precise terms come into use.

[186] Stearn (1966), p. 22.

[187] Rickett (1944), pp. 187–231, quoted by Stearn (1966), p. 34.

[188] Stearn (1966), p. 10.

[189] For information concerning Fitzherbert we are much indebted to the author of *SCC*, vol. 6, pt. 2, 'Agriculture', Francesca Bray, who has greatly helped our understanding of the construction of ploughs.

ploughing: 'Then is the plough the most necessary instrument that an husband can occupy, wherefore it is convenient to be known, how a plough should be made'. His aim is to describe the plough so accurately that it can be made from his instructions. This shift in motivation reflects the changed attitude to machinery which the Industrial Revolution was eventually to develop. What particularly distinguishes the writing of Fitzherbert from that of Pliny when describing ploughs is that he not merely states the function of a particular part of the implement, but sometimes gives rough measurements and tries to explain why it has a particular shape, or what would happen if it were different. For example, he says that the share-beam is 'four or five feet long, and it is broad and thin. And that is because the land [in Somerset] is very tough [i.e. sticky] and would suck the plough into the earth if the share-beam were not long, broad and thin.' In other words, the share-beam has something of the function of a ski in sliding over the earth. If the ploughwright understands the reason why it should be so, it will help him to build his plough efficiently.

Fitzherbert starts his description in a systematic fashion. He wishes his reader to be able to visualise (see Fig. 19) the four main pieces of wood which compose the framework of the plough. He gives their names as the plough-beam, the share-beam, the plough-sheath, and the plough-tail. It would have been convenient if he had been able to write somewhat as follows:

> The first two are horizontal timbers roughly parallel to each other, the former above the latter, kept securely in position by a piece of good quality oak, which is a vertical post, called the plough-sheath, morticed into the beams above and below. The fourth timber is morticed into the rear end of the share-beam, and comes up not parallel to the plough-sheath, but at an angle to meet the plough-beam, which passes through a slot in its upper end. The angle can be determined when one knows the dimensions of the other parts.

Unfortunately Fitzherbert was not able to write so precise a description. One reason was simply the lack of necessary terms. In 1523 when he was newly correcting his book, the word *parallel* had not yet been adopted into the vernacular language of England. It made its first recorded appearance in a mathematical context in 1549, but was not used to describe 'mechanical contrivances of which some parts are parallel' until 1594.[190] Similarly, the word *horizontal* did not enter the English language until 1555, and then only in the sense of belonging to the horizon, or occurring at it. Not until 1638 was it used in the sense of parallel to the plane of the horizon. *Vertical* made its first recorded appearance a little later, in 1559, and then chiefly in astronomical contexts. It is not known to have been used to describe mechanical appliances until 1825.

The word *parallel* was, of course, invented or adopted by the ancient Greek mathematicians, and was essential for their geometrical proofs. In mathematical contexts the word was familiar throughout classically educated Europe, but it was a

[190] *Shorter Oxford English Dictionary* (1972), p. 1429.

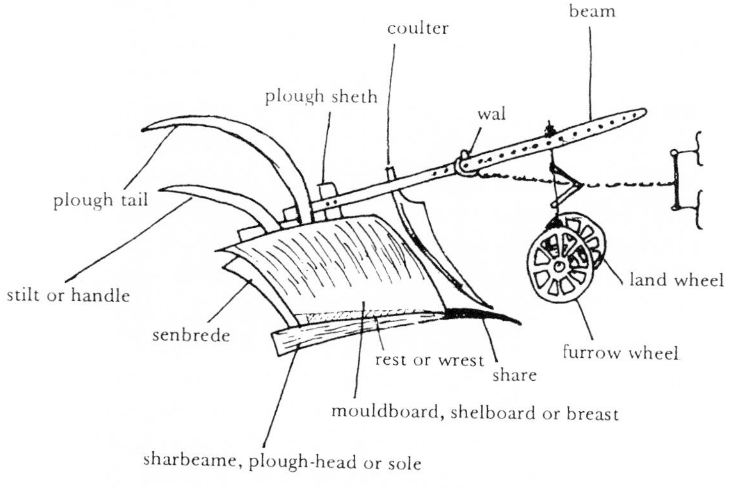

Main features of an English wooden plough

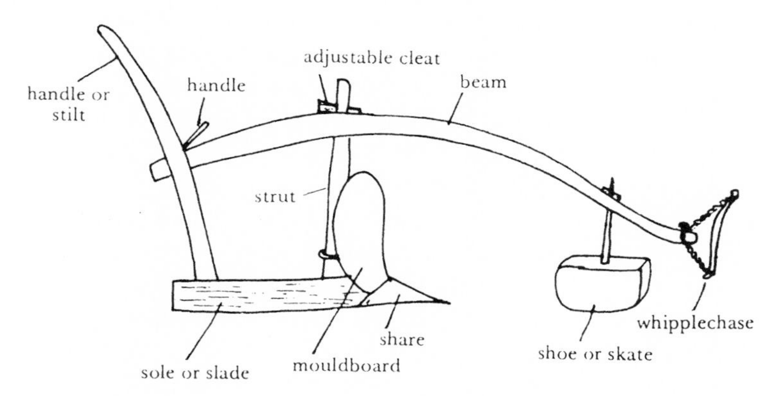

Figure 19. Diagram illustrating English and Chinese ploughs.

mathematician's term, and not used in the market place. With the rise of the vernacular languages of Europe and the decline of the classical language, such words as *parallel* naturally came to be used in the vernacular languages in mathematical contexts. It may seem surprising that the word parallel was not applied to 'mechanical contrivances' until some two generations after its first appearance. Yet this is only surprising if one forgets the social and intellectual gap which existed between educated mathematicians and unlettered carpenters or ploughwrights in the 16th century. The words *horizontal* and *vertical*, however, did not exist at all in classical Latin, but were Renaissance developments. To the Romans, horizons were merely lines of demarcation such as separate the sea from the sky, or were the outlines of buildings marked out on the ground.

The Greek word *horizein* also had the special sense of 'to define the meaning of words'. The horizon was not thought of as a line or direction at right angles to that which today we call the vertical. These words which have become so essential only began to be needed during the Renaissance as a result of the new attitude to mechanical contrivances. How then did Fitzherbert, using the vernacular language of England in the 16th century,[191] describe the plough for the benefit of those who knew what it looked like, but might want to make improvements in its design? He begins by defining the *plough-beam* as 'the long tree above, the which is a little bent', and the *share-beam* as 'the tree underneath, whereupon the share is set'. The word *tree* is not often used today in the sense of a beam, though it causes no real difficulty. But two other points may be noted. One is that no precise dimensions are given, or even proportions derived from one measured unit, as was the practice with early military engineers in classical times,[192] and the other is that, as so often in Chinese, the comparison of adjectives is implied but not made morphologically explicit.[193] As the plough-beam is 'the long tree above' it follows that the share-beam is not the equally long tree below, but that the upper beam is longer than the lower. Today it is the practice to make instructions explicit. There is, however, nothing misleading in what Fitzherbert has said in this particular, but when he says that the plough-beam is 'a little bent', his description is clearly defective, for we do not know in which plane it was bent. This defect is avoided by the skilled use of analogy in the description of a Chinese plough written by Lu Kuei-Meng 陸龜蒙 to be examined below.

The *plough-sheath* is the next part of the frame to be described. It is of cardinal importance, for if it did not hold the upper and lower beams rigidly together, the plough would soon become a rickety contraption. Fitzherbert says that the *plough-sheath* is 'a thin piece of dry [i.e. seasoned] wood made of oak, that is set fast in a mortice in the plough beam and also into the share beam, the which is the key and the chief band [bond] of all the plough'.

As no measurements are given, the ploughwright would have to judge how long and thick the plough-sheath should be from common sense and his knowledge of ploughs in general. But it is more serious that no indication is given as to whether this post, the plough-sheath, is vertical, meeting the share-beam at right angles, or whether it is morticed obliquely, as the plough-tail or handle undoubtedly is. The description says:

> The hind end of the plough beam is put in a long slit made in the same tail and not set fast, but it may rise by and go down, and is pinned behind, and the same plough tail is set fast in a mortice in the hinder end of the share beam.

[191] To make his description easier to follow, some modernisation of Fitzherbert's Tudor spelling has been made.

[192] Heron, in his *Artillery Manual*, gives the dimensions only of the base of the *gastraphetes* whose construction he describes (a machine first introduced in the early −4th century). Heron, however, was of the +1st century. Biton (−3rd century) describes the *gastraphetes* with precise measurements. See Marsden (1971), pp. 27 and 67.

[193] As Harbsmeier in *SCC*, vol. 7, pt. 1, p. 101, has put it: 'The notion "comparative construction" is systematically redundant in the grammatical system of Ancient Chinese. But this does not commit one to the obviously mistaken view that the ancient Chinese could not compare things, for example, for size and quantity!'

Fitzherbert continues to describe all the pieces of the plough which are made of wood with increasing lack of clarity, until he comes to the *sen-board*, of which he has this to say: 'To the other side of the stilt in the hind end, the sen board is a thin board, pinned or nailed most commonly to the left side of the sheath in the further end, and to the plough tail in the hinder end. And the said shelboard would come over the said sheath and sen board an inch, and to come past the midst of the share made with a sharp edge to receive and turn the earth when the coulter hath cut it'.

This description, lacking information about size, shape and the angle at which the board is set, leaves much to be desired. There is, however, rather more precision when he describes the metal parts of the plough, perhaps because rectifying mistakes in metal would be a more costly business. The share is described as 'a piece of iron sharp [i.e. angled] before and broad behind, a foot long'. This is at least more accurate than Pliny's description of the share in his fourth kind of plough whose 'spike is broader and sharper, ending off in a point'. Fitzherbert goes on to explain that it is 'made with a socket to be set on the further end of the share beam'.

It is scarcely possible accurately to reconstruct Fitzherbert's plough from his description alone.[194] He is handicapped by the undeveloped state of the English language in his day, and even more because the discipline of precise scientific description, which had begun long before with astronomy and mathematics, had not yet spread very far into the other sciences. What is lacking is measurement of angles, dimensions of timbers, relative positions of the parts, better description of shapes, or, if this is not possible, a more consistent explanation of the work the part is meant to do, and a better grouping of the parts to be described in such a sequence that those already described help the understanding of each of those in turn which are to follow – a Euclidean orderliness, as it were, developing from simple to complex. In Western Europe the discipline of mathematics and astronomy spread fairly quickly during the Renaissance into navigation, mechanics, and ballistics, and so hastened the birth of scientific method. If, as in China, the main national emphasis had been on agriculture, the rise of scientific method would certainly have taken longer, because of the greater complexity of organic matter and the difficulty of isolating the variables in botany, though mechanics could have been stimulated by a social need to develop agricultural machinery.

It is now time to consider Lu Kuei-Meng's 陸龜蒙 description of the Chinese plough. Francesca Bray says of the Chinese plough that already by Han times it had achieved a level of perfection and sophistication found nowhere else in the world. By Thang times it had developed a heavy share, a mouldboard and an adjustable strut. 'The adjustable strut is a fundamental development in the Chinese plough, for . . . it permits a precise regulation of ploughing depth . . . to suit different crops, soil types, seasons or weather conditions'.[195] Lu Kuei-Meng, the author of *A Classic of the Plough* (*Lei Ssu Ching* 耒耜經) written in +880, was a member of the official class who, during

[194] Nevertheless for an attempted reconstruction see Figure 19, reproduced from *SCC*, vol. 6, pt. 2, p. 140.
[195] *SCC*, vol. 6, pt. 2, p. 169.

the unpaid periods between official postings, supported himself by farming on his small family estate. So far from despising manual work he felt that in labouring on the land, or at least in supervising the labours of others, he was following the example of the Heavenly Husbandman, Shen Nung 神農.[196] His curiosity led him to enquire from a more practical farmer details concerning the terminology and structure of the plough, and these he recorded, not, as with Fitzherbert, for contemporary farmers to use, but from a characteristically Chinese love of recording useful information, 'thereby transmitting hope'. In judging his style it must be remembered that Lu Kuei-Meng was very much a member of the literati class, even if he occasionally got his hands dirty. He was familiar with the devices of an elegant style, and subject to the literary class pressures of his day.[197] Yet he handles the literary language with skill and clarity in his description of the plough. Even more striking is the disciplined and methodical way in which he sets about his description. This may now be summarised.

His introduction stresses the antiquity of the plough, and its importance for civilisation. He then states the source of his information, an old farmer who worked on his estate, whose clear explanation of the names and parts of the plough was a revelation to him. He next runs through the nomenclature, the two parts made of metal, and the nine parts made of wood, a total of eleven in all. These eleven parts are then grouped into three coherent features: (a) parts concerned with forming a ridge of earth; (b) parts concerned with regulating the depth of the furrow; and (c) parts concerned with the traction and steering of the implement.

He begins his explanation by defining the ridge, whose creation is the main purpose of the act of ploughing. Soil in the form of clods is thrown up by the share,[198] and turned over by the mouldboard. The reason for turning the earth over, in order to uproot the weeds, which does not occur in more primitive ploughs, is explained. Lu Kuei-Meng then uses the literary device of parallel sentence patterns to extract additional clarification from very few words. After pointing out the importance of uprooting the weeds he says:

That is why the ploughshare is elongated and remains below [i.e. on the bottom of the furrow], [and why] the mouldboard is swept backward and remains above. (*Ku chhan yin erh chü hsia, pi yen erh chü shang.* 故鑱引而居下,壁偃而居上)

The word *yen*, here translated as 'swept backward', is normally used to describe branches curving under the weight of snow, or grasses bending before the wind. In this case it is not wind, but the pressure of the earth as it is ripped by the ploughshare, which dictates the form that the mouldboard must assume in order to invert the ridge. It is an interesting question whether this manner of describing a piece of machinery,

[196] Shen Nung, the Heavenly Husbandman, was a minotaur-like deity, and the legendary inventor of agriculture. See *SCC*, vol. 1, p. 163.

[197] See Thilo (1980), who refers to Lu Kuei-Meng's Tea Book, and to his poetry.

[198] It is commonly believed that the Chinese used an iron share some centuries before it was introduced in Europe, but continued to do so long after the European iron share had been replaced by one of steel, which would be less brittle. But it has recently been discovered that the Chinese shares were in any case not brittle, being made of malleable cast iron, produced by a week or more of annealing, and this in the −2nd century for a process not known in Europe until the +18th.

in which the function is related to the shape, can be improved on. Certainly the very precise botanic terms do not really help. Lindley's botanical terminology,[199] for example, gives us *reflexus* (No. 411), 'suddenly bent backwards', *retrorsus* (No. 419), 'turned in a direction opposite to that of the apex of the body to which the part turned appertains', *introrsus* (No. 420), 'turned towards the axis to which it appertains', *extrorsus* (No. 421), 'turned away from the axis to which it appertains', *decumbens* (No. 424), 'reclining upon the earth, and rising again from it at the apex', but none of these, more accurate though they may be than Pliny's *resupinus*, fully describes the shape of the mouldboard in a Chinese plough whose graceful curve has something in common with a breaking wave driven obliquely onto a gently sloping beach. We may sympathise with Locke's powerful advocacy of pictographic script when he wrote:

Words standing for things that are known and distinguished by their outward shapes should be expressed by little draughts, i.e. outline-drawings and prints made of them. A vocabulary made after this fashion would perhaps, with more ease, and in less time, teach the signification of many more terms than all the large and laborious comments of learned critics.[200]

The word *yin* 引 on the other hand suggests the opposite, a line as straight and tense as a bowstring, or an arrow about to be discharged,[201] an excellent term for describing that part of the plough which leads the attack on the opposing earth. The vividness of Chinese similes and metaphors, and the skilful choice of words for this purpose, greatly contribute to the accuracy of Chinese scientific writing.

Further description of these two parts in a second parallel couplet is then provided.

The ploughshare shows its upper part angled. The mouldboard [is] formed [with its] lower part rounded. (*Chhan piao shang li. Pi hsing hsia yüan.* 鑱表上利,壁形下圜)

Writing of the Japanese language, Stanley Gerr has said that it 'has absorbed many of the virtues of the Chinese technique of expression: conciseness, convenience, accuracy, and frequently a vivid and picturesque phraseology'.[202] These are illustrated in the above passage.

In fairness it should be pointed out that Lu's use here of the word *yüan* 圜, round, leaves something to be desired. The word can mean circular, concave, convex or even dished. But it is extremely unlikely that so practical a man as Lu would have introduced it merely as a literary conceit for the sake of formal antithetical balance. In its context, in fact, it is quite clear, as Francesca Bray has pointed out,[203] that the mouldboard is 'curved at the base [to fit tightly against the share]', that is to say, a snug fit is required over that part of the share where the metal swells out to be socketed with the share-beam. So far Lu Kuei-Meng has succeeded in giving a clear description of the most important parts of his first feature as regards their functions. These are supplemented later with precise measurements. He goes on to describe

[199] Used with some modification in Stearn (1966), pp. 346–7.
[200] Quoted by Stearn (1966), p. 34, from Locke's *Essay Concerning Human Understanding*.
[201] Karlgren (1940), p. 371a, says of this character, 'Draw the bow, pull, draw, stretch, to lead, prolong. The graph has "bow" and a stroke which probably depicts the string.'
[202] Gerr (1944), p. 23. [203] *SCC*, vol. 6, pt. 2, p. 181.

the slade or share-beam, and two supporting struts. This part of the text is, however, somewhat corrupt, and contains curious technical terms like 'the turtle-flesh', *pieh-jou* 鼈肉, which are perhaps dialect words used by his rustic instructor, and represented by characters for the sake of their sound rather than for their meaning. We may therefore, without loss, proceed to his description of the second feature, the furrow depth mechanism.

It will be remembered that Fitzherbert described the vertical timber of his plough, which he named the 'plough-sheath', as 'the key and the chief bond of all the plough'. It was firmly mortised above and below. This rigidity was an essential characteristic of all European ploughs. To vary their depth of furrow was by no means easy. The essence of the Chinese plough, on the other hand, was lightness and flexibility. The equivalent of the 'plough-sheath' was in Chinese named 'the arrow', *chien* 箭. It was not mortised rigidly into the plough-beam, but passed through a slot in it, and could move up or down. The effect of this flexibility was that the angle at which the share entered the earth could be varied. This capacity for variation was provided because the 'arrow' or 'plough-sheath', though rigidly mortised below to the share-beam, to which the share was socketed, was free to move up and down through the slot in the plough-beam, where it could be locked with pins and wedges at the desired length. This length determined the angle of the plough-share's attack. Lu Kuei-Meng's description of this ingenious mechanism is as follows:

[The part which . . .] extends to the share-beam, and which is perpendicular and mortised to it, is called 'the arrow'. [The part which] to the front is curved like a carriage shaft is called the plough-beam. [The part] to the rear which is like a handle and rises up is called 'the rudder', [in English, the 'stilt']. (. . . *ta yü li ti tsung erh kuan chih yüeh chien. Chhien ju thing erh chin che yüeh yüan. Hou ju ping erh chhiao che yüeh shao.* 達於犁底縱而貫之曰箭。前如桯而樛者曰轅。後如柄而喬者曰稍。)

The least satisfactory part of this description is that of the handle which 'rises up'. Lu Kuei-Meng has failed to say to what other parts it is attached, though he does conclude his description by saying that it is taken in the hand, is used for controlling the plough, and forms what Fitzherbert called 'the key and the chief bond'.[204] He does not, however, have any difficulty in making it clear that the plough-sheath or 'arrow' is morticed into the share-beam vertically, i.e. at right angles, as is still the case with some modern ploughs in China.[205] Fortunately for him Literary Chinese had long had words for *vertical* and *horizontal*. It will also be noticed that whereas Fitzherbert described the curvature of the plough-beam as 'a little bent', Lu Kuei-Meng uses the neat simile of the carriage shaft whose characteristic curve would be perfectly familiar to his readers, and which gives a fairly precise idea of its shape. The mechanism itself is now described:

[204] Lu Kuei-Meng says (*Lei Ssu Ching*, p. 2), 'The "rudder" [or stilt] derives [its name] from the stern of a boat, and locks it in position'. (*Shao chhü chou chih wei chih yü tzhu hu.* 梢取舟之尾止於此乎)

[205] See Bray (1979), p. 235.

An extension of the 'arrow' protrudes from the plough-beam, and can be allowed up or drawn down (like an arrow in a crossbow). In this way, and only in this way, is it equivalent (to an arrow). (*Yüan yu yüeh chia chien, kho shih chang yen.* 轅有越加箭，可弛張焉。)

This excellent piece of description immediately becomes clear when it is remembered that the Chinese crossbow was loaded by a soldier either lying on his back or standing on his bow, and using both the thrust of his legs and the pull of his arms and back to develop the tension. In either position the arrow would be held vertically, and move up or down until locked in position by the trigger. The terms *shih* 弛 and *chang* 張 are used to mean 'slacken the tension on a bowstring' and 'draw the bow' respectively. Naturally the 'arrow' beam on a plough does not resemble a military arrow in any way except in its up and down movement during loading. Hence the use of the particle *yen* 焉.[206] Information missing is that the 'arrow' rides up or down through a *slot* in a plough-beam. Though obvious on reflection we would prefer it to be stated clearly.

Lu Kuei-Meng now describes the bolting part of the mechanism which, like the trigger on a crossbow, holds the 'arrow' at the properly adjusted height. This is done by the insertion through a slot in the 'arrow' at the height required to make a furrow of appropriate depth. This is what is done today. But here the text is subject to an apparent ambiguity, for this piece of wood, which is called the 'adjuster', can do its work equally well if it is aligned to the axis of the plough, or if it is aligned at right angles to the axis. The ambiguity is due to the fact that +9th-century Chinese evidently did not have terms for *laterally*, which is defined as 'acting or placed at right angles to the line of motion or of strain',[207] nor for *axially*, defined as 'aligned with the imaginary line that divides a regular shape into two equal parts with the same shape'.[208] This is hardly surprising, for the terms did not make their way into English in these senses until some five centuries after Western Europe began to give concentrated attention to the study of mechanics, namely in 1803 for the former, and by 1849 at the earliest for the latter.

Existing Chinese ploughs, and those for which clear illustrations exist, show the 'adjuster' as protruding through the 'arrow' above the plough-beam and laterally to it. Yet in spite of the lack of the two terms which could have made the placing of it unequivocal, there is reason to believe that the plough which Lu Kuei-Meng was describing had its 'adjuster' aligned axially and not laterally to it, and that, as often happens, lack of a technical term did not prevent an ingenious author from expressing his idea. He says:

On the upper [face] of the plough-beam there is also [i.e. in addition to the 'arrow' extension just mentioned] a groove formation. This also resembles an arrow, but only in respect to [the groove] [i.e. not in respect to the up and down movement of the 'arrow' just described]. (*Yüan chih shang yu yu ju tshao hsing, i ju chien yen.* 轅之上又有如槽形，亦如箭焉。)

[206] 'It is clear to the present writer that in a variety of contexts "*p yen q*" actually was intended to state what in our terminology is equivalential connnection.' Janusz Chmielewski, personal communication, 9 April 1981, MS, p. 21.

[207] *Shorter Oxford English Dictionary* (1972), p. 1111.

[208] *Ibid.*, p. 131.

In what way, we may ask, does a groove resemble an arrow? The up and down movements of loading and drawing the bow are precluded. Such qualities as sharpness, or of being feathered, are irrelevant. What remains is that an arrow has a certain length, and points in a certain direction. An arrow, by reason of its length, could only point along the axis of the plough-beam, and we are justified in thinking that this was the direction of the groove, for if it were cut into the narrow beam at right angles to this, the groove would be so short that any resemblance to an arrow would be lost. In spite of the lack of a necessary term Lu Kuei-Meng at this point has nevertheless succeeded in conveying the necessary information. It seems probable, therefore, that when he was writing this in the +9th century, what he was describing was an earlier form of plough in which the 'adjuster' was accommodated in a deep groove sunk into the top of the plough-beam along its axis. But such a groove, if cut deeply into the beam, would weaken it, and the making of it would involve quite unnecessary work, for an 'adjuster' entering the 'arrow' at right angles to the axis of the plough-beam, as is the practice in modern Chinese ploughs, requires no deep groove to contain it, but it can ride on the shoulders of the plough-beam. The earlier model would therefore become obsolete as soon as the advantages of the improved method had become apparent.

The text continues as follows:

[The part] which is cut to make steps high at the front and low at the rear, which can be moved backward and forward, is called 'the adjuster'. When it is pushed forward the 'arrow' goes down and [the share] enters the earth deeply. When it is pulled backward, the arrow comes up, and [the share] enters the earth shallowly. (*Kho wei chi chhien kao erh hou pei, so i chin thui yüeh phing. Chin chih tse chien hsia ju thu yeh shen, thui chih tse chien shang ju thu yeh chhien.* 刻為級前高而後卑,所以進退曰評。進之則箭下入土也深。退之則 箭上入土也淺。)

Lu then explains that the 'arrow' gets its name from its resemblance to the way in which an arrow moves when a crossbow is being loaded, and that the 'adjuster' is so called because it adjusts the depth of furrow. It is clear that the adjuster fits in the groove, for the groove is given no other function. The fact that the steps of the adjuster are said to be high 'at the front' and low 'at the back' reinforces the belief that the adjuster is aligned axially. The description then makes the following statement:

At the top of the adjuster [the part which runs or lies] athwart and horizontally is called the bolt. (*Phing chih shang chhü erh heng chih che yüeh chien.* 評之上曲而衡之者曰建..) With this the plough-beam and adjuster are pinned together. If one did not have this [bolt] the two parts would jump apart, and the 'arrow' would not be able to stay in position. (*So i ni chhi yüan yü phing. Wu shih tse erh wu, yüeh erh chhu, chien pu neng chih.* 所以柅其轅與評。無是則二物,躍而出,箭不能止..)

This is another indication that Lu is describing a more primitive type of plough in which the adjuster could easily jump out of its groove unless it was secured with a bolt or pin. That would not be possible on a later type, for the adjuster fits right through the 'arrow' in a slot, and could not jump out but only gradually get loose and fall out, for which reason it is secured by a pin on either side of the 'arrow'.

But in Lu's plough it makes good sense to say that the bolt goes 'horizontally and athwart', i.e. through the side of the plough-beam, through the adjuster sunk in its deep groove, and so out on the other side of the beam.[209]

This concludes the description of the second feature, and is enough to give a fair idea of the strengths and weaknesses of the writing as a whole. The description of the traction and steerage of the plough in the third feature may therefore be omitted. But what is to be noted is that at this early date Lu was concerned with precise measurements. Following the third feature, the main dimension of all the eleven parts is given. For some parts he gives more than one dimension, and perhaps further information about their shape, as for example that the mouldboard is 1 foot long, and 1 foot wide, and is slightly oval (*wei tho* 微橢). Measurements not given can readily be supplied by deduction or common sense. In sum the information is presented in so precise, orderly and vivid a fashion that it is possible to reconstruct this antique plough with few areas of doubt or ambiguity, no mean achievement for an amateur, and a tribute to the literary language in which he had received his education.

(5) Levels of Specification and Mathematical Terms

Modern science began with quantification, precise measurement and the invention of new terms for new concepts. Concentration on the noun as opposed to the adjective was all-important. It was not enough to say that something was moist. Scientists needed to know how much liquid was in it. A noun describes something which can perhaps be counted, weighed, measured, valued or listed. An adjective or verb does not. Moreover the use of nouns in apposition made possible neat captions for diagrams and graphs, such as 'Metal fatigue reduction chart', which would otherwise have to be expressed by some such verbal locution as 'Chart showing how metals are now less affected by continuous strain than they were before'. The Chinese language is particularly well-favoured in this regard since the function of a word in Chinese is to a large extent decided by its position in the sentence.

(i) *Mathematical terms*

To compare the terminology of both East and West is particularly interesting for mathematics, for without mathematics the sciences cannot develop. The vocabulary available to early Chinese geometers is of special interest. Take, for example, the terminology used to express the concepts of area and volume. In *SCC*, Volume 3, pp. 98–9, we listed terms that are used to express these concepts – twenty-nine for areas and thirty-eight for volumes. This list is by no means exhaustive. These terms derive from names of homely things which present a clear and often striking image

[209] For a useful description and measurements of a modern Chinese plough see Hommel (1937), p. 42. Additional reasons for thinking that the plough described by Lu Kuei-Meng was a clumsier and more primitive version of the refined modern Chinese plough are (a) its much greater overall length, 3.73 metres, as compared with the 2.33 metres of the plough measured by Hommel, and (b) the greater length of the adjuster, which, in Lu Kuei-Meng's plough was more than double the length of the 'stepped slide' described by Hommel.

to the eye, such as round and square fields, ramparts, pavilions, icehouses, awls, ponds, moats, and so on, and show the same facility in forming compounds as has been observed in other fields in Chinese science. A similar use of everyday words was adopted in ancient Greece in the formation of geometrical terms; there we find tables, spinning tops, sawn-off bits of wood and the like used to make terms signifying trapezoids, rhomboids and prisms. The Romans did the same, borrowing bows and bowstrings, wheel-spokes and finger rings to provide our modern vocabulary of arcs and chords, radii and annuli. But this was not an invariable rule in the West. Rather than find a local image to describe the shape of the pyramid, the Greeks probably borrowed the term intact from the Egyptian language. In China an image was found in the carpenter's awl and other objects. (See below, pp. 175 ff.) The Greeks also built up highly complex new words of the polysyllabic variety which Indo-European languages make so easily, such as parallelepiped from *para* (beside), *allelos* (one another), *epi* (on), *pedon* (the ground), i.e. having (like the ground floor of a house) a plane surface; a parallelepiped thus becomes a body with parallel surfaces. The Chinese did not attempt this sort of word-building.

Because of the ease with which the Chinese formed concise and picturesque compound words[210] it was not difficult for them to ensure that they had a particular term for every required shape. Because the Greeks could easily define shapes by using letters of the alphabet when tracing figures in the sand, it was not necessary for them to invent a special term for a figure which could just as well be described by letters at the points of terminus and intersection. An example of Greek letters so used is furnished by Apollonius of Perga (*c.* –225) who wrote: 'Then since the points Δ, E, H are in the plane through Δ, K, E, and are also in the plane through A, B, π therefore the points Δ, E, H, are on the common section of the planes; therefore HE is a straight line'.[211]

There are many reasons why Greek and Chinese geometry developed in different ways, but differences in language and script must surely be among them.[212] Chinese mathematical terminology had by the +5th century been developing over many hundred years, and in geometry was particularly rich, perhaps because it was pragmatically orientated and, like the founders of the Royal Society of London in the 17th century, preferred Works to Wit.[213]

[210] See *SCC*, vol. 3, p. 142, for picturesque terms describing unusual shapes, such as 'altar-step onions' and 'wine-kegs'.

[211] 'The Conic Sections', Proposition 9, tr. Thomas (1957), vol. 2, p. 303.

[212] Nevertheless the Chinese could have used the same device for indicating areas by symbols for points of angles and line intersections if they had wished. Indeed on occasion they did so as Karine Chemla has shown in her doctoral thesis (1982) on the work of the brilliant algebraist Li Yeh 李冶 of 1248. On pp. 2.5 to 2.8 she lists the Chinese characters used symbolically to indicate key points in a figure, these symbols being, however, related to Chinese cosmological thinking. Heaven (*thien* 天), the high point of the hypotenuse of a right-angled triangle, for example, descends by way of the sun (*jih* 日) and moon (yüeh 月) to earth (*ti* 地). For further details see Figure 20. It is thus possible to say, 'Heaven to sun equals sun to heart' (*hsin* 心), as the Greeks might have said, 'The line AB = the line BX'. Donald Wagner adds in a personal communication dated 9 June 1983: 'This may have been a Western influence, through Persian astronomers at the Mongol Court, but it shows at least that the Chinese could have used the same device if they had wanted to. Liu Hui used colours in his diagrams to do the same thing.'

[213] Sprat (1722), p. 116.

测圆海镜细草卷第九上

翰林学士知制诰　同修国史栾城李冶撰

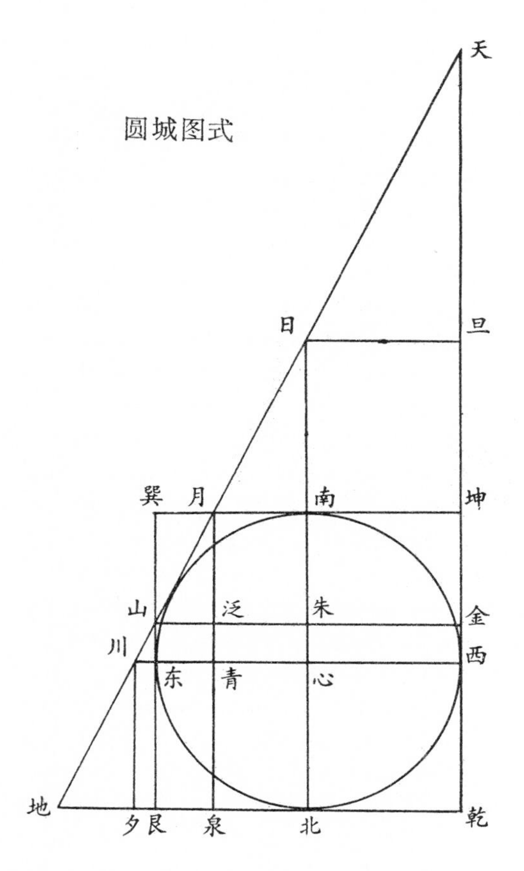

Figure 20. Example of Li Yeh's use of characters in geometrical constructions (1246). From Tse Yüan Hai Ching Chin I (*Tsinan: Shantung Chiao Yü Chhu Pan She* (1985), p. 583).

The close interest of Chinese geometers in expressing their thoughts in concrete terms is suggested by the use of blocks or cubes, perhaps made of wood or clay, which could be shaped and fitted together, and distinguished from each other by being of two colours, red and black. Wagner suggests[214] that there were sets of standard blocks which might have been part of 'some game or puzzle with a wider circulation than among mathematicians [alone]'.

[214] Wagner (1979), p. 168.

Unlike Greek geometry, the geometry of China was slow to separate itself from the concrete and the practical.[215] The over-riding obligation of a Confucian scholar was to foster the moral Way, and if he had time from this for other things, it should be devoted to public works which would promote the welfare of the people. The most obvious outlet for the mathematics of scholar administrators was in the calculation of taxes, for which see *CHS* 前漢書 24 and Swann (1950). As an illustration of this practical attitude, we find Liu Hui 劉徽 (+1st century) actually apologising for the fact that he is obliged to use a shape, which in itself had no practical use, namely the *pieh-nao* 鼈臑 mentioned below, p. 175, in order to investigate the volume of the *yang-ma* 陽馬, which itself must be used in order to investigate the volume of useful solids such as various types of cone and pyramid.[216] Nevertheless interest in theory for the development of efficient practice had long been present in China, and was by no means dormant in the early centuries of our era. This becomes very clear in the study of the volume of a sphere to be considered below.

An indication of an advance towards theory for its own sake in China is the fact that the sphere is dealt with at all in the *Chiu Chang Suan Shu* 九章算書. This book received its final form in the +1st century, though it undoubtedly contains earlier material. The original pragmatic bias of the book may be seen not only from the examples given, of which there are 246 in all, but even from the titles of some of the nine chapters, for example, 'Surveying the Land', 'Millet and Rice', 'Impartial Taxation'. The title of the fourth chapter, *Shao-kuang* 少廣 or 'Diminishing the Breadth', however, implies abstractions. The chapter is largely concerned with finding squares and cube roots. The diameter of a sphere, for example, is given as equal to the cube root of $\frac{1}{9}$ of its volume × 16. It would be surprising if a practical administrator in China in the +1st century needed to know the volume of a sphere, for it would have few applications.

Practical applications for the geometry of the sphere in the +1st century and earlier were rare in any civilisation. Its calculation may be taken as an indicator that mathematics was beginning to become abstract and theoretical. Because there was little chance that an error in the calculation of the volume of a sphere would be detected in practical work, as might well occur in the calculation of the amounts of material needed for building walls or embankments, the error could only be detected by logical thought and mathematical reasoning. It is this which the commentators on the *Chiu Chang Suan Shu* illustrate.

As in 17th-century London, 'the mathematicians seem to have been very plain practical men . . . their style of writing was quite unliterary'.[217] In this also they had something in common with the founders of the Royal Society who preferred 'the language of artizans, countrymen, and merchants, before that of wits or scholars'.[218]

[215] Gregory Blue has pointed out in a personal communication, however, that Greek geometry may be considered a relatively late development on the shorter time-scale of classical Greek society, with that society itself as a late flowering of civilisation in the East Mediterranean complex.

[216] See also *SCC*, vol. 3, pp. 98–9, where it will be noted that one term for a cone is *yüan-chui* 圓錐, 'circular awl', to be compared with the 'square awl', *fang-chui* 方錐 noted below p. 175.

[217] See *SCC*, vol. 3, p. 152, n. d, concerning the style of Taoist mathematicians, though verse was evidently not beyond them, as will be seen below from n. 234.

[218] Sprat (1722), p. 113.

The convenience of the brief Chinese geometrical terms, originally descriptive of some architectural feature, or borrowed from the vocabulary of artisans or earthwork engineers, may be seen from the list supplied by Wagner (1979), p. 166, which includes the following:

Chhien-tu: 塹堵 A right prism with right-triangular base. [Architect's term originally meaning a moat-wall.]

Yang-ma: 陽馬 A pyramid with rectangular base and with one lateral edge perpendicular to the base. [Literal meaning: 'male horse'.]

Pieh-nao: 鼈臑 A pyramid with right triangular base and with one lateral edge perpendicular to the base, this edge not being at the right-angled vertex of the base. [Literal meaning: 'turtle's shoulder-joint'.]

Fang-chui: 方錐 A right pyramid with square base. [Literally 'square awl'. The word awl (*chui*), perhaps suggested the tapering shape of a pyramid.]

Chhu-meng: 芻甍 A right wedge with rectangular base. [Perhaps an agricultural term originally meaning 'fodder loft'.]

The example of the pyramid is repeated in the nomenclature of many different shapes. In view of the wealth of terms in Classical Chinese for particular geometric shapes it is rather surprising to find that there is no specific word for volume as opposed to area.[219] The word *chi* 積 in fact covers both, and the context makes it clear which is intended. When the opening sentence of Problem 23 in Chapter 4 of the *Chiu Chang Suan Shu* began with the words: 'Consider a *chi* of 4,500 *chhih* [feet]. If it is a sphere, what is the diameter?', Liu Hui with the precision of a later age, made the comment: 'This in fact means cubic feet' (*I wei li-fang chih chhih yeh* 亦謂立方之尺也).

Nevertheless one can understand why in early times area and volume were not differentiated. The word *chi* is another agricultural term brought in to the aid of early mathematics, and originally meant, as did the word *area* to the ancient Roman farmer, a piece of ground cleared and levelled for a particular purpose. In the West this purpose was the threshing of corn. When 'the area' was not needed for threshing, it was used for ball games, wrestling and other pastimes, but in any case it was kept clear. It remained an 'area' and was in no sense voluminous. The Chinese 'area' on the other hand was the piece of ground on which the sheaves were piled, adding another dimension to the concept.[220] It has therefore a cluster of related meanings, 'to stack', 'to put in stack or store', 'to hoard', 'provisions', 'to accumulate'.[221] As each layer of sheaves accumulated, what had begun as an area would become a volume. In fact the notion that a volume consisted of layers of piled-up areas of non-negative thickness persisted for a long time. Tsu Keng-Chih 祖暅之, for example, using an assumption which is equivalent to Cavalieri's Theorem, says: 'If volumes are constructed of piled-up blocks, and corresponding areas are equal, then the

[219] See *SCC*, vol. 3, p. 98, n. b, where this is pointed out. It is even more surprising that Chinese technology advanced so far without the concept of degrees as a means of measuring angles for geometric purposes. For the Chinese centesimal division of degrees, see Libbrecht (1973), p. 76.

[220] The association of the area with cereals is suggested in the character by the radical *ho* 禾 meaning 'growing grain'.

[221] See Karlgren (1940), graph 868 (t).

volumes cannot be unequal'.[222] It has even been suggested[223] that the character for block, *chhi* 綦, should be emended to *mi* 冪, meaning a rectangular area. Naturally, as mathematics developed, it became necessary to distinguish the character *chi* 積 into separate terms for area and volume. Today *area* is referred to as *mien-chi* 面積 and *volume* as *thi-chi* 體積, *mien* and *thi* being surface and body respectively. There is also *jung-chi* 容積 meaning *capacity*, or the *chi* which contains. As so often in Chinese, verbal differentiation into compound terms accompanies the development of conceptual clarity.

In the early years of Chinese mathematics, then, the fact that the word *chi* could mean either area or volume was not found to be a handicap because the practical applications of mathematics would make it clear from the context which was intended. But by the +3rd century, with the growth of interest in mathematical theory, a commentator of Liu Hui's calibre would point out the need for greater exactness in expression, though naturally he would not presume to alter or emend in any way a revered text.

Another indication of increasing awareness of the importance of accurate terminology is the replacing of the old word for sphere – *li-yüan* 立圓, or 'circle made to stand upright', a term evidently analogous to that for a cube, *li-fang* 立方. In time it was replaced by the word *wan* 丸 which meant a ball or pill, as did *sphaera* in Greek and *pillula* in Latin. The term *li-yüan* may be assumed to have been becoming obsolete by the +3rd century, for if it had not, it would not have needed any explanatory comment from Liu Hui. The reason for its replacement by *wan* may have been that a circle made to stand upright could just as reasonably have been visualised as a cylinder. The word *wan* would be less ambiguous. Both terms occur in the same sentence of the *Chiu Chang Suan Shu*. Liu Hui is concerned to show that there is no difference in meaning. Clearly there was nothing inherently inhibiting about the language itself, and when the need for a more specific terminology was realised, it could easily be accomplished.

Greek mathematicians may have been led to determine the volume of a sphere by their desire for exact knowledge and their wish to explore the implications of new formulations regardless of any practical use they may have had. Chinese science on the other hand was intended to be used for the benefit of the people. But there was little practical use for spherical objects before the invention of cannon balls, footballs and gas balloons. Why was it, then, that Chinese mathematicians persisted for some centuries in their attempts to find an exact formula for determining the volume of a sphere?

[222] Wagner (1978), pp. 61 and 75 for this quotation, and p. 63 where he says: 'It seems possible, then, that Tsu Keng-Chih could have worked out some sort of derivation of Cavalieri's Theorem. The blocks in his statement of the theorem might then be cross-sections of non-zero thickness which, in his derivation, would be reduced in thickness to a limit.' Tsu Keng-Chih's words are: '*Fu tieh chhi chheng li chi yüan mi shih chi thung, tse chi pu jung i.* 夫疊棊成立積，緣冪勢既同，則積不容異。'. The correlation with Cavalieri's Theorem is also discussed by Li Wen-Lin (1982), p. 40.

[223] Wagner (1978), p. 61, who refers to the emendation by Li Huang 李潢, (*d.* 1812).

It seems likely that the calculations required for finding the diameter or volume of a sphere in the +1st century or earlier represent a rounding out of mathematical discipline, and perhaps a step towards a better understanding of spherical objects in the skies, rather than a useful training for dealing with man-made spheres.

This belief is supported by the nature of two problems to be found in Chapter 4 of the *Chiu Chang Suan Shu*. They represent the extremes of a scale. At the micro end, in Problem 23, the author supposes a sphere whose volume is only 4,500 cubic Chinese feet, or about the volume of a very small man-carrying gas balloon. At the macro end he supposes a sphere with a volume of 1,644,866,437,500 cubic Chinese feet, the equivalent of an asteroid of rather more than four kilometres diameter. The +1st century is clearly too early for any likely Chinese interest in either of these possible applications.

The problems are stated as follows:

> Problem 23. Consider a *chi* of 4,500 feet. If it is a sphere, what is the diameter? The answer is 20 feet. (*Chin yu chi ssu chhien wu pai chhih. Wen wei li yüan ching chi ho? Ta yüeh, erh shih chhih.* 今有積四千五百尺，問為立圓徑幾何？答曰:二十尺。)
>
> Problem 24. Also consider a *chi* of 1,644,866,437,500 feet. If it is a sphere, what is the diameter? The answer is 14,300 feet. (*Yu yu chi i wan liu chhien ssu pai ssu shih pa i liu chhien liu pai ssu shih san wan chhi chhien wu pai chhih, wen wei li yüan ching chi ho? Ta yüeh, i wan ssu chhien san pai chhih.* 又有積一萬六千四百四十八億六千六百四十三萬七千五百尺，問為立圓徑幾何？答曰:一萬四千三百尺。)

A literal translation is used here to bring out certain points. The fact that the word *chi* 積 could mean either area or volume was considered above. But first we may consider the background against which much mathematical thinking was carried on. In ancient and medieval Europe, as in ancient and medieval China, mathematics, music and astronomy were closely enmeshed. In Europe scales in music were ladders to heaven. In China rituals were held to be ineffective if instruments were tuned to the wrong pitch. Their correct tuning depended on mathematically precise measurement. In groping for formulae with which to express mathematical proportions, Chinese thinkers had conveniently to hand the theory of the 'generation' of notes from the orthodox scale derived according to the *lü-lü* 律呂 or standard pitches.[224] When there is no clear understanding of what is involved, an analogy is often helpful.

Just as the fundamental note represented by a string 81 inches in length generated other notes, so, as the *Chou Pi Suan Ching* 周髀算經 points out: 'The rectangle originates from (the fact that) $9 \times 9 = 81$'.[225] The rectangle was capable of generating other geometric shapes.

What is fascinating about the Chinese invention of a formula for calculating the volume of a sphere is that it was derived from an imaginary or at least unusual solid.

[224] See above, *SCC*, vol. 4, pt. 1, pp. 172 ff.

[225] These words occur at the very beginning of the *Chou Pi Suan Ching*, stabilised –1st century, but here possibly as early as –6th century. See above, *SCC*, vol. 3, p. 22. The number 81 was also used as the basis for the generation of the notes on the up-and-down principle. See above, *SCC*, vol. 4, pt. 1, p. 174.

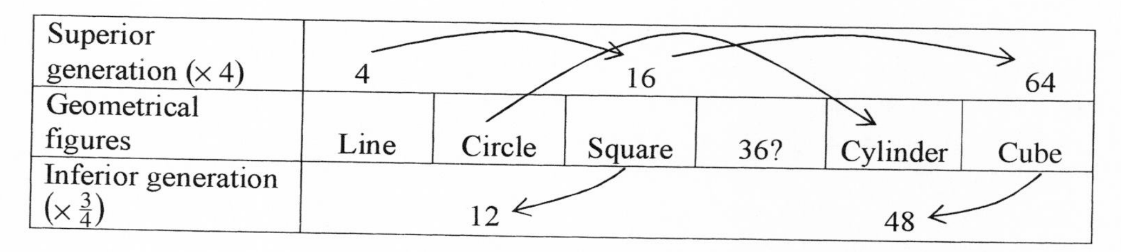

Figure 21. Geometric constructions produced by 'superior' and 'inferior' generation as in music.

Now why should they have done this? Every complete scholar had an understanding of music. We would like to suggest that in pursuing a musical analogy they became aware of a gap in the geometric scale and proceeded to fill it.

Let us now summarise the steps leading towards the finding of a formula for establishing the volume of a sphere from its diameter, or vice versa (see Fig. 21).

(1) The area of a square was to be found by multiplying two adjacent sides.

(2) The early Chinese method of finding the approximate area of a circle was to take $\frac{3}{4}$ the area of a circumscribing rectangle. The modern method is to multiply the radius of the circle squared by π. Today the figure for π is $\frac{22}{7}$. In the +1st century it was usually 3.

As the radius of the circle is the same length as half the length of the side of the square both methods are satisfactory, but when modern values of π are used results differ greatly from those obtained in the +1st century. For example:

	Radius	Area	Radius	Area	Radius	Area
+1st century	2″	12 sq.ins.	4″	48 sq.ins.	6″	108 sq.ins.
Modern	2″	12.7 sq.ins.	4″	50.2 sq.ins.	6″	113 sq.ins.

(3) The volume of a cube was to be found by multiplying a square by a third side of the same length. The resulting solid was termed a '*li-fang*' 立方, or 'square made to stand up'.

(4) From this it would follow that if the area of a circle was $\frac{3}{4}$ the area of its circumscribing square, the volume of a cylinder should be $\frac{3}{4}$ the volume of its circumscribing cube.

(5) The possibility then arises that perhaps a cylinder may by inferior generation give birth to some other solid shape, which would have the volume or number of 36, which is also that of a circle × 3. What more likely candidate to fill this gap than the sphere? The idea that Chinese mathematicians should have visualised one geometric figure 'giving birth' to another has not, as far as we are aware, been suggested before, but the possibility may be pursued a little further.[226]

[226] In the *Chiu Chang Suan Shu* of the +1st-century 'mother' and 'child' are normal terms for the denominator and numerator of a fraction respectively. See *SCC*, vol. 3, p. 81, n. (e).

The musical scale, as mentioned above, was generated by multiplying a fundamental number, which was usually 81, since the string could be conveniently divided by fractions of this number to obtain the desired intervals. This number 81 was processed by a sequence of fraction, either by $\frac{2}{3}$ (inferior generation), or by $\frac{4}{3}$ (superior generation).[227]

In generating a geometric scale one would multiply a given length of line, for example, 4 units, by the number 4 (superior generation), or by $\frac{3}{4}$ (inferior generation). Proceeding historically from simple to complex, once it had been discovered how to draw a straight line and to measure it, and to draw lines at right angles to it with the help of the carpenter's T-square, squares would be generated from lines. A line of measured length, therefore, when multiplied by its own length (superior generation), would produce its square, and this square when multiplied by the length of the original line (also superior generation) would produce its cube. The square when multiplied by $\frac{3}{4}$ (inferior generation) would produce what was believed to be the area of a circle circumscribed by the square, for as Shang Kao 商高 rather ambiguously expresses it in the second paragraph of the *Chou Pi Suan Ching*, 'The circle is derived from the square';[228] and the cube when multiplied by $\frac{3}{4}$ (inferior generation) would produce the volume of the cylinder. It must be remembered that the ratio between the diameter and the circumference of a circle was taken, in ancient China as elsewhere, to be 1:3, approximately until the beginning of our era, and in common practice for long after this.[229]

Just as the musical scale had five notes – *kung* 宫, *shang* 商, *chiao* 角, *chih* 徵 and *yü* 羽 – so the earliest geometrical figures also numbered five – the line, square, cube, circle and cylinder. Each of these may be given a number if the fundamental line is measured.[230] When given numbers they may be arranged from smallest to largest to make a scale. Let it be assumed that the line measures four units in length. Then the scale might be as follows:

Line	Circle	Square	[Sphere?]	Cylinder	Cube
= 4	$16 \times \frac{3}{4}$ = 12	4×4 = 16	[$48 \times \frac{3}{4}$] [= 36]	$64 \times \frac{3}{4}$ = 48	16×4 = 64

The 'generation' of this scale was shown in Figure 21 in which line, circle and square are on the left and volumes on the right.

[227] For which see *SCC*, vol. 4, pt. 1, p. 174.

[228] In *SCC*, vol. 3, p. 22, n. (c), however, we said, 'Presumably the writer was thinking of the diameter of a circle as equal to the diagonal of its inscribed square; perhaps also of exhaustion methods for getting'.

[229] Early approximations to π in China and elsewhere were considered above, *SCC*, vol. 3, pp. 99 ff. See also above, p. 178.

[230] As late as +3rd century a specific number was used to represent the more abstract concept of volume, for, as a solid could be measured, its dimensions, when appropriately multiplied, would yield a number. See also *SCC*, vol. 3, p. 22, where the belief that the properties of numbers must be known before one can work with geometrical figures is contrasted with the Euclidean method. It is not suggested that texts exist in which this analogy is made explicit, but rather that mathematics from –3rd century onward could not have been kept separate from music and numerology, for which see *SCC*, vol. 2, pp. 270 ff.

A mathematician with a feeling for music would perhaps be reminded of the pentatonic scale which has 'gaps' in it, gaps which may be filled with additional notes. Noticing that there are three area figures on the left of the scale, but only two volume figures on the right, and that there appears to be a numerical gap between 16 and 48, he could very well wonder whether there might not be an additional volume to restore the distressing lack of symmetry. An obvious lacuna is the sphere. If the volume of the cylinder were to be multiplied by $\frac{3}{4}$ according to 'inferior generation', he would, in the above example, get a number – 36 – and this, he might suppose, would be the number for the volume of the sphere, a figure to which there was as yet no approximation.

By the time the fourth chapter of the *Chiu Chang Suan Shu* had been written, i.e. not later than the +1st century, this is exactly what had happened, for we find there the formula for the diameter of a sphere in relation to its volume. This may be expressed as: 'The diameter of a sphere = the cube root of $\frac{16}{9}$ of the volume of a sphere'. If the diameter is known, this can be reversed to find the volume, i.e. $V = \frac{9}{16} \times d^3$, or in the above example, $\frac{9}{16} \times 64 = 36$.[231] By means of this formula the author of Problem 23 concluded that a sphere of 4,500 cubic feet must have a diameter of 20 feet. By modern reckoning, using the formula $V = 4r^3$, a sphere with a diameter of 20ft would have a volume of 4190.76 cubic ft. The formula used in Problem 23 would have been a useful approximation.

By the +3rd century the inaccuracy had been detected. Liu Hui realised that what we have called 'inferior generation' from the cube yields the cylinder, and that 'inferior generation' from the cylinder yields something bigger than the sphere. He went further, for he found a solid whose proportions matched the requirements of inferior generation from the cylinder. This he called the *ho-kai* 合蓋, literally 'close cover'. *Ho-kai* was the name he gave to a solid in which two cylinders intersect horizontally within the circumscribing cube. By further inferior generation from the *ho-kai* it was then possible to calculate new values for the volumes of spheres. The insertion of the *ho-kai* between the cylinder and the sphere was a remarkable piece of imaginative insight on the part of Liu Hui, comparable to the discovery by deduction of an invisible star. His reasoning was as follows: Having calculated the proportions of the *ho-kai* and found (a) that it was smaller than that of the cylinder circumscribed by the cube, and (b) that the sphere was smaller than the *ho-kai*, then if the ratio of the *ho-kai* to the cylinder was 3:4, this could not also be the ratio of the sphere to the cylinder. Liu Hui had shown that the ancient formula for the volume of a sphere could not be correct, but he confessed that he was unable to find the correct formula.

Liu Hui declared that though he realised the formula for the volume/diameter of the sphere as given in the *Chiu Chang Suan Shu* was incorrect, he could not himself give the correct one, or possibly, as Wagner points out, 'it may be that he knew the correct formula, $V = d^3$ but was unable to prove it.[232] The error which Liu Hui

[231] The author owes much to Donald Wagner whose article, Wagner (1978), was the original source for this subsection, and whose help in the writing of it was invaluable, and also to Catherine Jami for her patient help in its clarification.

[232] On the value d where d is the diameter of the sphere see above.

[233] Wagner (1978), 72.

discovered in the *Chiu Chang Suan Shu* was that if the ratio of the *ho-kai* to the sphere was 4:, and if the *ho-kai* was smaller than the cylinder, then the ratio of the cylinder to the sphere could not also be 4:. We are more inclined to think that he arrived at this inequality impressionistically by comparing the magnitudes of the two errors mentioned above'.[233] The correct formulation and its proof were first achieved two centuries later by Tsu Chhung-Chih 祖沖之 and later by his son Tsu Keng-Chih 祖暅之, who celebrated the achievement in verses which appear to 'give face' to Liu Hui, suggesting that he would have made the discovery if only he had had time. However, as Liu Hui was unlikely to have lacked time, the careful reader will perhaps interpret his verse as a courteous if posthumous stab in the back![234]

It would be instructive to compare the approach of Archimedes in –3rd-century Sicily with the writing of Tsu Chhung-Chih and his son in +5th-century China to the same problem concerning the relation of the volume to the diameter of a sphere. The logic of argument in each case is similar. The volume of A cannot be in the same proportion to B and C if the volumes of B and C are different, and: if the volume of A is four times greater than the volume of B, and six times greater than the volume of C, then B must be one and a half times greater than C. But the manner in which the arguments are presented is totally different. In Greece the essence of geometrical demonstration was the 'proof'. In China and India this was not at first considered important.[235]

The modern word 'proof' derives from the late Latin word *proba* which meant the probing and testing of the evidence of witnesses to see whether it merited the 'approbation' of judge or jury. Although the word is Latin the concept of seeking approbation by testing the evidence before a jury goes further back to Athens where litigation was a highly democratic process. 'Proofs' had to be terse and to the point, for only a limited time was allowed in court for each speech, checked by the clepsydra, the water-clock. In the Athenian law-courts, as no doubt in the early Christian assemblies, the supreme test of the truth of evidence given by free men was the solemnly sworn oath.[236] In the courts to which the geometers appealed it did not carry the same conviction to swear that the volume of a cylinder is $\frac{4}{3}$ times the volume of a sphere. Such evidence had to rely on reason alone, to the lasting benefit of mankind.

The social atmosphere in which the great Chinese mathematicians worked was utterly different.[237] In the writing of Tsu Chhung-Chih and his son, one does not sense the openness of Greek society, the challenge to the informed public to refute a proposition if they can, after the bases of the argument have been explained

[234] The verse is quoted in Wagner (1978), p. 76: 'The proportions are extremely precise, And my heart shines. / Chang Heng copied the ancient, / smiling on posterity; / Liu Hui followed the ancient, / Having no time to revise it. / Now what is so difficult about it? / One need only think'.

[235] Hindu mathematicians 'were not in the habit of preserving the proofs, so that the naked theorems and processes of operation are all that have come down to our time'. Cajori (1919), p. 83. But it should be noted that 'Euclid gave his division algorithm without strict proof. No one has criticised him for this.' Li Wen-Lin (1982), p. 10.

[236] See Nettleship & Sandys (1957), p. 332.

[237] This point has been discussed in Needham (1956), pp. 15–16, where reference is made to Vernant (1963) and (1964).

with almost legal precision in clear and simple language. Rather it is as if one were in the workshop of a craftsman who carves ivory balls and sees sphere within sphere, or in the studio of a scholar whose three-dimensional thinking is aided by the handling, separating and recomposing in different shapes of solid blocks. It seems not improbable, as Wagner suggests, that scholars played with blocks making geometric shapes, for the instructions in Tsu Keng-Chih's 'proof' imply a familiarity with their handling, rather as today a proof based on the behaviour of cubes within the rubik cube would imply some familiarity with their handling. The tensions are different. One misses in the Chinese the taut sense of urgency in the marshalling of the arguments. There is no water-clock dripping in the background to cut short a witness if he is too verbose. On the contrary there is time for literary allusions and the neat antithetical couplet. Tsu Keng-Chih gives a good example of his geometrical reasoning:

> Take a cubical block. Place the pivot [of a drawing compass] at the left back lower corner. From the arc [now assumed to have been drawn] remove the right upper side piece. Place [the pieces] together again, draw an arc at right angles, and remove the front upper side-piece. Now the cubical block is divided into four. There is one block inside the arcs called the inner block. There are three blocks outside the arcs [and these are] called the outer blocks.[238]

Tsu Keng-Chih then reassembles his four blocks and cuts them once again, horizontally, and develops his argument by the use of right triangles, i.e. the Theorem of Pythagoras, coming finally to his triumphant conclusion.

Tsu Keng-Chih's explanation would be hard for anyone to follow who had not had experience in handling the blocks or was not unusually gifted in being able to visualise plane figures stereoscopically. The operations he describes must of course be carried out in the imagination, and it may be that this form of imagination was highly developed in China, but the handling of three-dimensional shapes would have helped to develop this imagination. It is clear that he is writing for people who understand what he is doing, for he assumes that they are sufficiently familiar with his procedures to be able to bridge certain gaps. For example, the first time he mentions drawing an arc on the cube, he merely gives them the essential information as to where to put the point of the compasses, and assumes that it is obvious that the arc will be drawn between the only two available corners. Naturally he does not need to add anything so obviously explicit as 'draw it!' This was not because there would have been any difficulty in saying 'draw an arc' in Chinese. On the contrary, with the second arc he does just that, saying very concisely: *heng kuei chih* 橫規之, 'draw an arc at right angles'. The use of the word *heng* 橫 also indicates certain assumptions as to the intelligence of his reader. It assumes that he can visualise a compass point stuck into the corner of a cube. If this is done there are three surfaces on which the

[238] *Chhü li-fang chhi i mei, ling li shu yü tso hou chih hsia yü, tshung kuei chhü chhi yu shang chih lien. Yu ho erh heng kuei chih, chhü chhi chhien shang chih lien. Yü shih li-fang chih chhi, fen erh wei ssu. Kuei nei chhi i, wei chih nei chhi. Kuei wai chhi san, wei chih wai chhi.* 取立方棊一枚，令立樞於左後之下隅，從規去其右上之廉，又合而橫規之，去其前上之廉。於是立方之棊，分而為四。規內棊一，謂之內棊，規外棊三，謂之外棊。

compass can describe an arc. If two of these are regarded as having vertical planes, the third will be in the horizontal plane. *Heng* in Chinese means 'crosswise', 'at right angles to the vertical', or just 'at right angles'. As the compass point is to be inserted into the bottom left-hand corner of the cube, the cube itself must be resting on the only available horizontal plane. The second arc must therefore be drawn on the third face, at right angles to the first arc, and, as translated by Wagner, 'perpendicularly'.

As we saw in our description of the plough, Chinese authors had a clear sense of direction. Tsu Keng-Chih speaks of the 'left back lower corner', looking at the cube from the reader's chair, so to speak. This seems un-Greek because the Greeks, standing or sitting in a circle around the geometer, developed the convention of describing geometric positions by the use of letters.[239] But it must be remembered that originally even in Greece the *basis* or base of a triangle was the bottom, and the *hypsos* 'height' or apex was the top. The Chinese could easily have used the twenty-two characters of the 'celestial stems' and the 'horary characters' (or any other characters – see above, Fig. 20) in the same way as the Greeks used letters, but geometry did not develop in China in that way. A further example of the Tsu's assumption that their readers would be familiar with the handling of blocks is the way in which they state that after the arc has been drawn a portion of the cube can be removed as if it had been sawn through. Moreover, after describing two such arcs which the unversed reader might imagine to have separated the cube into three blocks, he says, 'Now the cubical block is divided into four'. In fact as the imaginary sawings are at right angles to each other, a fourth block is made separable by their intersection. No doubt adepts at the geometric art could visualise these blocks without difficulty, but if an author had wished to obtain a favourable verdict from the common reader, the Greek practice of indicating terminal and intersection points by symbols would undoubtedly have been helpful.

This discussion as to whether Literary Chinese was an efficient instrument for conveying mathematical ideas may now be summed up. Whether or not a language is influenced by the social climate of those who speak it may remain an open question, but it can scarcely be denied that new concepts and technological innovations are intimately associated with the form of society in which their inventors live. One of the greatest contributions of the ancient Greeks to world civilisation was the development of the concept of a proof to justify a contention. The method was elaborated by the mathematicians of Alexandria, who were themselves either formed or strongly influenced by the mathematics of Plato and the syllogistic reasoning of Aristotle. Euclid, for example, seems to have studied mathematics in Athens at the Academy, later (*c.* –300) migrating to the new city of Alexandria.

It is natural to ask, if the Greeks elaborated the idea of geometric proof, why was this not also done to the same extent in other civilisations? A better question is perhaps: If other civilisations, such as that of India and China, did not find it necessary to record the steps by which mathematicians arrived at their conclusions, and to arrange these

[239] The use of letters would be more convenient for a group of people sitting in a circle round a diagram.

steps in a concise and logical sequence, why did the Greeks do so?[240] A large part of the answer must surely lie in the form of society in which the mathematicians lived and worked. It has been suggested that factors which precipitated the Greeks into this mode of thinking were the openness of Greek public life, the institution of argument, and their habit of reaching decisions by secret ballot, as in the law courts and the dramatic festivals, or by the tumultuous one-man-one-vote system for the whole population of male citizens in the political assemblies, during the formative period of their civilisation. For any contention to survive in such an environment, it would need to be supported by cogent arguments briefly and clearly expressed.

In China there were different influences and other pressures. In the post-Han and pre-Sung periods, for example, less centralised government perhaps contributed to a more relaxed intellectual life, with freer discussion in which proof of a contention would be required. There was, on the other hand, the drive to ensure continuity, and related to this the traditional reverence for the past. It was not open to a minister to prove that the emperor's disgraceful conduct would inevitably lead to ruin, in the way that Nicias might have been tempted to speak of Alcibiades in the Athenian assembly. He could only suggest by allusion and historical parallels that perhaps all was not for the best!

Though Tsu Chhung-Chih's manner of proof would not have been familiar to the mathematicians of Alexandria, it would have been perfectly clear to the gentlemen scholars of China for whom it was intended. It is easy for the speaker of one language to regard the absence of a term in another language as a defect of that language. This may be so, but it is more likely that he is criticising the absence of the concept to which that term in his own language refers. This again may be a defect, but it may be that a given civilisation can manage perfectly well without it. It is, for example, astonishing that Chinese civilisation should have reached a high level of engineering without conceptualising the measuring of angles by parts of a circle expressed as degrees. The fact that the Chinese ecliptic circle was from early times known to have $365\frac{1}{4}$ days, rather than 360 as in early Babylonia,[241] may have delayed the subdivision of the circle into degrees; and the fact that for a long time the proportions of the measurable sides of right-angled triangles served as a substitute for the measurement of angles in degrees, explains why the use of measured angles was long delayed in China.[242] Similarly it is surprising that ancient Greek had no abstract term for volume, yet their use of the phrase 'the solid' *to stereon*, covered most of their needs, and in such a sentence as 'the volume of all these shapes is the same', where it is not possible to say 'the solid of all these shapes', it was still possible to express it another way, such as 'if the dimensions of these sides are multiplied the resulting number will be the same'. In short it would seem that any language will develop a way of expressing its speakers' concepts, once the concepts are clear, but that the

[240] See, however, Harbsmeier in *SCC*, vol. 7, pt. 1, on Chinese Buddhist logic and the role of public philosophical debate.

[241] For a detailed discussion of the origin of the division of the circle into 360°, see Heath (1921), vol. 2, pp. 215 ff.

[242] This may also explain why astronomical distances were sometimes measured in feet and inches instead of in degrees.

formulation of new concepts is largely a matter of the pressures exerted by society on inventive individuals.

(6) Concluding Reflections: Science and the Fate of Classical Languages in East and West

In the transition from medieval science to modern science, it seems probable that forging a suitable linguistic instrument is less important than creating an adequate descriptive base. Thus alchemy could not develop into modern chemistry until the table of elements had been worked out to form such a base. When thought is clear, scientific writing is usually clear. When thought is groping on the frontiers of knowledge, metaphors, analogies and very general terms take the place of precise terms and clear conceptions.[243]

Both Classical Latin, as it developed in the Middle Ages, and Literary Chinese would seem to have been magnificent instruments for the communication of scientific ideas. But it remains to be explained how it was that they failed to be developed for this purpose into the 20th century. In retrospect it now seems clear that given time and determination any classical language can adjust to changed conditions, and can even return to oust the vernacular language which has assumed its role, as has occurred with Hebrew in Israel. But for this to occur the classical language must return on a full tide of national sentiment.

On the other hand the diversion of the classical language for some specialised purpose is easier if it has already abandoned the main field. Latin does, in fact, afford an example of this phenomenon, for, having surrendered most of the field to the rising vernacular languages of Europe,[244] it then, in the late 18th century, made a limited comeback with the artificial creation of 'botanic Latin'. But this new specialised dialect illustrates the great extent of the modifications which had to be made, modifications which would have been resented and strenuously resisted not only by scholars but by all educated people, if the classical language had still been in possession of its full powers. In fact the botanic reformers were able to make the necessary changes from the classical form of the language only because the champions of Classical Latin had largely become indifferent to its fate as a world language, and were often unaware of the gathering momentum of modern science. They confined their interests to the ancient literature which was soon to become a province of specialists.

[243] For example, the plague at Athens in the year –439 as described by Thucydides cannot be identified, not because Thucydides was incapable of clear thought, or was having a lapse – on the contrary he was at pains to describe what he saw exactly – nor because the Greek language was an unsuitable language for describing diseases, but because Thucydides did not sufficiently understand the nature of the disease to know what to describe, or had not the technical means, such as a thermometer, to find out the precise aspects, such as temperature, which needed describing. See Thucydides, II, 47–55. Shrewsbury (1950) reviews the possibilities, but McNeill (1976), ch. 3, n. 34, points out that 'a virgin population exhibits symptoms far different from those manifest in a population already exposed'.

[244] New medical theses were, however, written in Latin in Britain till after the beginning of the 19th century, and Latin was needed by doctors and lawyers till much later. It is only recently that university public orators have ceased to orate in Latin.

If, on the other hand, the reform of a classical language needs to be made to meet the demands of a suddenly exploding new science, but these reforms are likely to be resisted and contested by conservative scholars supported by a less informed but equally conservative reading public, the attraction must seem irresistible to make a clean sweep. The temptation will then be to publish scientific works not in the old classical language at all, but in a newly current vernacular language in which less concern is felt about neologisms and innovations, and in which conservative attitudes have not yet become solidly established. This, however, is not the whole story in either China or Europe.

The classical languages of both Europe and China had many advantages. Latin was used internationally within its culture area until quite a late date. It continued as the main official language of France until 1539,[245] and much longer as the language of the universities, of international intercourse and of the church. The last scientific work of importance to be published in Latin in England was Newton's *Principia*, which appeared in 1687.[246] 'As a vehicle of scholarship it survived longest in the German Universities . . . In the German States between 1681 and 1690, more books were printed in Latin than in German, and Latin was still the medium of teaching in the German Universities.'[247] It continued to be the most useful lingua franca for educated people until well into the 18th century. Linnaeus, for example, used it for speaking when he was in Holland, and in correspondence with botanists. It was also used by Albrecht von Haller (1708–1777) in his correspondence, and by many others.[248] Its function in keeping scholars in touch with developments in other countries contributed greatly to the development of modern science in its early years, though the existence of a common language does not of itself guarantee the free flow of ideas nor that alien ideas shall be admitted to its network. Rufinus (13th/14th centuries), for example, did not know of the work of Albertus Magnus;[249] and the Chinese living in Peking in the 17th century did not feel moved to translate the useful work on glass-making, '*L'Arte Vetraria* by Antonio Neri (1612) which the Jesuit missionaries had brought with them to that city. As we said above in *SCC*, Volume 5, part 3, pp. 236–7, n. h, 'An up-to-date treatise on glass-making would have interested the Chinese'.[250]

[245] In which year Francis I decreed that all official publications should appear in French (*UNESCO History of Mankind*, vol. 4, p. 604). See also Sarton (1947), p. 1108.

[246] That is to say fifty years after Descartes had published his *Discours de la Méthode* in French.

[247] Bodmer (1944), p. 313.

[248] Stearn (1966), p. 7.

[249] Crombie (1952), p. 115, points out, 'Rufinus, . . . seems to have known nothing even of *De Vegetalibus et Plantis* of Albertus Magnus'. The high cost of books was probably also an impediment.

[250] The question of the transmission of Western chemical and metallurgical processes to China is considered in *SCC*, vol. 5, pt. 3, pp. 221–36. As we said there (p. 236): 'It was a tantalising twist of history that such works could get as far as Peking yet fail to penetrate the language barrier'. One should not, of course, consider a language to be a barrier in itself. It is only a barrier to those who are unable to learn it. As we said in note h, continued from p. 236, apart from some of the books brought by the Jesuits to Peking which were of little value to the Chinese, there were others 'which could have been very useful if there had been someone to expound them and take them through the language barrier'. Examples quoted were the *De Distillatione* of Geronimo Ross, the *Pirotechnia* of Biringuccio and *L'Arte Vetraria* of Antonio Neri mentioned above. The value of this book is also noted in *SCC*, vol. 5, pt. 2, p. 269.

Both Latin and Literary Chinese also had another advantage denied to the newly developing vernacular languages, which was that the great works of remote antiquity became available to the student who mastered the ancient medium. He was part of a continuing tradition, whereas there was a gulf between the ancient texts and the reader of the vernacular alone, which would not be bridged until translations had been made into the new medium. At first the vernacular languages of Europe could not compete with Latin.[251] As Stearn says, 'Latin survived through the Middle Ages and well into the 18th century as the one internationally used language of learning among the peoples of Europe, none of which then possessed a vernacular language sufficiently developed and widely enough known to challenge its supremacy in diplomatic, legal and ecclesiastical matters'.[252] It could, however, only be a matter of time before the challenge was made.

The reasons why the vernacular languages of Europe gradually ousted Latin are extremely complex. Perhaps the most important is that with the coming of the Renaissance there was a drawing together of the educated and the artisan classes, a rare situation, as Graham has observed in writing of the Mohist sciences: 'The sections on optics and mechanics surely reflect social conditions comparatively rare in history until the 16th century in Western Europe, where the Scientific Revolution soon followed, the explosive situation when men with speculative minds are in close contact with men who work with their hands'.[253] In Europe there was more to it than that. Not only did the two classes work closely together, but the need arose that the artisans should be able to read and write, record what they observed, make accurate measurements, and apply mathematics to the tests they made on their products.[254] The artisans had to be educated not in the ancient literary glories of Greece and Rome, but in the practical methods of increasing production and efficiency. Their education therefore had to be in the language which they spoke.[255] By the 17th century it had become quite clear that a form of education needed to be devised which would promote this aim. Such education would not stress the mere unthinking application of rules by menials, but, as Wilkins stated in his *Mathematicall Magick* of 1648, would be designed to create a form of education which would benefit common

The barrier to the transmission of ideas is made up of many things besides the difficulties of translation. Among them may be noted a contempt for the style in which the ideas are expressed; lack of enthusiasm or even interest among the parties who alone have opportunities for translation; the fact that sound ideas may be mingled with unacceptable theories; lack of funds to finance the translation project; political interference from rival parties; and, as was the case in the later stages of the Jesuit mission, distraction of attention by other over-riding pre-occupations, in this case Chinese concern with the imminence of the Manchu invasion. There was also the well-known distaste of Confucian scholars for artisan pursuits.

[251] Though Sarton (1947), vol. 3, p. 345, points out that even by +843 *langue d'oïl* was becoming accepted in its own right, the emancipation of that language had been possible only because Latin had reached an advanced state of disintegration. 'If the Latin language had remained alive, the Romance dialects would have had as much chance of developing as maggots on a living body.'

[252] Stearn (1966), p. 14.

[253] Graham (1978), p. 8.

[254] See above, Joseph Needham, 'Science and Society in East and West', and Joseph Needham and Huang Jen-Yü, 'The Nature of Chinese Society: A Technical Interpretation'.

[255] As Bishop Wilkins expressed it, books had to be written in the vernacular 'for which these mechanical arts of all other are most proper'. Quoted in Gillispie (1970), sub.nom. 'Wilkins' by Hans Aarsleff, vol. 14, p. 366.

artificers in gaining a right understanding of the grounds and theory of the arts they practise.[256]

Of great significance in Western Europe at this time was the fact that paper was known by the 13th century in Germany and by the 14th century in England, making possible the start of printing from moveable type in the 15th century and the rapid spread of printed books.[257] Standardised editions of important works form the very foundation of modern education, and perhaps the spread of relatively cheap printed books did more to undermine the ascendancy of the classical languages in Europe than anything else. Yet in China, where paper had been invented some fifteen centuries before it was used by Gutenberg, and where moveable pottery type had been invented[258] some four centuries before Gutenberg's first use of metal type, the invention of printing served to strengthen the classical language rather than the opposite. Had the writing of China represented the sounds of words as they were spoken, it can plausibly be argued that the effects of printing would have been similar to those in Europe, but the cheapening of books in logographic script at first only strengthened the hold of the logographic lingua franca.

Latin could not be ousted as a language of science in Europe, nor Literary Chinese in China, until the vernaculars were sufficiently developed to carry the load. In Europe it was not a matter of developing one vernacular only, but as many as there were nation-states, and sometimes more – Switzerland, for example, used four. In this there was a marked lack of co-ordination between contending states in a field where co-operation would seem desirable. The lack of an agreed nomenclature for the different distillations of petroleum in the various European languages even today (petrol, essence, benzine, kerosine, gasoline, paraffin) is an indication of this failure. In China the problem was rather different. It was somewhat as if Switzerland on a giant scale had selected French to be its language of the future, this language when written being expressed in traditional logographs. There would be nothing to prevent a German-speaking Swiss from reading the logographs in German if he wished to, though occasionally in China, and very much more often in Switzerland, such reading from the logograph would have an alien flavour, due to differences in idiom, and to differences in word order and syntax in general, which in Europe are very marked.[259] In China a further reform, now virtually complete, was to ensure that anyone reading characters would be able to speak them in that form of the Northern dialect which is now known as *phu-thung-hua* 普通話. It is as if Europe had made one of its original vernacular languages obligatory for all administrative, cultural and scientific purposes. The advantages of having a population of many

[256] Wilkins (1648), in his address 'To the Reader'.

[257] Carter (1925), pp. 245–50. The Chinese began to change from moveable wooden type to bronze type in the late 15th century, simultaneously with the development of moveable type in Europe. See Tsien Tsuin-Hsuin in *SCC*, vol. 5, pt. 1, p. 211.

[258] The earliest authoritative account of the use of moveable type is of that invented by Pi Sheng 畢昇 (*c.* +990 to +1051). See *SCC*, vol. 5, pt. 1, pp. 201 ff., and also pp. 1 ff. on the origin of paper.

[259] In *phu-thung-hua* a person wishing to say, for example, 'Are there any?' would say '*Yu mei yu* 有沒有?' A Cantonese speaker would expect to find the equivalent of '*yau mou* 有無', i.e. '*yu wu* 有無?'

hundreds of millions of people all able to pool their intellectual resources through the medium of a common language and script may be expected to become clear in the centuries ahead.

If the decline of the classical languages in Europe and China are to be properly compared, it is first necessary to be clear about the time dimension. China may reasonably be regarded as a geographical and political unit for this purpose, even though the initiative may sometimes have been taken in the coastal cities, sometimes in such a province as Hunan, with its progressive governor Chhen Pao-Chen 陳寶箴 (1831–1900), sometimes in the capital, as when the Jesuit mission arrived there towards the end of the 16th century. Europe, however, was not a geographical and political unit, and a decision taken concerning the use of a language in Italy, for example, would not necessarily apply in England or Germany. The comparison will therefore be drawn between China and England to avoid the distraction of time-lags in different European countries.

Both China and England experienced influences and needs which were similar in nature but different in timing. Among them may be mentioned the fermentation caused by ideas from an alien culture; the cheapening of books by the invention of printing; the demand for translations and their effect on the existing language; the need for new technology; the discovery that technology does not come alone, but demands a rethinking of fundamental concepts; the need for a new practical education enabling workmen as well as artisans to work with the new machines; the need for a simpler language which can be used by such people to raise their level of education, enabling them to study science; and finally the growth of democracy or of mass participation in politics. But whereas in England this was spread over some six and a half centuries, from say 1250 to 1900, in China it had to be compressed within three centuries, and the most difficult and important part of it within the fifty years from 1850 to 1900. That China was able to do it at a time when the Manchu dynasty was tottering to its fall was a remarkable achievement.

From about 1250 Arabic influence began to have strong effects in England, particularly on the Merton scholars at Oxford, in dynamics, for example. But European influence on Chinese mathematics and physics did not begin until after the arrival of the Jesuit mission in 1583. The closure of the mission in 1774 was followed by nearly a century without contacts with the West at a high intellectual level, a loss as great to the stimulation of China as Western Europe would have suffered if the flow of Greek manuscripts from Constantinople to Italy had dried up a hundred years before the city fell to the Turks in 1453. When contacts at a scientific level were reopened by Protestant missionaries in the second half in the 19th century, the days when China and Europe could roll back the frontiers of science in co-operation and mutual esteem had passed. For China to survive, drastic surgery was necessary.

Although many flinched from this operation, there was no shortage of great people in China who were prepared to see it through. Their actions may be briefly chronicled. Before the first stirrings of modern science began to appear in the writings of Arnald of Villanova (13th century), the Chinese had already been able to enjoy books printed from wooden blocks for over three hundred years.[260] But their use of logographic writing had one singular difference from alphabetic writing in Europe, namely that people begin to be literate from the time that they have mastered their first character, and as further characters are added to their memories, they are able to decode progressively more and more of what they see written. The time may come when they will be able to provide a character to match every word they use in the vernacular language, although they will still not be able to understand the more elaborate codification of Literary Chinese. What was written might therefore be found at several different levels of simplification. For example, it is known that 'Chang Po-hsing 張伯行, while governor of Fukien (1707–10), used three different versions of the *Sacred Edicts*: one for literati, "one illustrated with popular sayings for those of medium intelligence and scholarly ability, and one with memorable jingles for the simple country folk" '.[261]

Logographs therefore provided a ladder from one level of communication to another, from the open to the coded, whereas the alphabetic scripts of the West had a different function. A knowledge of the letters of the alphabet might help Europeans to decode that which was written in their own language. But if the writing was in a language which they did not know, they could only make approximations to the sounds, often without having any idea of their meaning. The decision to write and print books in European vernacular languages instead of Latin was, therefore, a revolutionary step, far more so than in China where thanks to the use of logographs, it was more a matter of style than change of language. This perhaps explains why it was possible for China to change the medium of administration from Literary Chinese to 'clear speech' (*pai-hua* 白話) so rapidly, in effect between the years 1897, when the 'Academy of Current Events' (*Shih-wu Hsüeh-thang* 時務學堂) was established in Hunan Province, and 1917, when *pai-hua* was formally declared the official medium during the Literary Revolution.[262] This compares very favourably with the three and a half centuries required in England for the displacing of Latin by English, as described above. Since in China the overwhelming demand, until the end of the 19th century, was for texts to be printed in the literary language, printing consolidated its hold there rather than the opposite, as in Europe.

As in Europe, the change of medium was preceded by several centuries of intense work on translation. In Europe it was a continuous process, beginning with the translation of Arabic and Greek works into Latin, and continuing from the 14th century with translations from Latin and many other languages into the appropriate vernacular. But in China the work of intensive translation begun by the Jesuits was not continuous. During the period from 1552 to the suppression of the mission in 1773 and the few years in which the remaining missionaries lingered on, a total of some 340 works were composed in Chinese on Western subjects or translated from Western languages into Literary Chinese, most of them before the year 1675.

[260] See *SCC*, vol. 5, pt. 1, p. 86, for details of the earliest complete paper book of +256 and pp. 146 ff. for block printing from the +6th to the +9th century.
[261] Rawski (1979), p. 15, quoting Spence (1968), 1.8:5.
[262] *UNESCO History of Mankind*, vol. 6 (2), p. 1292.

Of these, according to Pfister, those on scientific subjects equalled, if they did not surpass in number, those on religion and controversy.[263] In the 18th century this activity died down, but a residual effect was that many neologisms from the Jesuit books became accepted into the language.[264] Moreover, the translations which had been made, being written for their content rather than for style, were in a rather severe and simple form of Literary Chinese, which served as a model for scientific writing thereafter.[265]

There was, however, in 18th-century China, a changed attitude to science, which, because it had been used as an inducement by evangelists to obtain converts, became suspect. Nevertheless, by the late 18th century, the more practical aspects of Western science began to be clearly discernible.[266]

Attention was at first focused on the fact that 'political power grows out of the barrel of a gun'.[267] This lesson had been learnt by the nations of Western Europe after the walls of Constantinople had been breached by Turkish artillery in 1453, and the determination of each nation not to be outgunned by its neighbour led to a surge of interest in mining, metallurgy, ballistics, dynamics, mathematics, and those aspects of proto-chemistry which related to the manufacture of gunpowder, together with the production and translation of books and manuals concerning these sciences and technologies.

It was the arrival of the steam-engine propelling foreign ships which first compelled Chinese officials to begin a serious study of foreign science. Details of this may be found above.[268] It will suffice here simply to list the main dateable events:

In 1828, as reported in the *Hai Kuo Thu Chih* 海國圖志, the first fire-wheel boat arrived from Bengal.

In 1830 the East India Company's steamer *Forbes* arrived.

In 1835 the steam-vessel *Jardine* arrived, but the government caused its machinery to be dismantled.[269]

In 1841 diagrams of a model steamboat and locomotive were published in Ting Kung-Chhen's 丁拱辰 *Yen Phao Thu Shuo* 演炮圖說.

c. 1845 the first full-scale Chinese steam paddle-boat was built at Canton with the help of a foreign engineer.

In 1846 the first Chinese drawing of a steamship[270] appeared in Cheng Fu-Kuang's 鄭復光 (1847) *Huo Lun Chhuan Thu Shuo* 火輪船圖說.

In 1853 the first steamship service to China started.

[263] See Pfister (1932), p. xxiii. In the second half of the 19th century this proportion rose to four-fifths. (*UNESCO History of Mankind*, vol. 6 (2), p. 1059.) It must also be remembered that translations from Japanese to Chinese with many neologisms were also being made.

[264] Pfister shows the emphasis placed in the first century of Jesuit activity in China on making Western knowledge and 'Truth' available to the Chinese in their own language. From 1675 onwards the emphasis was on informing the people of Europe about the thought and technology of the Chinese.

[265] *UNESCO* (1976), vol. 5 (3), p. 1271. [266] *UNESCO* (1969), vol. 4 (2), p. 906.

[267] Mao Tse-Tung (1954), vol. 2, p. 224.

[268] *SCC*, vol. 4, pt. 2, pp. 387 ff. [269] Couling (1917), sub. 'Steam vessels'.

[270] See *SCC*, vol. 4, pt. 2, p. 389.

In 1868 the first two steamships of substantial size built in China were launched from the Kiangnan Arsenal and the Fuchow Dockyard. From then till 1874, fifteen further vessels were built at the dockyard. Such was the accelerating pace of innovation.

A leading figure in the attempt to make good the years lost at the end of the Ming period, when China's sea-power was cut to the bone, was Lin Tse-Hsü 林則徐 (1785–1850), who, in addition to his heroic attempts to prevent the importation of opium to his country, which could not be wholly successful in the face of British sea-power, assembled a staff of translators who were able to sift Western periodicals and books for information on sea-power, ship-building, maritime defence, and other military matters,[271] and make it available to China. He also encouraged experimental work and the testing of guns and ships.[272] The followers of Lin Tse-Hsü[273] specialised in a range of engineering techniques which included gun-casting, optical instruments, steam-engines, steam-boats and steam-locomotives, the making of gunpowder, mines, bombs, fuses and shells, and more generally all aspects of gunnery, the positioning of defence batteries and the mounting of artillery within defences. It will be noted that Literary Chinese, reinforced by a growing technical vocabulary, was still in use as the language of science.

All through the 18th century an awareness of the need for government to be able to communicate with the less literate part of the population was growing. There was a solid foundation on which to build, for the administration of China was based on an assumption of a certain level of literacy even in the most rural places. Tax obligations, for example, were listed in detail on a single document, and this document was passed from household to household, with penalties for failing to pass it on. For this to be workable every village must have had a reasonable number of people literate to the level required for tax demands.[274] Information for peasants concerning agricultural techniques, famine foods and medicines was also communicated by posters, as well as changes in rules and regulations. Nor in times of village uproar and commotion would the inhabitants fail to have recourse to the written word with the posting on village walls of denunciatory poems.[275]

Wang Hui-Tsu 汪輝祖 (1731–1807), a competent administrator and critical historian, spoke out powerfully on behalf of simplified language 'in all written public notices, since people usually find it difficult to understand the written language and soon tire of trying to read longwinded and verbose documents'.[276] At such a time the

[271] *UNESCO* (1976), vol. 5 (3), pp. 1264–5.

[272] In this he placed China ahead of Turkey, which in the 15th century had led the world in the use of heavy artillery, but thereafter entered a period of decline which was not arrested till 1923 with the advent to power of Kemal Ataturk.

[273] Memoranda by such men occupy chapters 84 to 95 of the 1852 edition of the *Hai Kuo Thu Chih* 海國圖志.

[274] Rawski (1979), p. 14.

[275] Rawski (1979), p. 16, which recounts such an incident in 1844 in Kwangtung. As Christoph Harbsmeier once observed: 'It is healthy to reflect that literacy is no lower and was no lower in China than in comparably developed other nations enjoying the benefits of alphabetic scripts. The linguist Bernhard Karlgren was certainly right when he emphasised that ultimately the difficulty of learning Chinese characters was not in fact the most difficult thing about learning Chinese.'

[276] Rawski (1979), p. 15, and see Balazs (1965), p. 56.

Western conception of a newspaper had only to be planted for it to strike root. This occurred in 1827 with the publishing in Canton of the English language newspaper *The Canton Register*. The idea of publishing news was by no means an innovation to China, for government information had been circulated by means of a gazette similar in content to the *London Gazette* (which was first published in Oxford in 1685), but anticipating this by six and a half centuries or more.[277] The *Peking Gazette* (*Ching Pao* 京報 or *Ti chhao* 邸鈔) was issued daily except at the New Year Festival (unlike *The London Gazette*, which only came out twice a week), and announced Court movements, lists of promotions, Imperial Rescripts, Edicts and the like. Although of restricted circulation it was avidly read in the provinces, and the copying of it in abridged form and sale at a reduced price constituted a minor local industry.[278]

The European newspaper differed from the previous gazettes in its aim, which was to enlighten, and in its content, which came to include informative articles on science, moral questions and popular progress as well as news of current events. Given the aim of such newspapers, and the fact that they depended for their success on attracting a wide circle of readers, it was inevitable that the language in which they were expressed should have approximated more and more to that of the largest possible number of readers capable of paying for a copy.

It is not possible to trace here the development of foreign newspapers in China from the beginning of the *Canton Register* in 1827 to the end of the century when there were twenty-five or thirty, excluding periodicals, nor of newspapers in Chinese, which by the end of the century numbered seventy or more.[279] It is often difficult to say when a missionary circular letter becomes a newspaper, and when an English-language newspaper which begins by having occasional articles in Chinese, and then a column or a page, can be described as having become a Chinese-language newspaper. But for a firm date we may note that John Fryer, of whom more will be said before long, assumed the editorship of the *Chiao-hui hsin-pao* 教會新報 (Mission News) in Shanghai in 1866, the paper having been started in 1862.

Newspapers certainly helped to spread education, if not literacy, for being written in Literary Chinese they could be read in all the provinces, and not merely on the coast where they originated. In 1895 there were nineteen Chinese newspapers published in China, Hongkong and Macao, but only three years later the number had increased to over seventy, an indication of the rate at which the pressure for change was building up. By the beginning of the 20th century there were seventeen newspapers in Chinese in Shanghai, some of which were said to have circulations of 10,000 a day.[280] Experiments were also being made in the language in which they were expressed. About the year 1900 a small monthly periodical was started in Canton which used a Romanised form of Cantonese, and in 1902 one of the

[277] On the *London Gazette* first edited by Joseph Williamson, and to include Sir Joseph Steele among its later editors, see the *Harmsworth Encyclopaedia*, vol. 5, p. 3874. An indication of the need for an international language, now that Latin was no longer acceptable, is the fact that the *London Gazette* was also published in French until 1696.

[278] See Couling (1917), p. 429.

[279] Ball (1904), pp. 478 ff.

[280] An interesting survey of the state of the press in China as it was at the turn of the century is given by Ball (1904), p. 479.

Hongkong Chinese newspapers began printing occasional pieces in colloquial, that is to say, using logographs to represent words as used in the vernacular, and not in the style of Literary Chinese.[281] This, however, is to anticipate the great effort which was being made, following the lead set by Lin Tse-Hsü 林則徐, in promoting modern technology, translating Western and Japanese books, and equipping Literary Chinese with an up-to-date scientific vocabulary.

Lin Tse-Hsü had died in 1850. But a young man of great intelligence was now preparing to bring to bear a comparable dynamism for the modernisation of China. This was Li Hung-Chang 李鴻章 (1823–1901). In 1863, when he was Acting Governor of Kiangsu Province, he obtained authorisation for a school of translators to be set up in Shanghai, similar to one which had just been established in Peking.[282] Two years later he was able to set up China's first Western-type arsenal with the help of Dr (Sir Halliday) Macartney, an officer seconded to him from special service with the British Army.[283] This was at first simply a machine shop for the production of weapons, but later it was found that the workmen it employed needed training, and a mechanical school was therefore added in 1867. In 1868 authorisation was requested for the addition of a translation department. Li Hung-Chang had already left the scene, becoming Acting Governor General in Nanking in the summer of 1865. The project had, however, by this time acquired momentum of its own, but Tseng Kuo-Fan 曾國藩, who had been appointed Governor General of Kiangnan and Kiangsi Provinces in South China, in which fighting he was desperately engaged, nevertheless found time to secure the authorisation of the department from Peking. The translation department, which had since 1863 been functioning independently in Shanghai, was therefore transferred to the Arsenal complex, in effect from early in 1868, though not with full authorisation until November of that year. In Tseng Kuo-Fan's memorial applying for this adjustment to be made, he also suggested and secured permission to establish a training school for linguists.[284] Even before the proposals had been officially approved, a Dr John Fryer, formerly mission school teacher, and later, as just noted, editor of the missionary newspaper *Chiao-hui hsin-pao*, was appointed as translator at the Arsenal, which now included four departments, (1) the arsenal and machine shop (1865), (2) the mechanical training school (1867), (3) the transferred translation department (1868), and (4) the school for training linguists (1868).

The Chinese bureaucracy of this period is often accused of endless procrastination, corruption and incompetence. Nevertheless, two facts emerge from this affair. One

[281] Ball (1904), p. 479. [282] Bennett (1967), pp. 18 ff. [283] Hummel (1943), vol. 1, p. 465.

[284] Bennett (1967), pp. 18–20. On the early introduction of modern chemistry to China, and the work of the Kiangnan Arsenal we have already written in *SCC*, vol. 5, pt. 3, pp. 250 ff. It would be wrong to give the impression that the staff of the Kiangnan Arsenal confined their interest to translation in and from English. Marianne Bastid, personal communication, 21 December 1985, points out that the Foochow Arsenal was built with the help of French technicians in 1866 and that the [French] Navy Archive report (Service Historique de la Marine BB4 1555, Rapport du sous-ingénieur de division Thibaudin, juin 1868) refers to glossaries of Chinese translations for technical terms compiled by the French engineers and teachers. She continues: 'the French director, Prosper Giquel, who himself could speak and write Chinese rather well, thought it necessary to teach French to the Chinese staff, including workers'. She continues: 'Among the early workers at the Arsenal was a small group from Hong Kong and those may have contributed something also because they had worked in the small British dockyards which had been established there'.

is that when the government wanted to push something through quickly, it could do it. Tseng Kuo-Fan made his proposals in a memorial to the throne 'in the [early] autumn of 1868'.[285] It was approved by 1 November of the same year, i.e. in a matter of weeks. There are few government departments today which could deal with the establishing of a new institution with such dispatch!

The second fact is the high level of delegation and of mutual trust between functionaries enabling the Arsenal to engage and start paying their foreign translator six months before the project in which he was working had been officially approved.

Now began a second period of intense translating activity comparable to that introduced by the Jesuits two centuries before. The number of scientific works published by Fryer and his colleagues in approximately thirty years appears to be 174, with another twenty or more translated but not published and excluding books which were more general than scientific.[286] This compares very favourably with the grand total of approximately 340 books and sundry documents on all subjects, translated over a period of more than two centuries into Literary Chinese by the Jesuit Fathers whose total number with dates was shown by Pfister as 442 and whose strength at its peak rose to 113, though it must be remembered that the Jesuits' main object was evangelisation, that they were translational pioneers, and in their second phase were also translating Chinese books into European languages. This was brought to a splendid climax by the heroic Amiot when the mission was under fire from many quarters. Nevertheless the work of the Kiangnan Arsenal was of critical importance in enabling Literary Chinese to adjust to the demands put on it by modern science which was developing in Europe with explosive rapidity.

If the influence which these translations had on the minds of the reading public and on the future development of the Chinese language is to be assessed, they must be looked at from several angles. To begin with, Fryer himself was disappointed that the very Arsenal which had commissioned the work sometimes made no use of the product, which he found it 'difficult to account for'.[287] But it must be remembered that a good source book does not necessarily make a good textbook for teaching purposes. On the other hand, more than 30,000 copies of books published between 1870 and 1880 were sold from the Arsenal itself. Many more were sold from other book centres.[288] There was a spurt in sales after 1890, and through all this period there is the indirect testimony of pirated editions selling vigorously at reduced price to indicate a healthy demand.[289]

The influence of these books on some of the younger generation who were beginning to play a major role in the modernisation of China is suggested by the fact that Khang Yu-Wei 康有為 (1856–1928), for example, bought copies of all the works which the Arsenal had published. A few years later Liang Chhi-Chhao 梁啓超 (1873–1929) and his brother also bought many of the published works.[290] Than Ssu-Thung

[285] See p. 194 above. [286] Bennett (1967), Appendix II, pp. 82 ff.
[287] *Ibid.*, p. 41, n. 67. [288] *Ibid.*
[289] *Ibid.* 'Copied by photolithography in small characters and sold at absurdly small prices', as Fryer observed.
[290] *Ibid.*, pp. 42–3.

譚嗣同 (1865–98), the first martyr of the 1898 reform, was a voracious reader of translations of Western books, and was certainly familiar with some of those from the Arsenal. It is evident that at the very least the style of Khang Yu-Wei was influenced by what he had read to the extent of absorbing much of the terminology which had been evolved in the course of Arsenal translation work.[291]

The manner in which the translations were made was for the foreign book to be taken passage by passage. First the reader of the foreign book who was fluent in colloquial Chinese would translate the passage into the colloquial. If there were points in doubt they would next be settled. Then the Chinese translator would turn the colloquial translation into Literary Chinese, and this when polished would appear in the Chinese version of the book. The main difficulties do not appear to have been in the nature of the literary language itself, but in the need to invent terms for words and concepts which had not till then made their appearance in Chinese natural philosophy. The procedure was

(1) to find an existing term which could be used without distortion of meaning beyond what was acceptable. This practice was very much used in English in the early years of the Scientific Revolution, as when the word *stroke*, meaning a blow with a weapon, came to be used in the sense of the movement of a piston in the cylinder, or of a certain type of brain damage.

(2) If no existing term came to mind Fryer and his collaborators would find an archaic or obsolete character which could be pressed into service, and would give it a new and precise meaning. This may be compared with the botanists of Europe taking ancient Latin terms and giving them a new botanic precision, or chemists taking the ancient word *aether* which had at one time indicated a fifth element, and using it in the precise sense of $(C_2H_5)_2O$.

(3) Sometimes, when the concept was new, it was necessary to coin a new character. There are many of these among Chinese terms used in modern chemistry. 'Oxygen', *yang* 氧, is an example. This may be compared with the new inventions of Europe, such as 'gas' or 'laser'.

(4) Sometimes what was invented was not a new character, but a new descriptive term formed from much used and well-established characters, some of which had changed meaning in the course of history. This type of invention had a long tradition in China, as is illustrated in the term 'wind-chasing gun';[292] a similar procedure in English may be seen in the invention of such terms as 'crocodile clips' and 'butterfly-nuts'.

(5) Transliteration of foreign words. When describing the sinicisation of modern chemistry above[293] we spoke of 'the age-old dilemma confronting translators from

[291] *Ibid.*, p. 43. Nathan Sivin has pointed out, however, in his review of Bennett (1967) in *Isis*, **61** (1970), p. 281, that the place of the Kiangnan Translation Bureau's output in the overall translation effort in China has not in that work been properly evaluated, and suggests Wright (1957), for a well-documented overview, in addition to Chinese sources.

[292] This term, *ta chui feng chhiang* 大追風鎗, is translated as 'large blowing-away-the enemy lance-gun' in *SCC*, vol. 5, pt. 7, p. 455. Other similar names for weapons are: the 'gripped-lightning musket' (*chhe tien chhung* 掣電銃) and the 'fast thunder musket' (*hsün lei chhung* 迅雷銃). See *SCC*, vol. 5, pt. 7, pp. 442, 448, 455.

[293] *SCC*, vol. 5, pt. 3, p. 255.

alphabetical languages into Chinese. . . . Should one employ an already existing . . . technical term, and risk a fatal distortion of one's meaning? . . . Or should one transliterate the foreign polysyllabic term or name into a string of monosyllables in meaningless juxtaposition, then define the ugly compound resulting?' This dilemma was generally resolved by using one or other of the first four methods just described, but a few transliterations seem to have made themselves a permanent niche in the language, such as *a-ssu-pi-lin* 阿司比林 (aspirin), and *ko-lo-fang* 哥羅仿 (chloroform). In general Chinese finds it far easier to accept new descriptive terms in which the characters give some indication of the meaning, as in *wei-sheng-su* 維生素 (life-maintaining quintessence) for vitamin, rather than a meaningless transliteration as in the earlier term *wei-tha-ming* 維他命.

Although much stress has been laid on the work of John Fryer and his colleagues, progress in the modernisation of Chinese science, and of the language with which to express it, was gathering momentum in many different centres at the same time, leading to problems in the standardisation of terms which eventually had to be resolved. What is so impressive in the work of the Kiangnan Arsenal is its scope. One cannot make an accurate translation of a technical term unless one understands the science behind it. Omitting fields of special applications of the sciences, the topics and sciences covered by Fryer alone include the following: acoustics; agriculture; algebra; analysis; anatomy; arithmetic; armaments; astronomy; botany; calculus; chemistry; coast defence; coinage and minting; conic sections; dynamics; economics; electricity; engineering; fluxions; fortifications; geography; geology; geometry; gunnery; heat; hydraulics; hydrodynamics; hydrographical surveying; hydrostatics; hygiene; international law; light; magnetism; *materia medica*; mechanics; medical jurisprudence; mensuration; meteorology; mineralogy; mining; model drawing; naval architecture; naval manoeuvring; naval regulations; navigation; palaeontology; photography; physiology; pneumatics; political economy; probability; prospecting; survey; therapeutics; trigonometry; X-ray; zoology.

The number of technical terms which this extremely wide range of topics implies is positively daunting. Apart from the gifts of the translators, their success can perhaps be attributed to three factors: first, the urgency of the situation – China's survival as an independent nation depended on it; secondly, the very friendly relations which existed between the team of translators; thirdly, the fact that the central government in Peking was prepared to delegate responsibility. After they had made clear what they wanted to be done, and had set the guidelines, they didn't interfere – the ancient Taoist principle of *wu wei* 無為 – but left the translators to get on with the job, even to the point of leaving it to Fryer to decide which books he would translate. One can imagine how much would have been produced under a heavy-handed bureaucracy.

By the end of the century chemistry and the other sciences were being widely taught in China. Whatever shortcomings the Kiangnan Arsenal translations may have had, they had brought China to the point of take-off for modern science. The old language of Literary Chinese had proved itself extremely adaptable and the adaptation had been accomplished at astonishing speed. Many hundreds of new terms had been created, and as Nathan Sivin has said, 'their success can easily be verified by the number of these terms which have survived into the Chinese literature of today'.[294] Fryer himself testified to the adaptability of Literary Chinese, and few Western writers have been in a better position than he to appreciate it. As early as 1862 he declared that 'the Chinese language like other languages is capable of growth', and believed that the Chinese language presented no serious problems for purposes of translation.[295] What is particularly interesting is that he even went so far as to say, in comparing Chinese with Western languages: 'our systems have no more right to universal use than the Chinese. Their ancient and wonderful language which for some reasons is more suited to become the universal language of the world than any other, must not be tampered or trifled with . . .'[296]

The situation at the turn of the century was therefore, as we pointed out above in *SCC*, Volume 5, part 3 (p. 262), that Literary Chinese had proved itself well capable of adapting to the needs of modern scientific writing, and yet, by 1917 it was formally dead. Many of the reasons are clearly the same as in the West – the growth of the modern nation-state with its stress on universal literacy, the time and expense needed to acquire a classical education, the economic pressure on newspapers to be published in a simple language related to the vernacular of the common people, and so on, together with one disability which was not shared with Latin, namely that it was not comprehensible when read aloud. There could be no equivalent of a public orator at a university addressing his audience in Literary Chinese, whereas in the West this was possible in Latin, and had been done for centuries, persisting into the present age until the fiction that at least the majority of listeners could understand what was being said finally broke down.

It is not necessary to describe the transition from Literary Chinese to written *pai-hua* 白話 and spoken *phu-thung-hua* 普通話 since they are events of the 20th century, and fall outside the range of these volumes. It will suffice to note that following the establishment of the Shih-wu Hsüeh-thang 時物學堂 (the Academy of Current Events) by Than Ssu-Thung 潭嗣同 in Changsha in 1897 the *pai-hua* movement developed rapidly. The abolition of the Civil Service Triennial Examinations in 1905[297] sounded the death knell for Literary Chinese as the language of administration, and the Literary Revolution of 1917 formally established *pai-hua* in its place. Yet it would be an exaggeration to say that Literary Chinese is dead. It is still a vehicle for poetry, and for as long as it forms even a small part of Middle School curricula, the gap between *pai-hua* and Literary Chinese will not be unbridgeable for a determined student. It may have a normative influence in other ways, not least on the development of languages for science in the future, in a way that Latin can no longer hope to do.

[294] Quoted from Sivin's review of Bennett, in *Technology Review* (March 1970), pp. 17–18.

[295] Bennett (1967), p. 30. See also *SCC*, vol. 5, pt. 3, p. 262.

[296] Bennett (1967), p. 31, quoting Fryer's 'Scientific Terminology: Present Discrepancies and Means of Securing Uniformity', 1891 in *Miscellaneous Pamphlets* (n.d.), vol. 1, no. 1, p. 11.

[297] The exhaustion of the type of language required for the Triennial Examinations had an adverse effect on the general use of Literary Chinese comparable to the adverse effect on the use of Latin caused by its misuse at the hands of the alchemists.

(*f*) CONCLUSIONS

Joseph Needham

(1) Science and Civilisation in China

The seed from which this whole series of some thirty volumes of *Science and Civilisation in China* would eventually sprout, was sown in 1937 when my future chief collaborator Lu Gwei-Djen 魯桂珍 came to Cambridge, and, with her two fellow-biochemists, acted as evocator of the series. I was then a research biochemist, specialising in the study of embryonic life. It is interesting to observe, as Francis Bacon might have said, the affiliation of events and people. I had been secretary of the Cornford-McLaurin Fund, set up to help the relatives of those who had been killed in the International Brigade in the Spanish Civil War, and this indication of my sympathies it was which had led the Chinese to Cambridge. Lu Gwei-Djen felt that Cambridge was the place where the laboratory atmosphere would be truly agreeable, and Needham was the man, and so it happened that they all came to Cambridge.

Lu Gwei-Djen worked with my first wife, Dorothy Needham, on aspects of muscle biochemistry; while Shen Shih-Chang 沈詩章 worked with me on the ultra-micro-manometers (the 'divers') introduced by Londerstrom-Lang in Denmark a short time before. Wang Ying-Lai 王應來, on the other hand, worked with the discoverer of cytochrome, David Keilin, at the Molteno Institute. After two or three years they all separated, Lu Gwei-Djen and Shen Shih-Chang to America, where she worked first in California and then at the Columbia Medical Centre in New York, while he found a permanent post at Yale University. Wang Ying-Lai returned to China, and spent most of his life as Director of the Shanghai Institute of Biochemistry. Lu Gwei-Djen spent the whole of the war in America, returning to China only at the end of it, in time to become Professor of Nutritional Biochemistry at Chinling 金陵 College, Nanking, but she was then called out again to UNESCO in which secretariat she worked for nine years. In 1957 she came back to Cambridge and never left it up to her death in 1991, except for relatively short visits with us to China and Japan or for conferences. In 1987 my first wife died, and two years afterwards Lu Gwei-Djen and I got married. But let me return now to 1937.

It was from working with these young biochemists that I found that their minds were exactly like my own. That raised in an acute form the problem of why modern science had not originated in China. They had such an influence on me (much more than Cambridge ever had on them) that I began to learn their language, which in its written form is particularly difficult for Europeans because it is ideographic, or, as we now say, 'logographic', and not alphabetic. I did not know a single Chinese character before I was thirty-seven years of age, but eventually I became quite fluent and I used to say that reading a page of Chinese was like going for a swim on a very hot day, because it got you completely out of the alphabet. But it is one thing to learn Chinese as part of the Oriental Studies Tripos in Cambridge, and quite another thing to learn it as a labour of love, which was my case.

It was from my friendship with these young biochemists that I first began to suspect that Chinese civilisation had from very early times made immense contributions to medieval science and technology, which have since flowed into the ocean of modern science. This is the story told by Volumes 3 to 6. Volumes 1 and 2 are essentially introductory, and Volume 7 attempts to explain why it was that modern science first developed in Western Europe, and only there, sketching out in a series of contributions some of that vast complex of factors which helped and hindered the development of modern science in different parts of the world.

My first collaborator was Wang Ling 王鈴 (Wang Ching-Ning 王靜寧), whom I had originally met when he was a Junior Fellow of Academia Sinica's History and Philology Institute, then evacuated to Lichuang 李莊 in Szechuan. He was then working on the history of gunpowder, and that was what made Volume 5, part 7, a crown of his efforts. We began the actual writing of Volume 1 in 1948 and it was eventually published in 1954.

I would like also to add here the name of Lu Gwei-Djen who collaborated with me in the writing of Volume 4, part 3, and other parts of the work to be listed below, but who was so much more to me than a collaborator.

Then there are many others whose work does not actually appear in the final publications, but also who nevertheless contributed greatly with their ideas. Among these I would mention Derk Bodde and Janusz Chmielewski.

With the publication of my 'General Conclusions' at the end of this Volume 7, the enterprise that we call *Science and Civilisation in China* may be considered as virtually completed. Only a few gaps in the overall plan, dealing with specific aspects of technology, remain to be filled. These will be published as the manuscripts become available, but I doubt whether these would seriously change the Conclusions we have drawn from the survey that my collaborators and I embarked on more than fifty years ago.

At this point, our readers may well say, 'You have written more than 20 volumes on *Science and Civilisation in China*. What precisely do you mean by "China", by "civilisation" and by "science"?' This question deserves to be asked. When we speak of the history of China we may be referring to the history of Chinese civilisation during different periods but we may also be referring to the history of the Chinese geographic space.[1] Discussion of China, if it is to be precise, must define which parts of East Asia are to be considered as China at any particular time; these have varied enormously. Next one should be able to say what the criteria are by which the civilisation which we call Chinese can be recognised. This we have attempted to do, and throughout the successive volumes we have related Chinese civilisation to neighbouring civilisations where possible.

[1] This is well illustrated in Chang Kwang-Chih (1963), pp. 299 ff.

The civilisations of very early times can be differentiated by fairly small sets of characteristics, though no single factor alone, as Chang Kwang-Chih 張光直 has pointed out, makes a civilisation appear.[2] If we wish to bring a civilisation into focus with more detail we must narrow the time span and the geographical area, and include more characteristics. As an example of this we may cite the eleven traits or characteristics selected by Chang Kwang-Chih as typifying the Chinese neolithic culture tradition.[3] They are as follows:

1. The cultivation of millet, rice and *kao-liang* 高粱 (and possibly the soybean).
2. The domestication of pig, cattle, sheep, dog, chicken and possibly horse.
3. The *hang-thu* 夯土 (stamped earth) structures and the lime-plastered house floors.
4. The domestication of silkworms and the loom(?) – weaving of silk and hemp.
5. Possible use of tailored garments.
6. Pottery with cord-mat-basket designs.
7. Pottery tripods (especially *ting* 鼎 and *li* 鬲) and pottery steamers (*tseng* 甑 and *yen* 甗) and the possible use of chopsticks.
8. Semilunar and rectangular stone knives.
9. The great development of ceremonial vessels.
10. The elaborate complex of jade artefacts; a possible wood-carving complex.
11. Scapulimancy.

This early culture could be readily distinguished from others of that time which cultivated rice but not millet, which domesticated the pig but not the sheep, which wove cloth from hemp or flax but not from silk, and so on. Similarly it can be distinguished from modern Chinese civilisation in that the Chinese among other things no longer practise scapulimancy, and now make vessels of metal and various plastic substances as well as clay. China geographically is a fluid conception, and the civilisation to be found there is to be distinguished from others by sets of characteristics which have changed from century to century.

We may also look at the word 'science' as it is used generally and as it is used in our *Science and Civilisation in China* series. Here again it is important to refer the word 'science' to the particular period in which it was used.

In *Science and Civilisation in China* we have viewed science at three levels, first what we call proto-science, such as is found in ancient Babylonia; then medieval science, such as is found in China before about the year 1700 and in Europe before 1500 approximately. Thirdly we refer to modern or international science. I disagree with those who call modern science 'Western', for though it began in Western Europe it has long ceased to be exclusively Western. Indeed I have pictured modern science as being like an ocean into which the rivers from all the world's civilisations have poured their waters.

[2] *Ibid.*, p. 55.
[3] But also see *Ancient China's Technology and Science*, Institute of the History of Natural Science (1983).

In discussions on this subject we have concluded that the growth of science may now be viewed in the following five stages:

(1) The primitive stage, when a body of techniques was developed. This may be referred to as proto-science.[4]
(2) The medieval stage, when bodies of sophisticated techniques were differentiated and general principles formulated, often related to some all-embracing theory.[5]
(3) Renaissance science, when universal laws were believed to apply to nature, but individual sciences were not thought of as components of one general Science.[6]
(4) 19th-century 'modern science' when each individual science is thought to exemplify a system of universal laws, and that the principles and procedures of rigorous description and explanation in physics may in time be fitted to the entire variety of the more complex sciences from chemistry to linguistics. This Science is thought of as a Unity.
(5) 20th-century attempts to establish a canon of accepted Unity of Science. Growing realisation that the 'Laws of Nature' are conjectural descriptions of the structural properties of Nature (e.g., Popper (1972), p. 190); and appreciation that science is radically human, that man is part of the context in which the conjectural descriptions of the properties of Nature have been formulated, that all disciplined enquiries are affected by this relationship, and that science has been brought forth by our cultural development from a multiplicity of historical structures. The idea of the Unity of Science is perhaps one of our mental constructions.

I believe in making a sharp distinction between modern science and that which preceded it, namely ancient and medieval science. Modern science, to my mind, consists in two things, the mathematisation of hypotheses about Nature on the one

[4] In the area of mechanics, for example, the Mohists took the initial steps to help progress along this path. The *Mo Ching* 墨經 (−5th to the −4th century) mentions the theory of the centre of gravity in steelyards, stress and deformation of solids in wooden beams and buoyancy. From the *Mo Ching* onwards many tools and machines were developed using the principles of mechanical motion. The *Khao Kung Chi* 考工記, for example, explains how the number of feathers on an arrow could affect the speed and accuracy of the shot. This was more advanced than the Aristotelian school of physics which thought that all missiles flew in a straight line.

[5] Lindberg (1990), pp. 13–15, describes how Pierre Duhem (1861–1916) argued the case of the rehabilitation of medieval science. He believed that the development of science was a steady, continuous progression, from primitive beginnings to a mature description of nature. Other scholars also arguing for medieval science were Lynn Thorndike (1882–1965) and Marshall Clagett. Duhem's assertion of 'continuity between medieval and early modern science was a revolutionary event in the historiography of science'. In his research into medieval science, he encountered figures such as Jordanus de Nemore, Albert of Saxony, Themo Judei, Jean Buridan and Nicole Oresme. The two major opponents of this view were Edwin E. Burtt (1892–1989) and Alexandre Koyré (1892–1964). Burtt would date the period of scientific change to between 1500 and 1700, basing his opinion on the work of Copernicus, Kepler, Galileo, Descartes, Boyle, Newton and others less well known. For an account of the differences between Chinese and Western science and a study of the effects of the reaction to the Jesuits, see Gernet (1980).

[6] McMullin (1990) explains the differing conceptions of science held by different philosophers, among them Isaac Newton (1642–1727) and John Locke (1632–1704). Newton was very much against basing his theories on hypothesis since this did not supply completely concrete proof, but instead used inductive and deductive methods. On the other hand, Locke saw hypothesis as allowable, since he recognised that many of the effects in nature are invisible and therefore require conjecture, and that hypothesis should be tested against experiment. See McMullin (1990), pp. 67–76. The contribution of Goodman and Russell (1991) is particularly valuable here.

hand, and on the other continuous and relentless experimentation.[7] In my opinion, there is no science outside modern international science.

In conferences one sometimes hears it said that, for example, 'this would be impossible in Muslim science'. I cannot agree that in the modern world there is such a thing as Muslim science, Japanese science, American or Russian science except as an abuse of the word science, unless one means Muslim, Japanese, American or Russian *use* of modern science.

Since the time of our first volume, attitudes towards Chinese civilisation among thinking people have greatly changed. Not only have great changes come about in the way in which Chinese civilisation is viewed and appreciated by Western nations, but enormous improvements have been made in the methods of studying Chinese civilisation. New tools have been created – dictionaries, indexes and translations, often of previously untranslated works, and an array of machines in the fast-developing world of computer technology. Archaeology has revealed not only artefacts and monuments previously unknown (such as the sky-maps on the ceilings of Han tombs) but has solved some of the enigmas of Chinese history and has given back to the world portions of a literature lost or forgotten, but now recovered from inscriptions on bronze vessels or written in ink on wooden tablets, bamboo strips and burial silks. The tombs at Ma-Wang-Tui of –168 are outstanding examples of these things. Collections of books have been discovered, including many hitherto unknown, for example the *Wu-Shih-Erh Ping Fang* 五十二病方 (Recipes for the Fifty-two Diseases), and the great advantage here is that such books have not been subject to imposition of intellectual trends in the name of orthodoxy. Finds such as these from Chinese antiquity may well be compared with the Egyptian papyri recovered from Oxyrrhincus. Among the recent discoveries of Chinese archaeology, we must mention the complete sets of bells and stone chimes, something we were unable to take account of in the section on physical acoustics in Volume 4, part 1.[8] Archaeology too has revolutionised our knowledge of Chinese music, and correspondingly acoustic studies have been pressed into increasingly specialised fields. When we published Volume 4, part 1, the splendours of a full Thang orchestra as reconstructed by Lawrence Picken had not yet been heard.

Again, in a number of our volumes we have now been overtaken by events. Far more is now known about Chinese mathematics than could be set forth when Volume 3 was published. Blemishes and omissions in our work there are bound to be. But we may console ourselves with the thought that *Science and Civilisation in China* has provided a model for similar studies of other great civilisations.

Up to 1945 sinology was concerned almost exclusively with pre-imperial times, while in the 19th century pre-imperial studies were left for civil servants, missionaries

[7] For a brilliant account of the history of modern science in the West, see Robinson (1993), chapter 1. With regard to our list of Chinese inventions and discoveries, his words are similarly very appropriate when he describes the Chinese mechanical power source and its applications (p. 192).

[8] See Chhen Cheng-Yih (1994). No account could then be taken of this vital book which was not published until 1994. Another essential book on this subject is that of von Falkenhausen (1993).

and journalists. In the 1930s China was widely regarded as a country of warlords, literati, industrious peasants and skilful artisans, but with no science and precious little mathematics. Indeed this was the reaction of some Cambridge scientists with whom I discussed my plans, 'You won't find any science at all *there*', they said. China was regarded as a stagnant country, where little had changed for 1,000 years, and where the written language encouraged artistic vagueness, and was inimical to precision and abstract thought.[9] And yet, as we were later to show, the first of all escapements of the mechanical clock was essentially Chinese. From China too came many more inventions than we were able to show in our first volume, which we shall list before long.

We shall soon draw attention to those areas of discovery or to those items of invention which we have previously noted as being of Chinese origin, or occurring earlier in China than elsewhere. For example, it is not yet generally realised that the Chinese pioneered the place-value system with decimal notation as used in modern mathematics,[10] or that they were the first to solve the problem in horology of how to construct a wheel that would revolve slowly and regularly.[11] Recognition is now commonly accorded to the Chinese invention of gunpowder, though it is often vitiated by the misinformation that the Chinese, being peace-loving people, used it only for the making of fireworks. This, nevertheless, is an advance in knowledge since the time of Francis Bacon, who was aware that gunpowder, printing and the magnet, i.e. the mariner's compass, were all inventions new to Europe, though where they had come from he could not say. In *SCC*, Volume 1, p. 19, I quoted from his *Novum Organum*, book I, aphorism 129 as follows:

> It is well to observe the force and virtue and consequences of discoveries. These are to be seen nowhere more conspicuously than in those three which were unknown to the ancients, and of which the origin, though recent, is obscure and inglorious; namely, printing, gunpowder, and the magnet. For these three have changed the whole face and state of things throughout the world, the first in literature, the second in warfare, the third in navigation; whence have followed innumerable changes; insomuch that no empire, no sect, no star, seems to have exerted greater power and influence in human affairs than these mechanical discoveries.

One would expect that after some forty years or more, some of the inventions and techniques which originated in China and were transmitted to the West as shown in

[9] It is hardly conceivable that anyone today should write a book entitled *Why China Has No Science* as Feng Yu-Lan 馮友蘭 did in 1922. But, this view survives to the modern day. See Chhien Wen-Yüan (1985). For an interesting comparison between this book and my own, see Gutmann (1992). A remarkable book by Kang Teng (1993) came out while this was in the press. He took the agriculturalist writers of *Nung-Shu* 農書 and shows that they were all very forward looking, giving important details about the roller mills and water mills. With regard to the question of why China in modern times fell behind Europe in terms of technology, he gives reasons such as the continuous uprisings and foreign wars as well as natural disasters such as flooding (pp. 168, 170). He argues the effect that these situations had on agricultural development in China. On the other hand, he also argues the strengths within the agricultural system in China, that is, the connection between the literati and the non-literati (p. 173).

[10] See George Gheverghese Joseph (1991). This is an interesting book on the non-European contributions to mathematics. Up to the 16th century, he instances (1) place value (the modern method); (2) magic squares (numbers which add up to the same amount whichever way you take them); (3) estimation of π; (4) solution of higher order equations and Pascal's triangle; and (5) indeterminate analysis as being of Chinese origin.

[11] In connection with horology, another example of the Chinese interest in this area is their incense seal clocks which were trails of powdered incense (in the form of a character perhaps) with markers at set intervals which thus marked the time it took for the incense to burn and therefore acted as a clock. See Bedini (1994) for a detailed account.

Table 8 of our Volume 1 would by now be recognised as part of China's contribution to world civilisation. Paper today is quite widely known to be of Chinese invention, though few are aware that our first wall papers came from China, or toilet papers where, Chhien Tshun-Hsün 錢存訓 (Tsien Tsuen-Hsuin) intriguingly informs us, 15,000 sheets 'thick but soft and perfumed' were manufactured in the year 1393 for the imperial family.[12]

But how many engineers are aware that Europe followed China in the use of interconversion of rotary to longitudinal reciprocating motion by the combination of eccentric, connecting-rod and piston-rod, and the opposite?[13] How many gardeners are aware that the first wheelbarrow was made in China, and that the design most commonly used in Europe is the least efficient of all, since much of the weight of the load is carried on the gardener's arms? The later Chinese designs carried all the weight on the wheel. How many people who take their pleasure in boats or on horses are aware that the centrally mounted rudder or the stirrups, on which their pleasure largely depends, were both developed in China so many centuries ago? How many are aware that modern astronomy is based on the adoption from Chinese astronomy of the system used in that country of polar equatorial co-ordinates or are aware of China's many contributions to mathematics listed above in Volume 3, pp. 146–7?

Indifference to the sources of our modern civilisation is likely to persist until the history of science becomes a regular school subject. Yet no subject has a better chance of cutting through national boundaries and giving children an international outlook. The Royal Society, founded by Charles II in 1660, set a splendid example in diffusing information. Though this was a period when the very popular 'books of secrets' came out, the Royal Society set its face against everything secretive – secrets of Nature or otherwise. Today, thanks to media programmes, very few such secrets remain unexplained. See Eamon (1994).

Many former beliefs about China have been swept away or have been drastically revised by our volumes, which, we believe, make clear the reason for this revision. But, it would indeed be surprising if half a century after the work was started, some of the opinions which we ourselves then held did not have to be modified.

In China also, many changes have been brought about, such as the simplification of characters and the use of *phin-yin* 拼音 (*pinyin*) as the national romanisation script. We, however, have continued to use the modified Wade-Giles system for the sake of consistency.

Our younger readers may find that other idiosyncrasies convey something of the atmosphere of yester-year. In 1938, Nanking 南京 (Nanjing) was the capital of China, and Peiphing 北平 (city of Northern Peace) was a northern provincial town. When, in 1945, it became the capital again, it was possible for us to refer to it as Peking, though it is now commonly called, in accordance with the way it is spoken, Beijing.

[12] *SCC*, vol. 5, pt. 1, p. 123.

[13] The interconversion of rotary and longitudinal motion may be seen in all the internal combustion engines of the present day, where the travel of the piston-rod is proportional to the distance between the eccentrics on the main wheel. The Chinese used it only for a water-wheel working on laundry, but in Europe the piston was moved by steam or vapour at the other end. And this system has continued until the present day. For details of this interconversion, see *SCC*, vol. 4, pt. 2, pp. 380 ff., 'Reciprocating Motion and the Steam Engine's Lineage'.

Peking was the correct name within the International Postal Service of China at the time. We continue to use the spelling Peking for the sake of consistency, just as anthropologists refer to Peking Man and not to Beijing Man.

If a single word was to be sought to describe the guiding thread which has run through all the volumes, I would be inclined to use the word 'justice'. When I started writing, justice was not being done in the West to the other great civilisations. The self-satisfaction of Europeans seemed to me more and more suffocating, and I should like to end this introduction to my 'Conclusion' with some words which I found in Polydore Virgil's *De Rerum Inventoribus* of 1512, words which I used at the beginning of Volume 4, part 2:

> And seeing that the Arts and Crafts with other like Feats, whose inventours be contained in this book, are in this Realm of England occupied and daily put in exercise to the profit of many and the use of all men, it were in mine opinion both a point of detestable unkindnesse and a part of extream inhumanity to defraud them of their praise and perpetual memory that were Authors of so great Benefits to the universal World.

Having reflected briefly on the small revolution which has taken place in Western thinking about China, let us now consider the tremendous revolutions which have occurred in Chinese thinking since the high days of Lord Macartney's embassy to Peking.

The political revolution of 1911 was only one of several occurring simultaneously in that tumultuous age. There was a revolution in thinking, in which the old philosophies had to adopt Western values and make room for modern science. There was the revolution in education which this necessitated and the abandonment of the traditional examinations for recruitment to the civil service, and the switch from Literary Chinese to written colloquial, for which see above [*SCC* 7.2.7.6 p. 27].[14] There was a revolution in the means of transport and the media of communication facilitating the rapid flow of ideas. There was also a change of attitude among the common people, now assiduously cultivated by those in authority, who wish to substitute scientific thinking for traditional patterns of thought. Age-old beliefs were subjected to observation and testing. Ancestral graveyards were destroyed, especially after the communist revolution of 1949, and village laboratories were established in which the local people could test their soils and conduct experiments relevant to their interests.

At the other end of the scale, we may notice the striking progress made in China in many modern sciences. Among the Nobel Prize winners, to take one example, there have been four Chinese; of these Li Cheng-Tao 李政道 (T. D. Lee) and Yang Chen-Ning 楊振寧 (C. N. Yang) were given their awards as long ago as 1957; Ting Chao-Chung 丁肇中 (Samuel C. C. Ting), and Li Yüan-Che 李遠哲 (Lee Yüan-Tseh) in the 1980s.[15] The first three were for physics and Lee Yuan Tseh was for chemistry.

[14] A foretaste of this work may be found in *Comparative Criticism*, **13**, pp. 17–30, 'The Decline of the Classical Languages in East & West', in Needham & Robinson (1991).

[15] Since the death of Needham in 1995 there has been a further Chinese Nobel Prize winner in the person of Chu Ti-Wen 朱棣文 in 1997, Physics, [Ed.]

T. D. Lee and C. N. Yang were from mainland China, while the other two came from Taiwan. Nor with such a history as was described in Volume 5, part 7, is it surprising that China has advanced so far and so fast in the field of rocketry and space exploration, as indicated by the launching of the Silkworm rocket and the placing of satellites in space around the earth. Indeed I ventured to say in that volume that the rocket may have been the greatest single invention ever made by man, because, if the sun cools or over-heats, and we have to go somewhere else in the universe, the only vehicle capable of travelling in outer space is in fact the rocket.

Nor has there been any loss of the momentum generated by centuries of Chinese scholarly experimenters seeking to improve the agriculture practised by their rural populations, and this has expressed itself particularly in the field of bio-chemistry. First came the total synthesis of a protein, insulin, in the 1960s, and then the synthesis of a nucleic acid, t-RNA, in the 1980s. Again in medicine, which began to be studied so early in China, the country which gave the world so many useful drugs and medicines, such as ephedrine and ginseng, and ideas, including variolation, the precursor of vaccination,[16] we find that Chuang Hsiao-Hui in the late 1930s made the interesting experiment of implanting pieces of adult organs into isolated ectodermal balls and finding that the types of induction produced by these implants were significantly different, for example, neural tubes, eye-cups, elongated notochords, nasal grooves, etc., and that the early research of Nieh Yen-Fu of Shantung (Shandong) University in developing nitrogen-fixing bacteria for cereals is being continued at the University of Sydney by A. M. Zeman, Chhan Yao-Tsheng (Tchan Yao-Tseng) and Ivan R. Kennedy. But the fusion point between modern-Western and Chinese-traditional medicine has not yet been reached. Today China is subjecting its traditional medicine to the rigours of modern scientific practice, not only for its own benefit but for other countries as well. One important point to be made is that though modern science was invented in Western Europe alone, due to a remarkable conjuncture of circumstances so well described by Immanuel Wallerstein (1992), yet once it had been invented, every country in the world was free to use it and has done so. But how successfully it could be used on a national scale has depended to a large extent on the quality of a nation's infrastructure. The dismantling of over-ambitious projects, and disasters such as that of Chernobyl, show that science cannot be divorced from problems of society. Yet every nation can now contribute something to the international pooling of ideas which is essential to the healthy development of modern science. Every nation can add to the list of great names which have created that development.

Let me now return to page 3 of Volume 1 of *Science and Civilisation in China*, which is the first page of the Preface as it was published in 1954:

What exactly did the Chinese contribute, in the various historical periods, ancient and medieval, to the development of Science, Scientific Thought and Technology? The question can still be asked for later periods, though after the coming of the Jesuits to Peking in

[16] See *SCC*, vol. 1, p. 135, and especially *SCC*, vol. 5, pt. 5, 'The Enchymoma in the Test-Tube; Medieval Preparations of Urinary Steroids and Protein Hormones', pp. 301 ff.

the early 17th century, Chinese science gradually fused into the universality of modern science. Why should the science of China have remained, broadly speaking, on a level continuously empirical, and restricted to theories of primitive or medieval type? How, if this was so, did the Chinese succeed in fore-stalling in many important matters the scientific and technical discoveries of the *dramatis personae* of the celebrated 'Greek miracle', in keeping pace with the Arabs (who had all the treasures of the ancient western world at their disposal), and in maintaining, between the 3rd and the 13th centuries, a level of scientific knowledge unapproached in the West? How could it have been that the weakness of China in theory and geometrical systematisation did not prevent the emergence of technological discoveries and inventions often far in advance (as we shall have little difficulty in showing) of contemporary Europe, especially up to the 15th century (see Vol. 1, Table 8, p. 242)? What were the inhibiting factors in Chinese civilisation which prevented a rise of modern science in Asia analogous to that which took place in Europe from the 16th century onwards, and which proved one of the basic factors in the moulding of modern world order? What, on the other hand, were the factors in Chinese society which were more favourable to the application of science in early times than Hellenistic or European medieval society? Lastly, how was it that Chinese backwardness in scientific theory co-existed with the growth of an organic philosophy of Nature, interpreted in many differing forms by different schools, but closely resembling that which modern science has been forced to adopt after three centuries of mechanical materialism?

These are some of the questions which the present work attempts to discuss. They have already been discussed, not only in our own pages, but frequently elsewhere. Although there are without doubt factors in any civilisation which inhibit certain developments, the idea of modern science being inhibited from its natural development which would otherwise have been inevitable, is not one which we would wish to defend today. The emphasis is now rather on why modern science should have arisen in Western Europe only, and not in Sri Lanka, India or Japan, all of which gave sets of factors differing from those of medieval China, and sometimes more closely resembling those of Western Europe.[17] In other words what was it that made Western Europe unique? This is precisely the question which Immanuel Wallerstein (1992) raised and answered, when he wrote of the rise of capitalism in Western Europe alone:

> In other words, all other known systems have 'contained' capitalist tendencies, in both senses of the word contain. They have had these tendencies; they have effectively constrained them. If so, the question then becomes what broke down in the historical system located in Western Europe such that the containment barrier was overwhelmed? This pushes us in the direction of exceptional circumstances, a rare coming together of processes, or what was referred to previously as a conjunctural explanation.

He then goes on to discuss four elements each formulated as a collapse – collapse of the seigniors, the States, the Church and the Mongol Empire – and to investigate

[17] Li Chih-Tsao 李之藻 (*d.* 1630) presented a memorial in 1613 in which he listed fourteen discoveries of Western science which had not been discussed in the writings of the ancient worthies. See Hummel (1943), vol. 1, pp. 452–3. Unfortunately, we have no idea what they were, but it certainly shows that the Chinese were interested in the inventions and discoveries of other civilisations.

the effect of the cumulative collapses. Out of this collapse arose the bourgeoisie, and with the bourgeoisie arose modern capitalism hand in glove with modern science.

In the Middle Ages, China was an Asian, 'feudal-bureaucratic' society, while Europe had military-aristocratic feudalism; much of what we are seeking must lie in this contrast. As we show in this volume, I think that a lot depended on the 'rise of the bourgeoisie', something that happened in no other civilisation in the world, neither India, nor South-East Asia, nor China. The new bourgeoisie was predisposed to look with favour on experimentation and its results, for exact knowledge meant greater profit. In the West, military-aristocratic feudalism was replaced by the bourgeois merchants. In China, on the other hand, this did not happen. Bureaucrats continued to operate as before, being inimical to anything which would change the age-old pattern of life, and opposing that which was fundamentally new.[18] Of course, we cannot explain the rise of modern science purely in terms of political revolutions. Other factors need to be considered, such as the invention of moveable-type printing.

Let me abbreviate slightly what I say in 'Science and Society in East and West':

> The study of great civilisations in which *modern* science and technology did not spontaneously develop obviously tends to raise the causal problem of how modern science did come into being at the European end of the Old World, and it does so in acute form. Indeed, the more brilliant the achievements of the ancient and medieval Asian civilisations turn out to have been, the more discomforting the problem becomes. During the past fifty years historians of science in western countries have tended to reject the sociological theories of the origin of modern science which had a considerable innings earlier in this century . . .
>
> However, the study of other civilisations places traditional historical thought in serious intellectual difficulty. For the most obvious and necessary kind of explanation which it demands is one which would demonstrate the fundamental difference in social and economic structure and mutability between Europe on the one hand and the great Asian civilisations on the other, differences which would account not only for the development of modern science in Europe alone, but also of capitalism in Europe alone, together with its typical accompaniments of Protestantism, nationalism, etc. not paralleled in any other part of the globe. They must in no way neglect the importance of a multitude of factors in the realm of ideas – language and logic, religion and philosophy, theology, music, humanitarianism, attitudes to time and change – but they will be most deeply concerned with the analysis of the society in question, its patterns, its urges, its needs, its transformations . . .
>
> If you reject the validity or even the relevance of sociological accounts of the 'scientific revolution' of the late Renaissance, which brought modern science into being, if you renounce them as too revolutionary for that revolution, and if at the same time you wish to explain why Europeans were able to do what Chinese and Indians were not, then you are driven back upon an inescapable dilemma. One of its horns is called pure chance, the other is racialism however disguised.

Neither of these alternatives is acceptable today. Chance declares the bankruptcy of history. Racialism would urge the possession by one particular group of peoples

[18] This chimes in well with the attitudes of Confucius to innovators. See Raphals (1992), pp. 53 ff. for a comparison of Mohist and Confucian attitudes to innovation.

of some intrinsic superiority to all other groups of peoples. Today this carries no conviction. I confidently anticipate, therefore, a great revival of interest in the relations of science and society during crucial European centuries, as well as a study ever more intense of the social structures of all the civilisations, and the delineation of how they differed in glory, one from another.

In sum, I believe that the analysable differences in social and economic patterns between China and Western Europe will in the end illuminate, as far as anything can ever throw light on it, both the earlier predominance of Chinese science and technology and also the later rise of modern science in Europe alone. Let us now, therefore, consider how it was that *China* got so far towards modern science and with what handicaps.

Let us first list some of the handicaps to the development of modern science which are found in the early and medieval civilisation of China. Of the greatest importance of all, perhaps, is the fact that the Chinese did not develop as far as the Greeks the idea of geometrical proof.[19] One reason why they emphasised algebraic mathematics and did not develop geometry may well have been that in China a circle was subdivided not into 360° as with the ancient Babylonians and through them Western Asia and Europe, but as with the number of days in the year, into 365.25°, a most inconvenient number. Chinese civilisation developed without trigonometry. And yet, at the technological level they managed very well by measuring the sides of triangles rather than angles.[20]

It is believed that geometric demonstration and formal proof grew up in Greece because of the public nature of Greek city life,[21] which was continued in the later city-states of Italy and elsewhere. This also was lacking in China where affairs were managed not by democratic assemblies but by the bureaucracy.

The development of modern science was undoubtedly assisted by the interest of the bourgeoisie in precise measurement and other commercial skills. The bourgeoisie itself was assisted into political existence by the long tradition of self-governing city-states. Neither of these political organisations were known in China.

Let me say again: 'All historians, no matter what their theoretical inclinations and prejudices, are necessarily constrained to admit that the rise of modern science occurred *pari passu* with the Renaissance, the Reformation and the rise of capitalism'.

Out of an unlikely military-aristocratic European milieu *modern* natural science could and did arise. When the merchants began to come out of their city-states in

[19] See the three-volume work of Crombie (1994) for an account of how Greek thinking on causality and proof, transmitted to European countries, led to the Scientific Revolution. The absence of this in China did not stop inventions and discoveries of every kind as will be shown below. Fang (1994) is of the opinion that it was China's lack of geometry which was the main reason for its lack of development of modern science. He shows how democratic discussion went on in the city-states of Greece, leading to the acceptance of geometrical proofs (see Fang (1994), pp. 56–61). Tyrants were not willing to discuss such things. It is impossible to imagine Chhin Shih Huang Ti 秦始皇帝 taking part in a discussion on natural phenomena in an age when the Greeks in the market place were shouting, 'He has proved it! He has proved it!' – after Thales had shown the equality of the angles at the base of an isosceles triangle using geometrical proof. For a perspicacious account of Greek proof, see Lloyd (1991), p. 292.

[20] Perhaps the Chinese developed as far as they did because they were so numerate. See Crump (1990), p. 40, for a table of stems and branches.

[21] See Fang (1994a) for an overview of Greek mathematics and the social setting of the city-states in Greece.

the 16th century,[22] first mercantile and then industrial capitalism arose, and *modern* natural science with it, in the time of Galileo and Torricelli.[23] This was the 'rise of the bourgeoisie', and though other factors were involved, such as the Protestant Reformation, it was this above all which happened in Western Europe, and in Western Europe alone. The point of view which has been adopted throughout these volumes is that modern science was that form of science at which the ancient and medieval sciences of all the countries of the world were aiming, but Europe alone was able to get there. Here the background of Greek logic and *mathesis universalis* was also important.

A good deal of work remains to be done on the exact nature of the tie-up between modern science and nascent capitalism. I have always pictured it as beginning with the exact specification of materials. If a merchant purchased a large quantity of oil from a Greek island he would need to know not only what its normal use was, but what it could also conceivably be used for; he would want to know its surface tension, its specific gravity, its refractive index, indeed all its properties, before he could decide who to sell it to. This would have been in the time of mercantile capitalism; in industrial capitalism there is less difficulty in imagining how intimately connected with it were science and technology. The accurate description of materials would have generated accuracy everywhere else, even in subjects like astronomy where there was no possibility of experimentation. And with exactness came the possibility of mathematisation. *Modern* science has been defined elsewhere[24] as the mathematisation of hypotheses about Nature, and the testing of them rigorously by persistent experimentation. Experiment was something rather new; the Greeks had done relatively little of it, and although the Chinese had been well acquainted with it, their purposes were primarily practical. Only the European Renaissance found out how to test mathematised hypotheses about Nature by relentless experimentation, and so to 'discover the best method of discovery'.

Of course military-aristocratic feudalism existed in other parts of the world besides Europe. I remember thinking when in Japan in 1986 how strange it was that modern science had not originated there as well. But then I reflected that the Japanese had not the tradition of the Greek city-state, which was so important for Europe. Athens gave rise, when the Renaissance came, to Venice and Genoa, to Pisa and Florence, and these in their turn to Rotterdam and Amsterdam, the cities of the Hanseatic League and finally London. In these cities, protected by their Lord Mayor or Burgomaster and their Aldermen, the merchants could shelter from interference by the feudal nobility of the surrounding countryside, until the day when they should come forth, and, after lending money to kings, princes and nobles, run the whole show.

[22] Goodman & Russell (1991), in the conclusion to their book, show how the Europeans developed modern science and explain the role of the city-states and merchants in the rise of modern science. See pp. 415–23. This book gives us the idea that Chinese science stimulated Western science when the time was ripe.

[23] The role of the old Chinese discovery of gunpowder in the dissolution of the city-states, and the role of the larger states in the sponsorship of the new science are both admirably explained by Dorn (1991), pp. 131–5. For a discussion of Hankow in recent centuries and its trade unions, see Rowe (1984), (1985) and (1989).

[24] See 'Science and Society in East and West', p. 1, n. 2.

As we said in our Foreword, page xlvii, it is worth while taking a look at the idea of the town or city in China compared with Europe. In China the town was simply a node in the administrative network,[25] held for the emperor by the civil governor, and (several bureaucratic ranks lower down) the military commander. It was the centre of the network of outlying villages, the people of which came in to market in the city. Compare with this an early 17th-century European painting, a group portrait of the 'Militia Company of Captain Frans Banning Cocq', known as 'The Night Watch', immensely proud of the city they were pledged to defend. Cities in Europe were really states within states, ready in the course of time to provide governments as alternatives (however much it might be glossed over in practice) to the medieval feudal-style governments which had preceded them.

China may have been the prime example of 'bureaucratic feudalism' but virtually all the other non-European parts of the world such as India, the South-East Asian countries and the whole Arab world may be said to have participated in it to some extent.[26]

In a word, if Chinese science, technology and medicine is to be understood, it must be related to the characteristics of Chinese civilisation. This is the point of Volume 7. Elsewhere we have explained how the bureaucratic ethos began by powerfully aiding Chinese science,[27] while only in the later stages did it inhibit any move towards modern science.[28] The examples of inventions we have shown illustrate how this came about. Such, at any rate, is our interpretation of the comparative developments in China and Europe.[29]

Yet in spite of these handicaps to the growth of science in China, scientific ideas comparable to those of medieval and Renaissance Europe nevertheless developed and technology flourished. Just how flourishing was technology in China and how wide ranging over the whole field of science was the Chinese genius for invention, may be seen from the following pages in which Chinese inventions, ideas and discoveries are listed in chronological order. This is a tentative list arrived at by an intense culling of our published volumes with some additions from our as yet unpublished

[25] This point is developed in more detail above, 'Science and Society in East and West', p. 8.

[26] See Wittfogel (1957) for details on how 'Oriental Despotism' was a phrase thought up by Karl Marx to designate those countries of East Asia which did not depend on rainfall for agriculture, but on hydraulic canals cleverly situated, i.e. the Asiatic Mode of Production. Wittfogel brought out well the connection between this and the bureaucratic government in China, Assyria, Babylonia and the like. He replies to his critics such as Toynbee (1958) in Wittfogel (1958).

[27] It must not be thought that the bureaucracy continued in power right until the end of the Chhing (1911). Rankin (1986) has shown that the commercialisation of the scholar-gentry had enormous consequences. For our purpose then, bureaucracy remained unchanged in the Ming and early Chhing state, but this was enough to spoil the origins of modern science in China. See Rankin (1986). Skinner (1977) points out the great importance of the city guilds' native-place associations which kept close touch with the bureaucracy and the city's élite in the places where they settled. These were important in the running of the city, setting up fire-brigades, hospitals and the like.

[28] See 'Science and Society in East and West', pp. 14, 17–18.

[29] Dorn (1991), pp. 157 ff., has a very interesting chapter which shows how close the Mormon community in the Salt Lake valley desert came to the Asiatic Mode of Production. They appointed their bishops as watermasters and formed a community whose job it was to open the sluice gates at need. The bureaucratic foundations of this and the setting up of a provisional government in 1849, with a society based on hydraulic agriculture, became more and more like the Asiatic Mode of Production.

volumes. We expect that the list will lengthen as we receive word from our friends and collaborators all over the world who have been invited and indeed have already begun to submit information on items about which they have special knowledge, and also, of course, to correct any errors in our present list.

In 1954 we were able to list no more than thirty-five mechanical and other techniques which had made their way from China to the West. (See Table III.) The number now becomes much greater if we include the many inventions of which we have become aware since 1954, some of which never left China, or which, like Su Sung's 蘇頌 giant astronomical clock,[30] developed no further and were forgotten. The number is also greater if we include not only mechanical and other techniques, but also ideas and theoretical practices.

It has sometimes been said that Needham 'tries to find a Chinese origin for everything'. This is not true. We have always tried to cast a balance sheet of giving and receiving. For example, on p. 240 of Volume 1, the subsection entitled 'The Westward Flow of Techniques' begins nevertheless by listing the techniques diffused in all directions from ancient Mesopotamia, Egypt and the Mediterranean Europeans. Then the Chinese techniques listed in Volume 1, Table 8 are given, and these are followed in Table 9 by mechanical techniques transmitted from the West to China, admittedly believed to be only four in number when Volume 1 was printed, up to the arrival of the Jesuits, after which there was a profusion of Western developments.[31] Nor have we failed to point out occasions when China could with advantage have accepted what the West had to offer, but did not do so either from lack of interest, or because it was too un-Chinese to be assimilable.[32] For example, the Jesuits were well able to raise the standard of glass-making in China by translating one of the books

[30] I can imagine one man saying to another, 'In the East people have invented a wheel that goes round in company with the stars, man's primary clock'. So the Europeans made their own escapement, the verge and foliot, instead of the waterwheel linkwork escapement. But it was characterised by a falling weight, sand running out of a holder, while the waterwheel was the origin of the escapement in China.

[31] Victor Mair, Professor of Chinese at the University of Pennsylvania, and Wang Ping-Hua, a Chinese archaeologist, have drawn attention to early European influences in China. They began searching in the north-east corner of Sinkiang province (Chinese Turkestan), between the Tianshan and the Taklimakan Desert on the edge of the Gobi Desert, for evidence of a Caucasian community dwelling there around −2000. Wang Ping-Hua excavated 113 bodies that were in such good condition that it was possible to see that they were North Europeans. In all, there were three burial sites. The technology of the cartwheels discovered, as well as the weave and pattern of the textiles, closely resemble that of people existing in the Ulevanian and Russian steppes about −2000. See Lowther (1994) for details. It is possible that they spoke a language related to Tocharian, the eastern-most branch of the European language family. Tocharian is now spoken in some parts of Germany and Scandinavia, the suggested place of origin of the people found in China.

It is not clear what the technology of the Caucasian people's wheels amounted to exactly. It cannot be dishing because that requires two wheels. We should like to know the origin of the Bronze Age in China. Iron was replacing bronze in about the −7th century. The ugly metal (iron) was replacing the lovely metal (bronze), though for a long while the two were used in conjunction. High tin bronzes for mirrors had already replaced iron by the end of the −11th century. It is possible that they brought it to the notice of the Chinese.

[32] In this connection, it is very interesting that the Chhien-Lung 乾隆 Emperor should have written to George III after he had been sent numerous elaborate astronomical apparatus and the like, that 'we have never valued ingenious articles, nor do we have the slightest need of your country's manufactures'. See Cranmer-Byng (1958), p. 137, for details. Alvares (1980) has shown that Chinese silk fibres were hundreds of metres long and that the Chinese never lost the art of making silk garments from them. Nor on the other hand did the Indians lose the art of weaving cotton of very short fibre length which needed ginning. The gin rollers running in contrary directions were connected by worm gearing, the oldest such examples in the world.

Table III. *Transmission of mechanical and other techniques from China to the West*

		Approximate lag in centuries
(a)	Square-pallet chain-pump	15
(b)	Edge-runner mill	13
	Edge-runner mill with application of water-power	9
(c)	Metallurgical blowing-engines, water-power	11
(d)	Rotary fan and rotary winnowing machine	14
(e)	Piston-bellows	*c.* 14
(f)	Draw-loom	4
(g)	Silk-handling machinery (a form of flyer for laying thread evenly on reels appears in the +11th century, and water-power is applied to spinning mills in the +14th)	3–13
(h)	Wheelbarrow	9–10
(i)	Sailing-carriage	11
(j)	Wagon-mill	12
(k)	Efficient harness for draught-animals: Breast-strap (position)	8
	Collar	6
(l)	Crossbow (as an individual arm)	13
(m)	Kite	*c.* 12
(n)	Helicopter top (spun by cord)	14
	Zeotrope (moved by ascending hot-air current)	*c.* 12
(o)	Deep drilling	11
(p)	Cast iron	10–12
(q)	'Cardan' suspension	8–9
(r)	Segmental arch bridge	7
(s)	Iron-chain suspension-bridge	10–13
(t)	Canal lock-gates	7–17
(u)	Nautical construction principles	>10
(v)	Stern-post rudder	*c.* 4
(w)	Gunpowder	5–6
	Gunpowder used as a war technique	4
(x)	Magnetic compass (lodestone spoon)	11
	Magnetic compass with needle	4
	Magnetic compass used for navigation	2
(y)	Paper	10
	Printing (block)	6
	Printing (moveable type)	4
	Printing (metal moveable type)	1
(z)	Porcelain	11–13

they had brought with them, entitled *L'Arte Vetraria distinta in libri sette* by Antonio Neri (1612). But for this there was no demand.

I have worked through all the published volumes of *Science and Civilisation in China* and through some of the volumes which at the time of writing are still in typescript, and find that my list has expanded from some thirty-five items to more than 250.[33] The mere fact of seeing them listed brings home to one the astonishing inventiveness of the Chinese people. They are printed here in alphabetical order for ease of reference. We hope to list them in a subsequent publication in chronological order, with further data.

This may well provoke further reflection. Our list begins from approximately the year −1500 by which time writing was well established in the Chinese culture area and continues down to the year +1700 by which time the Chinese medieval sciences were beginning to fuse with the new natural science which had originated in Western Europe.

Perhaps we should first list the outstanding inventions which had been shaping the destiny of the human race for the previous 30,000 years.[34] We must stress that definitions have changed, for example, the word 'Neolithic': less emphasis is now placed on technology and more on the economic aspect, as in 'self-sufficiency' or 'control of the environment'. 'Domestication' is now seen as a spectrum of phases from hunting, at one extreme, selective hunting, thinning, to increasing control of all phases of the animal's life – breeding being at the other extreme. By −30,000 mankind used fire, inhabited caves or had open-air shelters consisting of skin tents, and wore tailored clothing, for which eyed bone needles were used in the making of clothing and nets. The beginning of cave art dates from that time and has been found from France/Spain eastward to the Urals. In its production, lamps, ladders and scaffolding were used. From an even earlier date in Swaziland, South-East Africa, the human body was painted with red ochre (Fe_2O_3) mined for the purpose. Polished stone tools, possibly indicating the making of canoes, have been found in Borneo dating from this time. Weapons included lances and darts.

Communal houses made from the rib-bones of mammoths have been found from the period −29,000 to −21,000. Other inventions may be listed as follows:

−24,000

Carving of portable female figurines (e.g., the 'Venus of Willendorf'), suggesting the existence of religious cults.

−20,000

Paint platters, especially for black, red, white and browns but no green.

[33] From the beginning of my time surveying the vast subject, I have been interested in showing the Chinese inventions and discoveries that have been accepted in the West. See Dawson (1964), p. 234; Needham (1964), also reprinted in Needham (1969), pp. 55 ff.

[34] In the preparation of this preliminary list, I am grateful to Professor William C. Noble and to Dr Christopher Hallpike of McMaster University, Canada for their help and advice.

−18,000
Early use of harpoons and stylistically engraved antlers (Solutrean and early Magdalenian cultures).
−16,000
Development of log timber houses in Russia/Poland.
−14,000
A flowering of the use of burins and microblades (indicating a switch to the use of light-weight tools).
−13,000
Querns for grinding wild grains.
Circular houses with stone foundations (e.g., Ai'en Gen, Israel).
Use of sickle blades for reaping wild grasses.
−12,000 to −10,000
Initial taming/penning of gazelles (Near East) and red deer (South-East Europe), constituting the first stages in the process of the domestication of animals. Wheat (first of the cereal grains used by Man) domesticated *c.* −10,000.
−9,000
Use of clay for bricks (Natufian houses).
Early *fired* pottery, Jomon culture (Japan).
Rectangular houses with plastered floors.
Sheep domesticated in the Near East.
−8,000
Long-distance trade (Lake Van for obsidian) appears in Near East.
Evidences of irrigation (Jericho).
Use of ovens for baking.
−7,000
Copper working in North-East Turkey (Cayönü)
Megalithic architecture in town walls (e.g., Jericho).
Neolithic farming communities in Yang-shao culture area, North China.
Spindle whorls used for spinning thread in textile production.
Paddles for water craft (England, at Star Carr).
Bows and arrows in use.
−6,000
Cattle domesticated in Northern Greece.
Religious bull cults develop in Turkey (e.g., Catal Huyuk).
First true frescoes (paint on dry wall).
−5,000
Earliest writing (logography) in Northern Mesopotamia.
Development of true arch with keystone (Halaf) and stone vaults.
Rice-growing well established.
Looms in use.
−4,000
Formal writing (cuneiform appears −3,200).

−3,000
Large urban centres appear in Mesopotamia −2,900, India −2,700.
Dynasties established: Sumer; Egypt.
Egyptian hieroglyphic writing −3,000 to −2,900.
Use of facial cosmetics in Egypt (e.g., Narmer's palette).
First pyramids −2,650.
−2,500
Invention of the wheel for chariots – initially solid circle.
First use of papyrus for writing material – Egypt.
Cotton becomes a prime trade commodity from India
(Indus Valley civilisation).
First bronze casting, India −2,500.
Silk worms domesticated in China.
−1,500
Hittites discover iron-smelting and working.

The approximate dates given above are likely to be pushed back in time as archaeology proceeds. But what is striking is the immense tract of time which separates related developments. For example, more than 20,000 years passed between the invention of the needle for sewing and the spindle-wheel for making thread, a period longer than the whole history and neolithic pre-history of China together! But the inventive process has been speeding up continuously as our list of Chinese inventions will show. Use has been made of many sources in addition to those derived from our own researches, to be listed when the work is completed.

Such is the background culture from which specifically Chinese civilisation developed. Eleven specific characteristics of Chinese neolithic culture summarised by Chang Kwang-Chih were given above on p. 201. Invention is an ongoing process about whose history we are learning more and more. Here is my current list in alphabetical order for ease of reference, with approximate dates of earliest occurrence. It is proposed that a more detailed list with detailed references may be published separately at a later date.

(2) Chinese Inventions and Discoveries

Invention/discovery	*Date in list*
Abacus	+190
Acupuncture	−580
Advisory vessels	−3rd c.
Air-conditioning fan	+180
Alcohol made from grain by special fermentation process	−15th c.
Algorithm for extraction of square and cube roots	+1st c.
Anatomy	+11th c.
Anchor, non-fouling 'stockless'	+1st c.

Anemometer	+3rd c.
Anti-malarial drugs	−3rd c.
Arcuballista, multiple-bolt	−320
multiple-spring	+5th c.
Asbestos woven into cloth	−3rd c.
Astronomical clock-drive	+120
Axial rudder	+1st c.
Ball bearings	−2nd c.
Balloon principle	−2nd c.
Bean curd	+100
Bell, pottery	−3rd mill.
Bellows, double-acting piston-tuned bronze (see Piston-bellows)	−6th c.
Belt drive	−5th c.
Beri-beri, recognition of	+1330
Blast furnace	−3rd c.
Blood, distinction between arterial and venous	−2nd c.
theory of the circulation of	−2nd c.
Boats and ships, paddle-wheel	+418
Bomb, cast-iron	+1221
Bomb, thrown from a trebuchet	+1161
Book, printed, first to be dated	+868
Book, scientific printed	+847
Bookcase, vertical axis	+544
Bookworm repellent. See Fumigation	
Bowl, bronze water-spouting	−3rd c.
Bread, steamed. See Noodles	
Bridges, releasable	−4th c.
iron chain suspension	+6th c.
Li Chhun's segmental arch	+610
Bronze, high tin, for mirror production. See Mirror	
Bronze rainbow *teng* 鐙 (camphor still)	−1st c.
Callipers	+9
Camera Obscura, explanation of	+1086
'Cardan' suspension	−140
Cast iron	−5th c.
malleable	−4th c.
bombs. See Bomb, cast-iron	
Cereals, preservation of stored	−1st c.
Chain-drive	+976
Chess	−4th c.
Chimes, stone	−9th c.
Chhin 琴 and *se* 瑟 zithers. See Stringed instruments	
Chopsticks	−600
Clocks, sand	+1370
Su Sung's	+1088
Clockwork escapement of I-Hsing and Liang Ling-Tsan	+725

Coal, as a fuel	+1st c.
dust, briquettes from	+1st c.
Coinage	−9th c.
Collapsible umbrella and other items	−5th c.
Comet tails, observation of the direction of	+635
Compass, floating fish	+1027
magnetic needle	+1088
magnetic, used for navigation	+1111
Cooking-pots, heat economy in	−3rd mill.
Crank handle	−1st c.
Crop rotation	6th c.
Crossbow	−5th c.
bronze triggers	−300
gridsight for	+1st c.
magazine	+13th c.
Dating of trees by the number of rings	+12th c.
Decimal place value	−13th c.
Deep drilling and use of natural gas as fuel	−2nd c.
Diabetes, its association to sweet and fatty foods	−1st c.
Dial and pointer	+3rd c.
Differential pressure. See Ventilation	
Disease, diurnal rhythms in	−2nd c.
Diseases, deficiency	+3rd c.
Dishing of carriage wheels. See Wheel	
Distillation, of mercury	−3rd c.
Dominoes	+1120
Down draft (see also Reverberatory furnace)	−1st c.
Dragon kiln	+2nd c.
Draw loom	+1st c.
Drum carriage (see also Hodometer)	−110
Dyked/Poldered fields	−1st c.
Ephedrine	+2nd c.
Equal temperament, mathematical formulation of	+1584
Equilibrium, theory of	−4th c.
Erosion and sedimentary deposition, knowledge of	+1070
Esculentist Movement (edible plants for time of famine)	+1406
Ever-normal granary system	+9
Fertilisers	−2nd c.
Firecrackers	+290
Firelance	+950
First Chinese Natural Science Congress. See Natural Science Congress	
First dated printed book	+868
First scientific printed book	+847
Flame test. See Potassium	
Flame-thrower (double-acting force pump for liquids)	+919
Folding chairs	+3rd c.

Free reed	−1000
Fumigation	−7th c.
Furnace, reverberatory	−1st c.
Gabions	−3rd c.
Gauges, rain and snow	+1247
Gear wheels, chevron-toothed	+50
Ginning machine, hand-cranked and treadle	+17th c.
Gluten from wheat	+530
Gold, purple-sheen	−200
Grafting	+806
Gravimetry	+712
Great Wall of China	−3rd c.
Grid technique, quantitative, used in cartography	+130
Gunpowder, formula for	+9th c.
firecrackers and fireworks	+12th c.
government's gunpowder department and monopoly on	+14th c.
used in mining	Ming
Hand-carts	−681
Hand-gun	+1128
Harness, breast-strap	−250
Harness, collar	+477
Helicopter top	+320
High temperatures, firing of clay at	−2nd mill.
Hodometer (see also Drum carriage)	−110
Holing-irons	+584
'Hot streak' test	+1596
Hygrometers	−120
Indeterminate analysis	+4th c.
Interconversion of rotary and longitudinal motion	+31
standard parts for the	+1313
Kite	−4th c.
Knife, rotary disc, for cutting jade	+12th c.
Kuan hsien system	−240
Lacquer	−13th c.
Ladders, extendable	−4th c.
Leeboards and centreboards	+751
Lodestone, south-pointing ladle	+83
Magic mirrors	+5th c.
Magic squares	+190
Magnetic declination noted	+1040
Magnetic thermo-remanance and induction	+1044
Magnetic variation observed	+1436
Magnetism, used in medicine	+970
Malt sugar, *i*, production of	−1st mill.
Mangonel (see also Trebuchet, simple)	−4th c.
Maps, relief	+1086
topographical, unearthed from Mawangtui	−3rd c.
Masts, multiple	+3rd c.

‘Matches’ (non-striking)	+577
Melodic composition	+475
Mercury, distillation of. See Distillation	
Metal amalgams used to fill cavities	+659
Metals, to oxides, burning of	−5th c.
densities of	+3rd c.
Mill, wagon	+340
Mills, edge-runner	−200
edge-runner with water-power applied	+4th c.
Mining, square sets for	−5th c.
differential pressure mine ventilation	−5th c.
Mirror with ‘light penetration’ surface	−11th c.
Mirror, magic. See Magic mirrors	
Mouldboard	−2nd c.
Mountings, vertical and horizontal	+1st c.
Mouth-organs	−9th c.
Moxibustion	−3rd c.
Multiple-spindle silk-twisting frame	+1313
Natural Science Congress	+5th c.
Negative numbers, operations using	+1st c.
Noodles (filamentous), including bread	+100
Nova, recorded observation of	−13th c.
Numerical equations of higher order, solution of	+13th c.
Oil-lamps, economic	+9th c.
Paktong, 白銅 (cupro-nickel)	+230
Pandects of Natural History (*Pen Tshao* 本草 literature)	−5th c.
Paper, invention of	−300
money	+9th c.
toilet	+589
wall	+16th c.
wrapping	−2nd c.
Parachute principle	+8th c.
‘Pascal’ triangle of binomial co-efficients	+1100
Pasteurisation of wine	+1117
Pearl fishing conservancy	+2nd c.
Pearls in oysters, artificial induction of	+1086
‘Pi’ (π), accurate estimation of	+3rd c.
Piece moulding for casting bronze	−2nd mill.
Piston-bellows, double-acting	−4th c.
Place-value	−13th c.
Placenta used as source of oestrogen	+725
Planispheres	+940
Plant protection, biological	+304
Planting in rows	−3rd c.
Playing cards	+969
Polar-equatorial co-ordinates	−1st c.
mounting of astronomical sighting instruments	+1270
Porcelain	+3rd c.

Potassium, flame test used in identifying	+3rd c.
Pound-lock canal-gates	+984
Preservation of corpses, achieved by Taoists	−166
Printing, bronze type	+1403
with earthenware moveable type on paper	+11th c.
multi-colour	+12th c.
with woodblocks	+7th c.
Propeller oar, self-feathering	+100
Prospecting, bio-geochemical	+6th c.
geological	−4th c.
Rainbow *teng* (camphor still). See also Bronze rainbow *teng*	−1st c.
Recording of sun halves, parhelic spectres and Lowitz arcs	+635
Reel on fishing rod	+3rd/+4th c.
Refraction	−4th c.
Refractive clay. See High temperatures, firing of clay	
Reverberatory furnace. (see also Furnace)	−1st c.
Rocket arrow, invention of	+13th c.
launchers	+1367
winged	+14th c.
Rockets, two-staged	+1360
Roller bearings. See Ball bearings	−2nd c.
Roller-harrows	+880
Rotary ballista	+240
Rotary fan. See Winnowing-machine	−1st c.
Sailing carriage	+16th c.
Sails, mat-and-batten	+1st c.
Salvage, underwater	+1064–7
Seawalls	+80
Seed, pre-treatment of	−1st c.
Seed drill, multiple-tube	+155
'Seedling horse'	+11th c.
Seismograph	+132
Ships, construction principles of	−1st c.
paddle-wheel. See Boats and ships	
watertight compartments in	+5th c.
Silk, earliest spinning of	−2850 to −2650 (Liang Chu culture)
Silk reeling machine	+1090
Silk twisting frame, multiple spindle	+10th c.
Silk twisting frame, multiple spindle, water-powered	+1313
Silk warp doubling and throwing frame	+10th c.
Sluices	−3rd c.
riffles added to	+11th c.
Smallpox, inoculation against	+10th c.
Smoke-screens	+178

Snow crystals, six-sided symmetry of	−135
Soil science (oecology)	−5th c.
South-pointing carriage	+120
Soybean, fermented (*shih* 豉), paste (*chiang* 醬) and sauce (*shih yu* 豉油)	−2nd c.
Sprouts, for medicinal and nutritional purposes	−2nd c.
Spindle wheel	−5th c.
multiple-spindle	+11th c.
treadle-operated	+1st c.
Spooling frame	+1313
Square pallet chain-pump	+186
Stalactites and stalagmites, records of	−4th c.
Star map. See planispheres	+940
Stars, proper motion of	+725
Steamers, made of pottery	−5th mill.
Steel production, co-fusion method of	+6th c.
Sterilisation by steaming	+980
Steroids, urinary	+1025
Still, Chinese-type (see also Alcohol, distilled)	+7th c.
Stirrup	+300
Stringed instruments, *chhin* and *se* zithers q.v.	−9th c.
Su Sung's clock. See Clocks	
Sugar. See Malt sugar	
Tea, as a drink	−2nd c.
Teng. See Bronze rainbow *teng*	
Thien Yuan algebraic notation	+1248
Thyroid treatment	−1st c.
Tilt-hammer, water-powered spoon	+1145
Toothbrush	+9th c.
Trebuchet, simple (see also Mangonel)	−4th c.
Trip-hammers	−2nd c.
water-powered	+20
Umbrella, collapsible	−5th c.
Ventilation, using differential pressure	Warring States
Vinegar	−2nd c.
Wallpaper. See Paper	
'Water-bound macadam'	−178
Water-powered multiple spindle frames	+1313
Water-wheel, horizontal	+31
Watermills, geared	+3rd c.
Weather vane	−120
Wet copper method (precipitation of copper from iron)	+11th c.
Wheat gluten. See Gluten	
Wheel, dishing of	−4th c.
Wheelbarrow, centrally mounted	−30
with sails	+6th c.
Windlass, well	−120
Windows, revolving	−5th c.

Winnowing-machine, rotary-fan	−1st c.
Wu-thung 烏銅 (black patinated copper)	+15th c.
Zoetrope	+180

(3) Modern Science: Why from Europe?

As can be seen from the above list, 'China has the longest unbroken history of progress in science and technology (over 4,000 years) of any nation in the world'.[35]

The inventions listed above at present number over 250. This list is provisional. The word invention itself is ambiguous. To what extent does an existing idea need to be altered to constitute a new invention? On the other hand it may be shown that some of the inventions listed were not truly Chinese in origin and will have to be withdrawn. But others will certainly need to be added to the list. Many of the planned total of volumes in *Science and Civilisation in China* are yet to be published and when published will certainly add their quota. When it is published there may well, for example, be inventions from the volume on ceramics to be added to those listed here, for this was an art in which the Chinese excelled.[36]

It may also be worth mentioning that the Chinese had no idea of patenting anything,[37] a process which was important in the West where life was competitive.[38] The textile inventions would have been particularly suited to this system. It was not until 1449 that the first 'letters patent' were granted by the City of London.

While using this list, we should also bear in mind that the date of an invention is often that of the first literary reference, or is given according to the earliest archaeological remains; it is rarely the date of the actual invention. Some may have been around for centuries before the dating on the list; and in some cases mistaken literary references may have misled us into giving too early a date for an entry.

Nevertheless our list is already sufficiently workable to provoke us into asking a number of questions. For example, we have a total of approximately 250 inventions over the last seventeen centuries; this means an average of nearly fifteen new

[35] Finniston *et al.* (1992), p. 72.

[36] In contrast to invention, we take discovery to mean the extension of the knowledge of natural phenomena. Thus, paper, printing and gunpowder (the first chemical explosive known to man) would be considered inventions, whereas the magnetic compass and all that flowed from it would be a discovery. In this way, the entries in the above list can be divided.

[37] It would have been a good thing if the Chinese had patented gunpowder, paper and printing and so on, so that the inventors' names would have been better known in the West than they are today. Another example is the silk twisting and doubling machines which could twist silk thread 700–900 m long (*SCC*, vol. 5, pt. 9, p. 346) as opposed to the cotton ginning machines which could only twist threads of a half-metre in length.

[38] Basalla's (1988) *The Evolution of Technology*, pp. 119–24, gives an interesting account of the patent system in Western countries, especially the countries which did not have it (e.g., The Netherlands and Switzerland). Isambard Kingdom Brunel, designer of the Great Western Railway and the Great Eastern Steamship, opposed the patent system, saying that it stifled innovation. To the Select Committee of the House of Lords on the Patent Laws in 1851, he wrote: 'I believe that the most useful and novel inventions and improvements of the present day are mere progressive steps in a highly wrought and highly advanced system suggested by, and dependent on, other previous steps, their whole value and the means of their application probably dependent on the success of some or many other inventions, some old, some new'. See Petroski (1993), pp. 44–5. Many of the Chinese inventions can be considered in this light, for example, the helicopter top which was the first step towards the development of the rotor blade used in modern helicopters in the West.

inventions every hundred years. How does this compare with the inventiveness of the other great early civilisations – Babylonia, Egypt and India?

Again it is worth classifying inventions into categories. Chinese inventiveness seems to have been particularly fruitful in the production of machines and instruments; these served five main fields of interest, namely agriculture, metallurgy, transport and travel, war and the study of the heavens. Ploughshares, bellows, paddlewheels, crossbows and rain gauges offer an example of these in each field. Chinese inventiveness had a particularly practical bent. But there were more abstract fields of interest also. The human body and the medicines needed to treat its diseases were also backed up by theoretical interest into, for example, the circulation of the blood. Some inventions seem to have arisen merely from a whimsical curiosity, such as the 'hot air balloons' made from eggshells which did not lead to any aeronautical uses or aerodynamic discoveries, or the zoetrope which did not lead on to the kinematograph, or the helicopter top which did not lead to the helicopter. The Chinese had in fact so many inventions to do with flight, crowned by the man-carrying kite, that one is surprised that they did not invent a glider at least as good as that which broke the legs of the Saxon monk Aethelmaer at Malmesbury early in the +11th century. Or perhaps did Yüan Huang-Thou 元黃頭 in +559 glide even further on his kite shaped like an owl and made of paper than Aethelmaer's furlong? (See *SCC*, Volume 4, part 2, pp. 587–8.)

Inventions in themselves, however, do not foster or inhibit science. It is necessary to study the social background, the ideas and practices which generate them and ultimately science itself. Our task is to study those ideas and practices which are recorded and seem either to have fostered or to have impeded the growth of science and technology in China. Let me speak of those social and economic factors which appear to have most powerfully advanced or hindered the growth of science in China from the medieval to the modern stage. The inventiveness of the Chinese people cannot be gainsaid, nor can their curiosity about natural phenomena. And yet they did not discover the supreme method for making scientific discoveries. Why was this? Perhaps it is now time for me to write my last few words in concluding this long series of volumes.

These volumes began slim enough, but already by the third they were developing a middle-age spread. I felt that if a volume were too large to read comfortably in one's bath it was time to subdivide it. *SCC*, Volume 4 was accordingly divided into three parts or 'physical volumes', but such was the wealth of Chinese science and technology which opened up before us that these too became unwieldy. *SCC*, Volumes 5 and 6 had to be subdivided into so many parts that it is doubtful whether I can see their completion within my lifetime. Now at the age of ninety-three, I should not wait any longer to summarise why I think it was that China, after such a promising start, did not go the whole way to developing modern science, whereas Western Europe, for special reasons which have been set out above, succeeded in doing so.

I would like to begin with why, in my opinion, China made what I have called 'a promising start'. In the preceding pages of this concluding volume I have made it clear that I cannot go along with any theory that the Chinese were better or less well-endowed with brains or moral qualities than Westerners. Racism is not for me. Nor can I subscribe to the notion that it was just a matter of luck that Europe happened to be the place where modern science was born. I am convinced that it was due to the different types of society generated at opposite ends of the Eurasian land mass, and of course differences in geography and in economics had a big hand in shaping those societies. But what are the qualities that make a good scientist?

First of all, perhaps, curiosity is needed, combined with persistence. Curiosity alone is not enough. The many-wiled Odysseus did not lack curiosity, but was in no way a scientist. The early Taoists, not only curious about what they saw, but observing nature patiently and persistently, were proto-scientists. The scientists of modern China are heirs to an immensely long tradition of careful and persistent observation. In ancient times the Chinese tradition of keeping records was rivalled only by that of the ancient Babylonians. Indeed, in the recording of eclipses and many other celestial phenomena the Chinese excelled, as a glance at our *SCC*, Volume 3, pp. 409 ff. will make clear.

Other important qualities come to mind in which the Chinese were by no means lacking. They had the ability not only to observe and record, but to make accurate comparisons between things observed; and having observed and compared, to place things in groups and to classify. Chinese skill and interest in classification is amply demonstrated in *SCC*, Volume 6, part 1, where we wrote of Chinese botany and the *pen-tshao* 本草 literature, what we have called pandects of natural history. Ability to classify is one of the green shoots of budding science.[39]

Today we think of science very largely in the context of experiment aimed at the isolation of variables in order to test a possible theory. This is a fairly modern development. In medieval China there was a tradition of experimentation, sometimes misconceived, yet by sheer persistence yielding rich rewards.[40] The gunpowder epic described in *SCC*, Volume 5, part 7, is one example of this. Some of the requirements for modern experimentation were there, but not all.

There were other abilities which the Chinese possessed in good measure, such as the art of logical disputation, and the ability to hypothesise and to make counterfactual hypotheses. The logical furnishings of the Chinese mind have been well described by Christoph Harbsmeier in *SCC*, Volume 7, part 1. Nor did the Chinese lack scepticism, though there were some areas where scepticism was not acceptable, just as in medieval Europe to challenge the verity of the Holy Ghost would not have

[39] Also see Gutmann (1992), p. 231, for an appreciation of the merits of taxonomy.

[40] What a contrast with the Roman Empire for the period from the −2nd century to the +2nd century, when Marius' decision to shorten the sword and extend the frontiers was virtually the last spark of Roman inventiveness. In China in the same period there were thirty-four *scientific* discoveries out of a total of sixty-seven discoveries and inventions. Later, in the Sung period (+11th to +14th centuries), there were twenty-three scientific discoveries out of a total of forty-seven discoveries and inventions.

been acceptable. All these were valuable aspects of Chinese intellectual attitudes highly relevant to the birth of science.

Encompassing Chinese thinking was the overall predilection for action at a distance, as opposed to the Western predilection for atomism. This may well have handicapped the Chinese in elaborating a theory of dynamics, but was helpful in discovering such things as the nature of the tides (*SCC*, Volume 3, pp. 483 ff.), of musical resonance (*SCC*, Volume 4, part 1) and of magnetic phenomena (*SCC*, Volume 4, part 1, pp. 229 ff.).

No less important were the social structures of ancient and medieval China. Of those which did not help China on the road to modern science I shall speak in a moment. But among those which gave China a good start I would mention the fact that chattel-slavery was never a basis for Chinese society, as it was in the Roman Empire, and that those who were willing to apply labour-saving inventions did not have to fight the cut-throat competition of slave labour working the fields in gangs, as was normal in the Roman *latifundia*. This is perhaps the main reason why Chinese technology got away to a flying start in the centuries when Rome was in decline and Western Europe was descending to the Dark Ages.

Next I would mention that, as in the West, the Chinese had the idea of linear time, but along with a reverence for times past was the idea of progress. Whereas classical Europe viewed time as descending from a golden age long past to a leaden age of dullness and immobility yet to come, the Chinese, with every century producing its crop of inventions and technological improvements, could scarcely view the future with such bleak pessimism. Calendars also, with each succeeding dynasty, were revised and recalculated with a view to improvement.

Another factor which gave China a good start for science was the educational structure. It had an astonishing resilience, surviving at village level when central government had broken down, and returning to the administrative centres from the villages where scholars had taken refuge, once central government was re-established. The examination system helped to keep the school system in being, and was itself kept in being by the need for a bureaucracy capable of running the country. Though the examination system was not designed to select scientists, or even to encourage an interest in science, it did nevertheless select educated people for positions of power, something that was not even contemplated under feudalism in Europe. Though Confucians were often indifferent to nature and science, well-educated administrators would nevertheless be in a better position to appreciate the value of new inventions and technological developments than the illiterate feudal lords of ancestral acres in Europe whose main prowess lay in jousting.

I have referred to certain disadvantages of the Chinese script, but it may be said here that the Chinese empire, like the Roman, had a common language by which educated people in all the provinces could communicate and exchange ideas. The fact that in China it was more often common script than common speech explains why characters continued to be used for this purpose long after Latin had ceased to be the *lingua franca* of Western Europe. Nor should one overlook the fact that the

Chinese were singularly well equipped to deal with mathematics, having had from their earliest times a numerical system based on place-value which was infinitely more satisfactory than the Greek and Roman systems based on letters of the alphabet. Only lacking was the zero which, as I have said in *SCC*, Volume 3, pp. 10 ff. probably evolved in Cambodia and was laid like an Indian garland on the empty space on a Chinese counting-board (*SCC*, Volume 3, pp. 148). It is an interesting fact that whereas Chinese characters in some measure inhibited the writing of speech but aided computation, alphabetic letters aided the writing of speech but inhibited computation. Perhaps the culminating achievement of indigenous Chinese mathematics was the *Thien-yüan* 天元 matrix algebra of the Sung dynasty which enabled Chinese mathematicians to solve equations of high degree. (See *SCC*, Volume 3, p. 129).[41]

There is also the matter of the bureaucracy itself. The negative aspects of bureaucracy in relation to science have been considered in earlier sections. But there were many positive aspects. For example the tremendous effort made in establishing a meridian line 3,500 kilometres in length under the direction of Nan-Kung Yüeh 南宮說 and I-Hsing 一行 during the years +721 to +725 in order to obtain more precise data for calendrical astronomy can find no comparable effort in Europe until post-Renaissance times. For an equivalent one must go back to Eratosthenes' attempts to calculate the circumference of the Earth by the use of gnomons placed at Alexandria and Syene in the –3rd century. (See *SCC*, Volume 3, pp. 225 and 292 ff.).

From the qualities required in a scientist, and from such social structures as schools and universities necessary for the development of science, we may turn to the materials needed to record scientific information. The earliest Chinese writing materials – bone, tortoise shells and slivers of bamboo – were less helpful to writers than the Babylonian clay tablets or Egyptian papyrus sheets. But certainly by the –6th century ink brushes were being used for writing on sheets of silk, and by the +2nd century the manufacture of paper in quantity had been organised. Paper is renewable, but when the papyrus reed-beds of Egypt became exhausted, the Roman bureaucracy was forced to turn to parchment and vellum prepared slowly and expensively from animal skins. From then on until the introduction of paper and printing to Europe, China had the advantage in the dissemination of ideas. Nor was there any lack of books; the Chinese were themselves pioneers in printing on paper. Like the Greeks they had also learnt to bring words together in dictionaries, ideas together in encyclopaedias and books together in libraries for the use of scholars; and like the Greeks they had learnt how great a loss the world of learning can suffer when books are destroyed in conflagrations.

I should draw attention also to the fact that the Chinese had invented many instruments for precise measurements which are necessary for the onward march

[41] Alvares (1980), Joseph (1991) and Bhanu Murthy (1992) show how far ancient peoples could get in mathematics without the intervention of modern methods.

of science, instruments for map-making (the hodometer, the south-pointing carriage and the concept of the rectangular grid, for example), and for the study of climate (rain gauges and snow gauges), to name but a few. Many of their inventions, though utilitarian at the outset, stimulated scientific investigation. The magnetic compass, for instance, provoked an understanding of magnetic polarity and declination before even polarity was known in Europe. This cannot be dismissed as mere technology. It was early science. Again, the recognition of fossils was an example of early science and not merely technology.

Finally I would like to say that just as the great ocean voyages stimulated the men of the Renaissance to rethink their science when confronted by unimaginable beasts and plants never seen before, so also the Chinese had their age of great ocean voyages when Admiral Cheng Ho 鄭和 explored the South Seas and the western limits of the Indian Ocean. The giraffe was at least as great a surprise to the people of China as was the armadillo or the duck-billed platypus to the people of the West, and required no less revision and expansion of classification systems.

Such then are some of the factors in China and the West which it seems to me were of cardinal importance for the growth of modern science. On the whole China was well endowed, and we can understand how it is that they obtained an early lead in technology and a promising start in some of the sciences. Yet it was Western Europe which discovered the method of developing the new natural science which was to transform the world. What were the other factors?

One factor which stands out above all others is the rise of the bourgeoisie for the first time in history. We may review the factors which led to it, and we may discuss the results which followed from it, but it is the rise of the bourgeoisie in Western Europe from the 15th century onwards which decided that Europe would not continue trying to emulate its classical past, nor evolve in Chinese fashion, improving its technology and making small advances in various sciences without actually coming to grips with scientific method, but would strike out in a new direction, developing capitalism, scientific method and the industrial revolution one after the other.

So complex was the process that we cannot flatter ourselves that in our Volume 7 we have done more than hint at the main movements. In speaking of the rise of the bourgeoisie I am mindful that its origin can be traced back to the city-states of Greece, the Roman concept of law, and to an old passion for individual freedom and self-government. I am very much aware that the geography of Europe contributed to these things – Europe was in fact a nursery for merchant princes – and that the ability of merchants to amass wealth and retain it made it possible for them to overthrow governments, something which the stable political system of China could not countenance until the West made it impossible to avoid.[42]

We have wished to show in *SCC*, Volume 7 that the bourgeoisie succeeded in gaining and keeping power because of a unique situation following what Immanuel Wallerstein has called 'the four collapses' – the collapse of the Seigneurs or feudal

[42] Also see Dorn (1991).

order, of the States, of the Church and of the Mongol Empire. The importance of the Black Death in weakening these four orders and so enabling the bourgeoisie to gain control can hardly be exaggerated.[43] The Black Death struck not only Western Europe but also the Mongol Empire, including China where it had originated. But in China no bourgeoisie arose to take control. The case of Western Europe was unique.

Under bourgeois dominance modern science came to birth. I have described how propitious was the mercantile atmosphere for careful and accurate measuring, recording and testing – first this factor and then that, to see which was the more profitable. In that way the elimination of variables became a method, and scientific method was born.

But the great step during the bourgeois revolution was the mathematisation of hypotheses, and this was something which was never achieved in China before the 19th century. It became possible in Europe because scholars and artisans at last found themselves communicating not fitfully but regularly in the course of business and daily work. And this became possible because a new type of education opened for the artisans. For centuries the sons of artisans and merchants had been able to get an education if they were sufficiently lucky or gifted. Cardinal Wolsey, for example, was the son of a butcher. Such educated persons, however, did not make their way into commerce, but were absorbed into the Church or the old classically educated orders. The new education began when for the first time books were printed in the vernacular languages of Europe. The importance of printing for the Renaissance was brought out in 1979 in such books as *The Printing Press as an Agent of Social Change*, by Elizabeth L. Eisenstein. Although the Chinese invented moveable-type printing (at the time of Pi Sheng 畢昇, +11th century) they could not achieve the same information revolution as the Europeans. They were hamstrung by the complexities of Chinese characters. Douglas S. Robinson has recently stressed to me in a personal communication the importance of the printed word for a technological revolution as follows: 'a technological society requires a certain minimum quantity of information . . . a quantity that is essentially unachievable without the technology of printing . . . the absence of an alphabetic writing system handicapped its development there [in China] to some degree, perhaps to the point that the necessary level of information production was never achieved, or at least was much delayed'.

Before the introduction of printing, books were too expensive for the common people. There would therefore have been little point in translating texts from Latin into modern European languages. Most of those who could afford books could read Latin. But with printing in the vernacular, written information was at last within the reach of Western European artisans. A child could then begin to read in its mother tongue. It was no longer necessary for a boy or girl first to learn Latin in order to be able to read a serious book, unlike China where mastery of the old classical language was essential for this purpose. Armed with no more than an understanding of the alphabet a man could tear an education for himself out of books printed in his mother

[43] The role of gunpowder in the changing of European political structure is mentioned above, pp. 53, 211, n. 23.

tongue. William Cobbett is an example of one who did this as late as in the 19th century. Thus it was only when the old Chinese invention of moveable-type printing met with the alphabetical languages of the West that an 'information revolution' came about.[44]

Even with only a rudimentary education artisans became more useful to their employers, and educated employers needed the help of artisans who could read and write. The gulf between artisans and scholars had at last been bridged. The fruit of this union was modern science and the industrial revolution. To explain why this did not happen in China one needs to be able to explain why in China there was no bourgeois revolution. Many of the reasons for this may be found in the previous volumes – the establishment of the bureaucracy, the stability of the Chinese system of government, deep-rooted opposition to a mercantile ethos, the failure to develop what Huang Jen-Yü 黄仁宇 calls 'a mathematically managed economy', all these have been discussed. One may also speculate on whether Western Europe would have elaborated scientific method without Euclid, or whether the bourgeoisie would have established their position if it had not been for the Black Death. These are questions comparable to that old favourite, 'Why did the Roman Empire fall?' or, for that matter, 'Why did the Western Empire not last as long as the Eastern Empire?' These questions are stimulating, and sometimes provoke fresh thinking, but they have no definitive answers.

My concluding thought is that by an extraordinary series of events modern science was born, and swept across the world like a forest fire. All nations are now using it, and in some measure contributing to its development. We can only pray that those who control its use will develop it for the good not only of mankind but of the whole planet.

[44] See Cardwell (1994), p. 55.

APPENDIX
JOSEPH NEEDHAM: A SOLILOQUY*

Gonville and Caius College Chapel, Cambridge.
Funeral of
Noel Joseph Terence Montgomery Needham
Friday 31 March 1995

Let us begin and carry up this corpse
  Singing together,
Leave we the common crofts, the vulgar thorps
  Each in its tether.

Kenneth: I wonder how we begin. At omega of course, and proceed to alpha.

Service: As soon as thou scatterest them they are even as asleep: and fade away suddenly like the grass.

Kenneth: Tell me, Joseph, have you any fears about an after-life?

Joseph: My greatest fear is that after death one might find that the medieval teaching is true, and that there is a hell, a Buddhist hell with demons! I prefer the Taoist belief that after death we are resolved into our elements, received as it were into the Great Inner Chamber, where no one should pursue us with shouting and bawling. That is what I wished for my first wife Dophi, and my second wife Gwei-Djen.

Service: Let us pray.

Kenneth: Do you pray, Joseph?

Joseph: Sometimes.

Kenneth: Are your prayers answered?

Joseph: Only if they are in accordance with the Tao.

Kenneth: How do you know if they are in accordance with the Tao?

Joseph: Ah! That is the big question.

Kenneth: I think you claim to be a Christian, a Marxist and a Taoist. You may be each of these at times, but surely not all three simultaneously? I know you like to build bridges between contradictory opinions, but isn't this a bridge too far?

Joseph: They all have a lot in common. Perhaps what the world needs is a synthesis of the best in all three.

Service: Man that is born of a woman hath but a short time to live. . . . He cometh up, and is cut down, like a flower – he fleeth as it were a shadow, and never continueth in one stay.

* This was written by Kenneth Robinson, shortly after Needham's funeral. The soliloquy takes the form of an imaginary conversation with Needham during the funeral service, but is based on actual conversations which had taken place between the years 1949 and 1995.

Kenneth: What sort of woman were you born of, Joseph?

Joseph: My mother, Alicia? She was a very exciting person, Scots-Irish, musical – she composed the melody for the song 'Nelly Dean' – full of life, endlessly at war with my father, who was a pioneer anaesthetist. I was always trying to mediate, to build bridges of understanding between them, but I never succeeded.

Kenneth: That was a bridge you failed to build, but what a bridge you built between the civilisations of China and the West! Till then these two civilisations had thought of each other as rivals, but now, thanks to your life's work, we all regard them as complementary.

I think you once characterised the civilisation of the West as a 'slap in the face' civilisation. That is to say, when there is a confrontation, Westerners tend to call a spade a spade and a liar a liar. When a person is given a slap in the face and blatantly insulted, dialogue comes to an end; China, on the other hand, you described as having a 'stab in the back' civilisation. When there is a confrontation, though they may stab their opponent in the back, they keep smiling and dialogue can continue.

Joseph: Yes, that is so. Although neither of these forms of behaviour is desirable, I prefer to keep talking. You find this contrast even in their logic. Whereas in the West we say: all men are mortal. Socrates is a man. Therefore [even a fool can see that] Socrates is mortal; the Chinese end this syllogism with a polite question and a smile as they put the knife in: How [may I ask the learned gentleman] could Socrates not be mortal?

Kenneth: Tell me, Joseph, how did your interest in different civilisations begin?

Joseph: Well, I began as a classicist, not as a scientist, but at Oundle, where I went to school, they put considerable stress in the curriculum on working with one's hands, and on the making of models and machinery. That is perhaps where I got my interest in all sorts of machines and mechanisms, and in comparing the mechanical inventions of China and the West, which came to fruition – or should I say fructification? – in *Science and Civilisation in China*, Volume 4. This volume dealt with physics and its practical applications.

Service: So teach us to number our days that we may apply our hands unto wisdom.
Prosper thou the work of our hands upon us, O prosper thou our handywork.

Joseph: And then I lived in France to acquire French and to broaden my education.

Kenneth: I think you spoke French very well, and of course had no trouble in reading it.

Hélas!

Ne vais-tu pas hier, Brinon,
Parlant et faisant bonne chère,
Qui n'est aujourd'hui, sinon
Qu'un peu de poudre en une bière,
Et dont ne reste rien que le nom?

[Was it not only yesterday, Brinon, that you were getting around, chatting and living it up, who today are nothing at all except perhaps a handful of dust in a casket, and of whom nothing is left but the name?] *(Ronsard. Trans. K. R.)*

Kenneth: You got caught up in the First World War a little bit, didn't you?

Joseph: Yes, in 1918 I was seventeen. I was born on 9 December 1900. It means I never had any difficulty in working out my age. In fact I prefer the Chinese way of calculating age, which states the number of years you are completing.

Kenneth: So in 1995 you are in your ninety-fifth year?

Joseph: Yes, that's it. So for a few months I was a surgeon sub-lieutenant. I went up to Caius in 1918 intending to become a surgeon. But the saw-bones side of the business was too mechanical. It was too much like carpentry. I needed something more intellectually challenging. So I switched to biology and worked on the newly developing science of biochemistry with Sir Frederick Gowland Hopkins in his lab. We called him Hoppy. That was a time when we made up limericks, clerihews and other poems on scientific subjects. There was one by J. B. S. Haldane I remember, who preceded me as Sir William Dunn Reader in biochemistry. It went like this:

You cannot synthesise a bun
by simply sitting in the sun.
I do not answer – Yes, yes, yes
When I am offered meals of S.
But readers, rhizostomes and rats
Are fairly good at making fats.
So let us firmly stick to this,
Our most effective synthesis.

Kenneth: So that was when you started analysing the metabolic changes in the embryonic development of chickens from the egg. How did you do it?

Joseph: I had three eggs in an incubator, and every day I took one out, churned it to a pulp and then carried out a chemical analysis of the liquid. It soon became clear that a chicken embryo was a small chemical factory, exchanging some chemicals for others.

Kenneth: And after that you got down to the writing of your *Chemical Embryology* in three volumes which established your reputation as a leading scientist. It must have taken an immense effort. Gwei-Djen told me you had a breakdown after it.

Joseph: Yes, that was in 1931. For two weeks after it was finished I had difficulty in sleeping because I was incessantly hearing music in my head.

Kenneth: You were at that time, and for some years to come, a scientist, but one with an extraordinarily wide range of interests. Yet, in the end, your life turned into a sort of pilgrimage in pursuit of knowledge, justice and freedom. How did that come about?

Service:

No goblin nor foul fiend
Can daunt his spirit;
He knows he at the end
Shall life inherit.
Then fancies, fly away;
He'll not fear what men say;
He'll labour night and day
To be a pilgrim.

Joseph: It came about because four Chinese graduate research students came from Shanghai in 1937 to study biochemistry with Dophi and me in our labs. One of these was Lu Gwei-Djen whom I married in 1989. From them I learnt that China had developed mathematics, astronomy and several other sciences to a level comparable to that achieved in medieval Europe. Till then people in the West believed they had a monopoly in scientific thinking.

I began to teach myself Chinese. When in 1942 our government felt it should make a friendly gesture towards the Chinese government by sending a scientist to join the Embassy in Chungking they sent me, not because I was likely to do much good, but because I was sufficiently maverick as to be clearly expendable.

Kenneth: And then followed your four years of extraordinary exertions on behalf of Chinese university education and research. You visited the remotest parts of the country with what transport you could get, or on foot, learning all the time from Chinese scientists about the history of science in China, and gradually forming the nucleus of the library which is today housed in the Institute which bears your name.

Joseph: That is correct. And already there was forming in my mind the plan for my book *Science and Civilisation in China*. This book, I was determined, would do justice to the hundred or more generations of thinkers and experimenters who had been measuring and recording phenomena in China for so long.

Kenneth: I imagine it was your dedication to justice that led you to stick your neck out over the controversy in the Korean War about germ warfare. Did you ever test the exhibits?

Joseph: At the end of the Second World War, when I was still in China, the American Air Force dropped anthrax-impregnated cattle cakes in South China, apparently to test their effectiveness. I myself carried out tests on these cakes and found that the allegations were true. I therefore had no reason to doubt the truth of what my informants told me about a similar occurrence in Korea.

Kenneth: I remember you telling me about this not so long after it happened, and when you were blacklisted from entry to the United States. I suppose now that we know more about how war was waged in Vietnam, and about the use of nerve gases in Britain, it does not seem so improbable.

Joseph: I think there has been a certain swing in public opinion since those days.

Kenneth: So much for justice, but what about freedom? I do think, Joseph, that you have been pretty blinkered about freedom in the People's Republic. I remember you saying in a radio talk that there was freedom of movement in China, way back in the 1970s. A person had only to go to the railway station, you said, and buy a ticket, and he could go anywhere in China that he liked. But could he get permission to be absent from work, or to spend two or three days queuing to buy a ticket?

Joseph: The first freedom is freedom to eat. People in China were at last being fed.

Kenneth: What about personal freedom in this country? Many people used to believe that you were an extreme Marxist, if not a member of the Communist Party.

Joseph: Rubbish. I was a member of the Labour Party.

Kenneth: Certainly you were a champion for justice and freedom when you made it possible for Professor Jacobsen to escape from Germany and the threat of the Holocaust. You brought him to Cambridge in 1933, didn't you?

Joseph: I think that would be the year.

Kenneth: And what about personal freedom for individuals here, and sexual freedom for students? I remember when you were Master of Caius, you did not agree to the installing of vending machines for condoms in the college.

Joseph: I had no objection to the free association of students, and to their buying packets of condoms in the town if they wished to, but if these machines were installed in the college it would mean that someone wanted a condom instantly. I was not in favour of instant sex.

Kenneth: But you were keen on other sorts of freedom. For instance on Sunday afternoons you would take a group of friends out to Quy where I now live, and find your way to Quy pool, and there everyone except Gwei-Djen would strip to the bare buff and go prancing off into the water. But Gwei-Djen would sit modestly by herself, her gaze averted!

Joseph: Have you been out to the pool then?

Kenneth: No. It's no longer used for swimming. It's clogged with weed, and someone was drowned in it. But your description of it which Laurence and Helen Fowler reproduced in their book *Cambridge Commemorated* catches the spirit of it. Here it is:

> Happily the country around Cambridge, especially towards the north-east where the fens are, is wonderfully provided with wild pools. . . . One of these pools lies very difficult to find in what is practically a piece of untouched fen. It is almost surrounded with reeds whose stems are almost white, but bear at the top their long green blades which stream unanimously out in one direction if there is a little wind. If you lie flat on the bank of the diving place and look along the pool, you see a picture of reeds in the best Chinese manner. Then if you dive in, you find the water absolutely clear and beautifully brown, contrasting marvellously with the brilliant blue of the surface, reflecting the sun. Or at midnight, on one of those few nights of the year when the air and water are warm enough, it is lovely to swim through the moon and stars glittering on the water . . . *(History Is on Our Side, 1946)*

Kenneth: Cambridge must have been a tremendous free-thinking place in the 1930s, and it was still very exhilarating when I first knew you. You used to drive your great Armstrong-Siddley car which had previously belonged to my schoolboy hero Sir Malcolm Campbell, the racing driver who set up the world's land-speed record at over 300 mph in 1935.

I first met you, I think, at Oxford in 1949. You had read the first draft of my thesis on the evolution of equal temperament in China, and you said that you were engaged on Volume 4, *Physics*, of your *Science and Civilisation in China*, but that acoustics were your 'blind spot'. Would I write you about twenty pages? I said I couldn't possibly do that. I didn't know enough about it. You said, 'Of course you can', with such conviction that no further argument was possible. And so, in due course, I wrote 103 pages for Volume 4.

Joseph: Yes, I remember that very well. You used to bring the manuscript down in instalments to Caius, and we would work through it.

Kenneth: I don't think you have any idea of the strain I experienced when I was working with you then. So little was known about acoustics in China in those days. The great sets of bronze bells so wonderfully described by Chen Cheng-Yih had not yet been discovered, and we were often skating on thin ice. I would try to explain some difficulty I had encountered in a text, and would be sweating to hold together ideas which wanted to fly apart, and would be floundering with my words, breaking off, starting again. . . . You sat there like Patience on a monument, eyes closed, silent, waiting, waiting. Finally you would say: 'I haven't the slightest idea what on earth it is you are trying to tell me'. And then, in two highly distilled lines, you would resolve all the difficulties. Later I was taught by Gwei-Djen how to deal with you. She often expressed herself in the most exotic English, which you understood perfectly well, but affected not to. When you said, 'I haven't the slightest idea what it is you are trying to say', she would reply: 'You silly old man, you're not even trying!'

You used to take me to dinner in Hall, and after it we'd work through my manuscript with incredible speed and concentration, and then we'd talk of other things. You were always forthright and unconventional. I remember in the 1980s, when you were beginning to find movement painful, you used to make your way up to the dining hall via the kitchens, getting the kitchen staff to hoist us up on the vegetable lift! You were a wonderful and generous host, and invited me to Caius many times during the years when I was working abroad, if I happened to be in England. I enjoyed those visits very much.

In 1979 you invited me to leave my post with the UNESCO Institute in Hamburg and 'come over to Macedonia and help us'. How could I refuse such an offer? I started working with you again in 1980. My task, editing the final volumes of your work, was particularly difficult because you had sometimes been unable to tell old friends who pressed their writings on you for inclusion in the series that they were not in harmony with your thinking. It was easier to say that Kenneth would sort out any problems arising from different points of view.

Joseph: When a really eminent person devotes a lot of his time to writing a contribution, and then presents it to you as a gift, how can you refuse it? As you know, I never subscribed to the slap in the face approach.

Kenneth: As the *Tao Te Ching* makes clear, if things go wrong you must correct them early, otherwise you are in real trouble. Field Marshal Montgomery used to say the same thing to his officers: 'You must not get into battle on the wrong foot, gentlemen. You must not get into battle on the wrong foot.'

Joseph: The trouble was that when I embarked on the writing of *Science and Civilisation in China* I had no idea of the immense wealth of ideas and information that we would discover. People in the Western world used to think that Chinese civilisation was one of pretty paintings, of rather unconventional poems, of unspeakable music, of wonderful results in handicrafts – silks, porcelains and bronzes – and of the accidental discovery of gunpowder, which they used for fireworks, but no real science. I myself visualised a single slim volume on Chinese science in the days before the Second World War, when I was first becoming interested in the possibility. During my time in China I realised that one volume would not be enough, and that it would probably have to be seven. Volume 1 would sketch in the general background. Volume 2 would describe the philosophies and religions of China as they affected the development of science. Then in Volumes 3 to 6 we would describe what the Chinese knew of mathematics and astronomy, with related studies, of physics, chemistry and biology. That left Volume 7 for describing the social background and coming to some general conclusions.

As we got deeper into our work the volumes increased in size. More and more fascinating stuff kept coming out as we ransacked the treasure house of Chinese literature. Volume 3 was already too big for comfortable reading in one's bath. Volume 4 burst at the seams and had to be subdivided into three fat volumes. Volumes 5 and 6 are still in progress, still expanding. I had at one time hoped to be still with you when the last volume was published, but I promised that even if I was not on the bridge the ship would sail safely into port with others in command. I am sure that day will come. My dilemma was this: when I discovered the immense wealth of Chinese studies in science, technology and medicine, was I to ignore the riches which were being unearthed, and concentrate on a mere outline in just seven volumes, or was I to publish everything I could as it arrived? I chose the latter course. I published no less than four volumes, for example, on alchemy in China, and with Gwei-Djen decided on a change in policy by which we would co-opt other writers to take charge of the volumes we couldn't handle ourselves. Till then your contribution on acoustics was the only one not to be written by myself.

Kenneth: Until you changed your policy you had tackled every subject, every field of science, head on, almost from scratch. I remember when you were just starting the first volume on botany, I asked you how it was shaping and whether you already knew much about Chinese botany. You replied: 'No. At present it's just a green blur!' But two years later it was no longer a green blur. You had got all the great Chinese botanists and their systems of classification of plants in sharp focus. That would have been a life's work for many people.

There are many things about your writing which I find most admirable. For example, even on the most erudite subject there is a colloquial friendliness clothing a sentence structure of classical rigour, and with the most precise use of words chosen for their accuracy. This often requires a high level of education in the reader, and it will be some time before the full impact of your work is felt all the way through the education systems of Western countries, filtering into school textbooks, though that process has already begun. But I also admire your dogged persistence, brushing aside superficial criticisms and not wasting time defending your reputation. As you used to say: 'The dogs bark, but the caravan moves on'. Working with you I found it most comforting to know that your broad shoulders were there to carry the weight.

Service:

God be in my heart
And in my thinking.
God be at my end
And in my departing.

Kenneth: You loved rituals, didn't you, Joseph, and all ritualised dancing, as with the Morris dancers. What was the rhyme that Francesca Bray illustrated for you?

Dr Joseph Needham
Dances with philosophic freedom.
You'd better watch your toes if
You dance with Joseph.

And you loved poetry and amusing songs, and would often sing them at parties and even at radio interviews. Hélas, hélas . . .

Ne vais-tu pas hier, Joseph,
parlant et faisant bonne chère?

So we gather you up for your last great ritual – greater than all the honours you received during your long life; more memorable than the day we escorted you to Buckingham Palace.

Service: The coffin is carried round Caius Court, followed by the Fellows walking two by two, and then to the Gate of Honour . . . The choir then sing the Nunc Dimittis. The coffin is carried up the Senate House Passage, and the congregation disperses.

In the room marked K2 in Caius College, where Joseph Needham did so much of his work, four Chinese characters may be seen on the wall:

Jen chhü 人去
Liu ying 留影.

The man departs.
There remains his shadow (his umbra or influence).

Kenneth Robinson
Cambridge Review
November 1995

BIBLIOGRAPHIES

A. CHINESE BOOKS BEFORE +1800

Chih Wu Ming Shih Thu Khao (and *Chhang Phien*).
See Wu Chhi-Chün in Bibliography B.

Ching Ching Ling Chhih 鏡鏡詅癡.
Cheng Fu-Kuang 鄭復光 +1845
Shanghai, 1936.

Ching Shih Chih Yao 經世指要.
Important Principles in the Cosmological Chronology.
Southern Sung.
Tshai Chhen 蔡沉.

Ching Shih 鏡史.
History of Optick Glasses. (A portfolio of MSS) *c.* +1660.
Sun Yün-Chhiu 孫雲球.

Chiu Chang Suan Shu 九章算術.
Nine Chapters on the Mathematical Art.
H/Han, +1st century (containing much material from C. Han and perhaps Chhin).
Writer unknown.

Chou Li 周禮.
Record of the Rites of [the] Chou [Dynasty] [descriptions of all government official posts and their duties].
C/Han, perhaps containing some material from late Chou.
Compilers unknown.
Tr. E. Biot (1).

Chou Pi Suan Ching 周髀算經.
The Arithmetical Classic of the Gnomon and the Circular Paths [of Heaven].
Chou, Chhin and Han. Text stabilised about the −1st century, but including parts which must be of the late Warring States period (*c.* −4th century) and some even pre-Confucian (−6th century).
Writers unknown.

Chu Phu 竹譜.
A Treatise on Bamboos [and their Economic Uses; in verse and prose] [probably the first monograph on a specific class of plants].
L/Sung, *c.* +460.
Tai Khai-Chih 戴凱之.
Tr. Hagerty (2).

Chu Ping Yüan Hou Lun 諸病源候論.
Discourses on the Origin of Diseases [systematic pathology].
Sui, *c.* +607.
Chhao Yüan-Fang 巢元方.

Chu Tzu Chhüan Shu 朱子全書.
Collected Works of Chu Hsi.
Sung (ed. Ming; *editio princeps* +1713).
Chu Hsi 朱熹.
Ed. Li Kuang-Ti 李光地 (Chhing).
Partial trs. Bruce (1); le Gall (1).

Chhi Min Yao Shu 齊民要術.
Important Arts for the People's Welfare [lit. equality].
N/Wei (and E/Wei or W/Wei), between +533 and +544.
Chia Ssu-Hsieh 賈思勰.
See des Rotours (1), p.c.; Shih Sheng-Han (1).

Chhien Han Shu 前漢書.
History of the Former Han Dynasty [−206 to +24].
H/Han, *c.* +100.
Pan Ku 班固, and after his death in +92 his sister Pan Chao 班昭.
Partial trs. Dubs (2), Pfizmaier (32–4, 37–51), Wylie (2, 3, 10), Swann (1), etc.
Yin-Te Index no. 36.

Chhien Tzu Wen 千字文.
The Thousand-Character Primer.
Ascr. S/Chhi, +520.
Attrib. Chou Hsing-Ssu.
Tr. St Julien (10).

Erh Ya 爾雅.
Literary Expositor [dictionary].
Chou material, stabilised in Chhin and C/Han.
Compiler unknown.
Enlarged and commented on *c.* +300 by Kuo Phu 郭璞.
Yin-Te Index no. (Suppl.) 18.

Hai Kuo Thu Chih 海國圖志.
See Wei Yüan and Lin Tse-Hsü (1).

Honan Chheng Shih I Shu 河南程氏遺書.
Collected Sayings of the Chheng Brothers of Honan [Neo-Confucian philosophers].
Sung, +1168.
Ed. Chu Hsi 朱熹.

Hou Han Shu 後漢書.
History of the Later Han Dynasty [+25 to +220].
L/Sung, +450.
Fan Yeh 范曄. The monograph chapters by Ssu-Ma Piao 司馬彪.
A few chs. tr. Chavannes (6, 16); Pfizmaier (52, 53).
Yin-Te Index no. 41.

Huai Nan Tzu 淮南子.
[= Huai Nan Hung Li Chieh 淮南鴻列解.]
The Book of [the Prince of] Huai Nan [compendium of natural philosophy].
C/Han, *c.* −120.
Written by the group of scholars gathered by Liu An (prince of Huai Nan) 劉安.

Partial trs. Morgan (1); Erkes (1); Hughes (1); Chatley (1); Wieger (2).
Chung-Fa Index no. 5.
TT/1170.

Hsi Ming 西銘.
The Inscription on the Western Wall [of his lecture-theatre].
Sung, *c.* +1066.
Chang Tsai 張載.
Tr. Eichorn (3).

Hsi Yüan Lu 洗冤錄.
The Washing Away of Wrongs [treatise on forensic medicine].
Sung, +1247.
Sung Tzhu 宋慈.
Partial tr. H. A. Giles (7).

Hsiao Ching 孝經.
Filial Piety Classic.
Chhin and C/Han.
Attrib. Tseng Shen (pupil of Confucius) 曾參.
Tr. De Rosny (2); Legge (1).

Hsing Li Ching I 性理精義.
Essential Ideas of the Hsing-Li [Neo-Confucian] School of Philosophers [a condensation of the Hsing Li Ta Chhüan, q.v.].
Chhing, +1715.
Li Kuang-Ti 李光地

Hsing Li Ta Chhüan [*Shu*] 性理大全[書].
Collected Works of [120] Philosophers of the Hsing-Li [Neo-Confucian] School [Hsing = Human Nature; Li = the Principle of Organisation in all Nature].
Ming, +1415.
Ed. Hu Kuang 胡廣 *et al.*

Huang Chi Ching Shih Shu 皇極經世書.
Book of the Sublime Principle which Governs All Things Within the World.
Sung, *c.* +1060.
Shao Yung 邵雍.
TT/1028. Abridged in *Hsing Li Ta Chhüan* and *Hsing Li Ching I.*

Huang Ti Nei Ching, Ling Shu 黃帝內經靈樞.
The Yellow Emporer's Manual of Corporeal [Medicine], the Vital Axis [medical physiology and anatomy].
Probably C/Han, *c.* −1st century.
Writers unknown.
Edited Thank, +762, by Wang Ping 王冰.
Analysis by Huang Wen (1).
Tr. Chamfrault & Ung Kang-Sam (1).
Commentaries by Ma Shih 馬蒔 (Ming) and Chang Chih-Tshung 張志聰 (Chhing) in *TSCC, I shu tien*, chs. 67 to 88.

Huo Lun Chhuan Thu Shuo 火輪船圖說. +1846.
Illustrated Explanations of the Fire Wheel Ship by Cheng Fu-Kuang 鄭復光.
Appended to the *Ching Ching Ling Chhih.*

I Ching 易經.
The Classic of Changes [Book of Changes].
Chou with C/Han additions.
Compiler unknown.
See Li Ching-Chhih (I, 2); Wu Shih-Chhang (1).
Tr. R. Wilhelm (2), Legge (9), de Harlez (1).
Yin-Te Index no. (Suppl.) 10.

Khao Kung Chi 考工記.
The Artificers' Record [a section of the *Chou Li*].
Chou and Han, perhaps originally an official document of Chhi State, incorporated *c.* −140.
Tr. E. Biot (1).

Kuan Tzu 管子.
The Book of Master Kuan.
Chou and C/Han. Perhaps mainly compiled in the Chi-Hsia Academy (late −4th century) in part from older materials.
Attrib. Kuan Chung 管仲.
Partial trs. Haloun (2, 5); Than Po-Fu *et al.*

Kuan Wu Phien 觀物篇.
Treatise on the Observation of Things.
Sung, *c.* +1060.
Shao Yung 邵雍.

Kuan Yin Tzu 關尹子.
[= *Wen Shih Chen Ching.*]
The Book of Master Kuan Yin.
Thang, +742 (may be Later Thang or Wu Tai). A work with this title existed in the Han, but the text is lost.
Prob. Thien Thung-Hsiu 田同秀.

Lei Ssu Ching 耒耜經.
The Classic of the Plough.
Thang, *c.* +880.
Lu Kuei-Meng 陸龜蒙.

Li Chi 禮記.
[= *Hsiao Tai Li Chi.*]
Record of Rites [compiled by Tai the Younger].
(Cf. *Ta Tai Li Chi.*)
Ascr. C/Han, *c.* −70 to −50, but really H/Han, between +80 and +105, though the earliest pieces included may date from the time of the *Analects* (*c.* −465 to −450).
Attrib. ed. Tai Sheng 戴聖
Actual ed. Tshao Po 曹褒
Trs. Legge (7); Couvreur (3); R. Wilhelm (6).
Yin-Te Index no. 27.

Loyang Chhieh Lan Chi 洛陽伽藍記.
(or *Loyang Ka-Lan Chi; seng ka-lan* transliterating *sāngharāma*).
Description of the Buddhist Temples and Monasteries at Loyang.
N/Wei, *c.* +547.
Yang Hsüan-Chih 楊衒之.

Lü Hsüeh Hsin Shuo 律學新說.
A New Account of the Science of the Pitch-Pipes.
Ming, +1584.
Chu Tsai-Yü (prince of the Ming) 朱載堉.

Lü Shih Chhun Chhiu 呂氏春秋.
Master Lü's Spring and Autumn Annals [compendium of natural philosophy].
Chou (Chhin), −239.
Written by the group of scholars gathered by Lü Pu-Wei 呂不韋.
Tr. R. Wilhelm (3).
Chung-Fa Index no. 2.

Lun Heng 論衡.
Discourses Weighed in the Balance.
H/Han, +82 or +83.
Wang Chhung 王充.
Tr. Forke (4).
Chung-Fa Index no. 1.

Lun Yü 論語.
Conversations and Discourses (of Confucius) [perhaps Discussed Sayings, Normative Sayings, or Selected Sayings]; Analects.
Chou (Lu), *c.* −465 to −450.
Compiled by disciples of Confucius
(chs. 16, 17, 18 and 20 are later interpolations).
Tr. Legge (2); Lyall (2); Waley (5); Ku Hung-Ming (1).
Yin-Te Index no. (Suppl.) 16.

Meng Chhi Pi Than 夢溪筆談.
Dream Pool Essays.
Sung, +1086; last supplement dated +1091.
Shen Kua 沈括.

Meng Tzu 孟子.
The Book of Master Meng [Mencius].
Chou, *c.* −290.
Meng Kho 孟軻.
Tr. Legge (3); Lyall (1).
Yin-Te Index no. (Suppl.) 17.

Mo Ching 墨經.
See *Mo Tzu*.

Mo Tzu (incl. *Mo Ching*) 墨子.
The Book of Master Mo.
Chou, −4th century.
Mo Ti 墨翟 and disciples.
Tr. Mei Yi-Pao (1); Forke (3).
Yin-Te Index no. (Suppl.) 21.
TT/1162.

Nan Shih 南史.
History of the Southern Dynasties.
+659.
Li Yen-Shou 李延壽.
Peking: Chung Hua Shu Chü, 1975 edition.

Nung Sang Chi Yao 農桑輯要.
Fundamentals of Agriculture and Sericulture.
Yuan, +1273. Preface by Wang Phan (b) 王磐.
Imperially commissioned, and produced by the Agriculture Extension Bureau (Ssu Nung Ssu) 司農司.
Probable editor, Meng Chhi 孟祺. Probable later editors, Chhang Shih-Wen.
暢師文 (*c.* +1286). Miao Hao-Chhien 苗好謙 (*c.* 1318).
Cf. Liu Yu-Chhien (1).

Nung Shu 農書.
Treatise on Agriculture.
Yuan, +1313.
Wang Chen 王禎.

Pen Tshao Ching Chi Chu 本草經集注.
Collected Commentaries on the *Classical Pharmacopoeia [of the Heavenly Husbandman]*.
S/Chhi, +492.
Thao Hung-Ching 陶弘景.
Now extant only in fragmentary form as a Tunhuang or Turfan MS, apart from the many quotations in the pharmaceutical natural histories, under Thao Hung-Ching's name.

Pen Tshao Kang Mu 本草綱目.
The Great Pharmacopoeia; or, The Pandects of Natural History [Minerology, Metallurgy, Botany, Zoology etc.], Arrayed in their Headings and Sub-headings.
Ming, +1596.
Li Shih-Chen 李時珍.
Paraphased and abridged tr. Read & collaborators (2–7) and Read & Pak (1) with indexes. Tabulation of plants in Read (1) with Liu Ju-Chhiang.
Cf. Swingle (7).

Pen Tshao Phin Hui Ching Yao 本草品彙精要.
Essentials of the Pharmacopoeia Ranked According to Nature and Efficacity (Imperially Commissioned).
Ming, +1505.
Liu Wen-Thai 劉文泰, Wang Phan 王槃 & Kao Thing-Ho 高廷和.

Pen Tshao Yen I 本草衍義.
Dilations upon Pharmaceutical Natural History.
Sung, pref. +1116, pr. +1119, repr. +1185, +1195.
Khou Tsung-Shih 寇宗奭.
See also *Thu Ching Yen I Pen Tshao* (*TT*/761).

Shen Nung Pen Tshao Ching 神農本草經.
Classical Pharmacopoeia of the Heavenly Husbandman.
C/Han, based on Chou and Chhin material, but not reaching final form before the +2nd century.
Writers unknown.
Lost as a separate work, but the basis of all subsequent compendia of pharmaceutical natural history, in which it is constantly quoted.
Reconstituted and annotated by many scholars; see Lung Po-Chien (*1*), pp. 2ff., 12ff.
Best reconstructions by Mori Tateyuki 森立之 (1845), Liu Fu 劉復 (1942).

Shih Pen 世本.
Book of Origins [imperial genealogies, family names, and legendary inventors].
C/Han (incorporating Chou material). −2nd century.
Ed. Sung Chung (H/Han) 宋衷.

Shih Yao Erh Ya 石藥爾雅.
The Literary Expositor of Chemical Physic; or, Synonymic Dictionary of Minerals and Drugs.
Thang, +806.
Mei Piao 梅彪.
TT/894.

Shu Ching 書經.
Historical Classic [Book of Documents].
The 29 'Chin Wen' chapters mainly Chou (a few pieces possibly Shang); the 21 'Ku Wen' chapters a 'forgery' by Mei Tse 梅賾, *c.* +320, using fragments of genuine antiquity. Of the former, 13 are considered to go back to the −10th century, 10 to the −8th, and 6 not before the −5th. Some scholars accept only 16 or 17 as pre-Confucian.
Writers unknown.
See Wu Shih-Chhang (1); Creel (4).
Tr. Medhurst (1); Legge (I, 10); Karlgren (12).

Shuo Fu 說郛.
Outer Ramparts of Reportage.
+1652 (expanded edition).
Thao Tsung-I 陶宗儀 and Thao Thing 陶廷 eds.
On editions, see S. King P'ei-yuan, *Etude comparative des diverses éditions du* Chuou Fou. Peiping: Centre franco-chinois d'études sinologiques, 1946.

Shuo Wen.
See *Shuo Wen Chieh Tzu*.

Shuo Wen Chieh Tzu 說文解字.
Analytical Dictionary of Characters [lit. Explanations of Simple Characters and Analyses of Composite Ones].
H/Han, +121.
Hsü Shen 許慎.

Sun Tzu Ping Fa 孫子兵法.
Master Sun's Art of War.
Chou (Chhi), *c.* −345.
Attrib. Sun Wu 孫武, more probably by Sun Pin 孫臏.

Tao Te Ching 道德經.
Canon of the Tao and its Virtue.
Chou, before −300.
Attrib. Li Erh (Lao Tzu) 李耳(老子).
Tr. Waley (4); Chhu Ta-Kao (2); Lin Yü – Thang (1); Wieger (7); Duyvendak (18); and very many others.

Tao Tsang 道藏.
The Taoist Patrology [containing 1464 Taoist works].
All periods, but first collected in the Thang about +730, then again about +870 and definitively in +1019. First printed in the Sung (+1111−1117). Also printed in J/Chin (+1168−1191), Yuan (+1244) and Ming (+1445, +1598) and +1607.
Writers numerous.
Indexes by Wieger (6), on which see Pelliot's review (58); and Ong Tu-Chien (Yin-Te Index, no. 25).

Thien Kung Khai Wu 天工開物.
The Exploitation of the Works of Nature.
Ming, +1637.
Sung Ying-Hsing 宋應星.
Tr. Sun Jen I-Tu & Sun Hsüeh-Chuan (1).

Thung Chien Kang Mu 通鑑綱目.
Essential Mirror of Universal History [the *Tzu Chih Thung Chien* condensed].
Sung, +1189.
Chu Hsi 朱熹 and his school.
With later continuations.
Partial tr. Wieger (1).

Tso Chuan 左傳.
Master Tso-Chhiu's Enlargement of the *Chhun Chhiu [Spring and Autumn Annals]*.
Chou, compiled between −430 and −250, but with additions and changes by Confucian scholars of the Chhin and Han, especially Liu Hsin. Deals with the period −722 to −453. Greatest of the three commentaries on the *Chhun Chhiu*, and others being the *Kungyang Chuan* and the *Kuliang Chuan*, but unlike them, probably itself originally an independent book of history.
Attrib. Tso-Chhiu Ming 左邱明.

Tung Ching Meng Hua Lu 東京夢華錄.
Dreams of the Glories of the Eastern Capital.
S/Sung, +1148, first printed +1187.
Meng Yüan-Lao 孟元老.
Peking: Shang Wu, 1959 edition.

Tzu Hsia Chi 資暇集.
Providing for Leisure Time.
Late Thang +9th century.
Li Khuang-I 李匡義.
Chhin Ting Ssu Khu Chhüan Shu 欽定四庫全書 edition. [Taipei: 1983], vol. 850.

Wu-Shih-Erh Ping Fang 五十二病方.
Recipes for the Fifty-two Diseases.
−2nd century.
Manuscript excavated from a tomb dated −168 at Ma-Wang-Tui 馬王堆, Changsha 長沙 in 1973.

Wu Tsa Tsu 五雜組 [or 俎].
Fivefold Miscellany.
+1608.
Hsieh Chao-Che 謝肇淛.
Taipei: Hsin Hsing Shu Chü, 1971 edition.

Yen Thieh Lun 鹽鐵論.
Discourses on Salt and Iron [record of the debate of −81 on State control of commerce and industry].
C/Han, *c.* −80 to −60.
Huan Khuan 桓寬.
Partial tr. Gale (1); Gale, Boodberg and Lin (1).

Yü Kung 禹貢.
The Tribute of Yü.
(A chapter of the *Shu Ching*, q.v.).

B. CHINESE AND JAPANESE BOOKS AND JOURNAL ARTICLES SINCE +1800

Chhien Pao-Tshung 錢寶琮 ed. (1963).
Suan Ching Shih Shu: II *Chiu Chang Suan Shu*
算經十書. II: 九章算術.
Ten Mathematical Classics. II: Nine Chapters on the Mathematical Art.
Peking: Chung Hua.

Kuo Mo-Jo (郭沫若) *et al.*
Kuan Tzu Chi Chiao 管子集校.
Collected Commentary on the Book of Master Kuan.
Hsin Hua Shu-Tien, Shanghai, 1956.

Kuo Shuang-Lin 郭雙林 & Hsiao Mei-Hua 肖梅花 (1995).
Chung-Kuo Tu Po Shih 中國賭博史.
A History of Gambling in China.
Peking: Chung-Kuo She Hui Kho Hsüeh.

Lo Hsin-Pen 羅新本 & Hsüeh Jung-Sheng 許蓉生 (1994).
Chung-Kuo Ku Tai Tu Po Hsi Su
中國古代賭博習俗.
Gambling Customs in Ancient China.
Sian: Shan-hsi Jen Min Chhu Pan She.

Pai Shang-Shu 白尚恕 (1983).
Chiu Chang Suan Shu Chu Shih 九章算術注釋.
Annotated Explanations of Nine Chapters on the Mathematical Art.
Peking: Kho Hsüeh.

Teng-chhuan Chou Chih 鄧川州志.
Gazetteer of Teng-chhuan Department.
+1854/5.
Comp. and rev. by Hou Yün-Chhin 侯允欽.
Repr. Taipei: Cheng Wen, 1968.

Wang Chien-Chhiu 王建秋 (1965).
Sung Tai Thai Hsüeh Yü Thai Hsüeh Sheng
宋代太學與太學生.
The Great School and its Students in the Sung Dynasty.
Taipei: Thai-wan Shang Wu Yin Shu Kuan.

Wei Yüan 魏源 & Lin Tse-Hsü 林則徐 (1).
Hai Kuo Thu Chih 海國圖志.
Illustrated Record of the Maritime [Occidental] Nations.
1844; enlarged ed. 1847; further enlarged ed. 1852

Wu Chhi-Chün 吳其濬.
Chih Wu Ming Shih Thu Khao 植物名實圖考.
Illustrated Investigation of the Names and Natures of Plants.
1848. See SCC, vol. 7, pt. 2, 4, pp. 7–8.

Wu Wen-Chün 吳文俊 ed. (1998).
Chung-Kuo Shu Hsüeh Shih Ta Hsi 中國數學史大系.
Compendium on Chinese Mathematics.
Peking: Shih Fan Ta Hsüeh.
Vol. 2: Shen Khang-Shen 沈康身 ed.
Chung-Kuo Ku Tai Shu Hsüeh Ming Chu Chiu Chang Suan Shu. 中國古代數學名著九章算術.
Nine Chapters on the Mathematical Art: A Masterpiece of Ancient Chinese Mathematics.

Yabuuti Kiyoshi 藪内清 ed. (1970).
So Gan Jidai no Kagaku Gijutsushi.
[Essays] On the History of Science and Technology in the Sung and Yuan Periods.
Kyoto: Jimbun Kagaku Kenkyusho.
Repr. 1970.

C. BOOKS AND JOURNAL ARTICLES IN WESTERN LANGUAGES

ADANSON, MICHEL (1763). *Familles des plants*, 2 vols. Vincent, Paris, 1761. 2nd ed. of Vol. 1 only, posthumous; with many additions: *Familles naturelles; Premiere partie, comprenant l'histoire de la botanique*, ed. A. Adanson & J. Payer, Paris, 1847, 1864.

AGASSI, J. (1963). *Towards an Historiography of Science*. Mouton, s-Gravenhage, 1963. (History and Theory; Studies in the Philosophy of History, Beiheft no. 2.)

AJDUKIEWICZ, KAZEMIERZ (1948). 'Zmiana i Sprzecznosc' (Change and Contradiction), *Mysl Wspolczesna* (7-8), 35; reprinted in his *Jezyk i Poznanie* (Language and Cognition), Vol. 1: 'Selected Papers (1920–1939)' (1960), Vol. 2: 'Selected Papers (1945–1963)' (1965). Panstwowe Wydawnictwo Naukowe, Warsaw.

ALBERTUS MAGNUS (1517). *De Vegetalibus et Plantis*. See WEISHEIPL (1980).

ALVARES, CLAUDE (1980). *Homo Faber – Technology and Culture in India, China and the West from 1500 to the Present Day*. Martinus Nijhoff, The Hague.

ANGSTRÖM, A. J. (1868). *Recherches sur le spectre solaire*. W. Schultz, Uppsala.

ANON. (1962). *Chinese Therapeutical Methods of Acupuncture and Moxibustion*. Foreign Languages Press, Peking. An authoritative statement prepared by the National Academy and Research Institute of Traditional Medicine in Peking.

ANON. (1972). *How Things Work: The Universal Encyclopedia of Machines*. 2 vols. Granada, in Paladin Books, London, 1967; German original 1963; repr. 1972.

ANON. (1974). 'Genetics and the Quality of Life', *Study Encounter*, **10** (1); *Study Encounter Report no. 53*, WCC, 160 Route de Ferney, Geneva, 20, 1.

ANON. (1975). 'International News' (re Berg Conference at Asilomar), *N*, **254**, 6.

APPULEIUS, LUCIUS (*c*. +15th cent., ascribed). *De Herbarum Virtutibus*. Rome, *c*. +1480.

ARBER, A. (1912). *Herbals, their Origin and Evolution: A Chapter in the History of Botany, 1470 to 1670*. Cambridge University Press; 2nd ed. 1938; repr. 1953, 1986.

BACON, FRANCIS (1620). *Instauratio Magna [Sive] Novum Organum*, Lib. I. II. and the *Parasceve ad historiam naturalem et experimentalem*. Joannes Billius, London.

BAILEY, ANNE M. & LLOBERA, JOSEP R. (eds.) (1981). *The Asiatic Mode of Production*. Routledge and Kegan Paul, London, Boston and Henley.

BALAZS, E. (1965). *Political Theory and Administrative Reality in Traditional China*. SOAS, London.

BALL, J. DYER (1904). *Things Chinese or Notes Connected with China*. Murray, London.

BARATOUX, J. (1942). *Précis elémentaire d'acuponcture; avec repérage anatomique des points et leurs applications thérapeutiques*. Le François, Paris.

BARRETT, T. (1998). *Needham Research Institute Newsletter*, no. 17, December.

BASALLA, GEORGE (1988). *The Evolution of Technology*. Cambridge University Press.

BAUDELAIRE, CHARLES (1857). 'L'Albatros', *Les Fleurs du mal: The Complete Text of the Flowers of Evil*; in a new translation by Richard Howerd. Harvester, Brighton, 1982.

BAUER, WOLFGANG (1966). 'The Encyclopaedia in China', *JWH*, **9**, 667.

BAYLIN, J. tr. (1929). *Extraits des Carnets de Lin K'ing [Wanyen Lin-Chhing]; Sites de Pekin et des environs vus par un lettré chinois*. Lim. ech., Nachbaur, Peiping. Reproduction of 26 illustrations and descriptions from the *Hung Hsueh Yin Yuan Thu Shuo*.

BEALE, LIONEL L. (1878). *The Microscope and its Application to Clinical Medicine*. Highley's Library of Science and Art, London, 1st ed. 1854; 4th ed. 1878.

BEAU, G. (1965). *Le Médecine chinoise*. Editions du Seuil, Paris. (Le Rayon de la Science series, no. 23.)

BEDINI, SILVIO A. (1994). *The Trail of Time – Time Measurement with Incense in East Asia*. Cambridge University Press.

BEER, A. ed. (1955). *Vistas of Astronomy*. (Stratton Presentation Volume) Pergamon, London.

BEER, A., HO PING-YÜ, LU GWEI-DJEN, NEEDHAM, JOSEPH, PULLEYBANK, E. G. & THOMPSON, G. I. (1964). 'An Eighth-Century Meridian Line: I-Hsing's Chain of Gnomons and the Pre-History of the Metric System', *VA*, **4**, 3.

BEER, JOHN (1968). *Blake's Humanism*, Barnes & Noble, New York; Manchester University Press, Manchester.

BENNETT, A. A. (1967). *John Fryer: The Introduction of Western Science and Technology into Nineteenth Century China*. Harvard University Press, Cambridge, Mass. (East Asian Monograph no. 24.)

BENTLEY, G. E. (1963). *Vala, or the Four Zoas*, Clarendon, Oxford.

BERNIER, FRANÇOIS (1909). *Bernier's Voyage to the East Indies: containing The History of the Late Revolution of the Empire of the Great Mogul; together with the most considerable passages for five years following in that Empire; in which is added A Letter to the*

Lord Colbert, touching the extent of Hindustan, the Circulation of the Gold and Silver of the world, to discharge itself there, as also the Riches, Forces and Justice of the Same, and the principal Cause of the Decay of the States of Asia – with an Exact Description of Delhi and Agra; together with (1) Some Particulars and Customs of the Heathens of Hindustan, (2) The Emperor of Mogul's Voyage to the Kingdom of Kashmere, in 1664, called the Paradise of the Indies . . . Dass (for SPCK), Calcutta, 1909. [Substantially the same title-page as the editions of 1671 and 1672.]

BETTLEHEIM, CHARLES (1975). 'Economics and Ideology', *CNOW*, **52**, 9.

BEVAN, EDWYN ROBERT (1940). *Holy Images: An Inquiry into Idolatry and Image-Worship in Ancient Paganism and in Christianity*. Allen & Unwin, London.

BILLETER, J. F. *et al.* (1993). 'Florilège des *Notes du Ruisseau des Reves* (Mengqi bitan) de Shen Gua (1031–1095)', *Etudes Asiatiques*, **47** (3), 389–451.

BILLETER, J. F. (2000). *Chine: Trois fois muette*. Allia, Paris.

BIOT, E. tr. (1851). *Le Tcheou-Li ou rites des Tcheou*. 3 vols. Imp. Nat., Paris. (Photographically reproduced, Wentienko, Peking, 1930.)

BIRINGUCCIO, *De la Pirotechnia* (1540). See SMITH & GNUDI (1959).

BJÖRK, LARS E. (1975). 'An Experiment in Work Satisfaction', *SAM*, **232** (3), 17.

BLACKBURN, T. R. (1971). 'Sensuous-Intellectual Complementarity in Science', *SC*, **172**, 1003.

BLOFELD, JOHN (1965). *The Book of Change*. Allen and Unwin, London.

BLOOM, ALFRED (1981). *The Linguistic Shaping of Thought: A Study in the Impact of Language on Thinking in China and the West*. Lawrence Erlbaum Associates, Hillsdale, NJ.

BLUE, GREGORY (1979). 'A Brief Account of the 1931 Leningrad Conference on the Asiatic Mode of Production'. Unpublished paper prepared for the Social and Political Sciences Committee, University of Cambridge.

BLUNDEN, C. & ELVIN, M. (1998). *A Cultural Atlas of China*. Phaidon, Oxford, 1983; rev. ed., Facts on File, New York, 1998.

BODMER, FREDERICK (1944). *The Loom of Language: A Guide to Foreign Languages for the Home Student*, ed. and arranged by Lancelot Hogben 2nd imp. Allen & Unwin, London. (Primers for the Age of Plenty no. 3.)

BOYM, MICHAEL (1656). *Flora Sinensis, Fructus Floresque humillime porrigens serenissimo ac potentissimo Principi, ac Domino Leopoldo Ignatio, Hungariae Regi florentissimo, etc. Fructus saeculo promittenti Augustissimos, emissa in publicum a R. P. Michaele Boym Societatis Iesu Sacerdote, et a domo professa ejusdem Societatis Viennae Majestati Suae una cum faelicissimi Anni apprecatione oblata Anno Salutis MDCLVI*. Richter, Vienna, 1656; repr. in M. Thevenot's *Voyages*, vol. 2. Langlois, Paris, 1730.

BRANDT, C., SCHWARTZ, B. & FAIRBANK, J. K. (1952). *A Documentary History of Chinese Communism*. Harvard University Press, Cambridge, Mass.

BRAY, FRANCESCA (1979). 'The Evolution of the Mouldboard Plough in China', *TT*, **3** (4), 227–40.

BRAY, FRANCESCA (1984). *Science and Civilisation in China*, Vol. 6, pt. 2: *Agriculture*. Cambridge University Press.

BRAY, FRANCESCA (1986). *The Rice Economies: Technology and Development in Asian Societies*. Blackwell, Oxford.

BRAY, FRANCESCA (1989). 'The Classification of Crop Plants in China', *CSCI*, **9**, 1–3.

BRETSCHNEIDER, E. (1871). 'On the Study and Value of Chinese Botanical Works; with Notes on the History of Plants and Geographical Botany from Chinese Sources.' Rozario & Marcal, Fuchow, 1871. First published in *CRR*, 1870, **3**, 157, 172, 218, 241, 264, 281, 290, 296. Chinese tr. by Shih Sheng-Han (down to p. 24, omitting the discussion on Palmae, but with the addition of critical notes) Chung-Kuo Chih-Wu-Hsüeh Wen-Hsien Phing-Lun. Nat. Compilation & Transl. Bureau, Shanghai, 1935; repr. Com. Press Shanghai, 1957.

BRITISH MUSEUM. See MUSEUM, THE BRITISH

BRITTON, ROSWELL S. (1935). *The Cowling-Chalfont Collection of Inscribed Oracle Bones* [drawn by: Frank H. Chalfont]. Commercial Press, Shanghai, 1935.

BROCK, SEBASTIAN (1974). 'World and Sacrament in the Writings of the Syrian Fathers', *SOB*, **6** (10), 685.

BROCKBANK, Y. (1954). *Ancient Therapeutic Arts*. Heinemann, London. (Fitzpatrick Lectures, Royal College of Physicians, 1950–1.)

BURY, J. B. (1924). *The Idea of Progress: An Inquiry into its Origin and Growth*. Macmillan, London.

BUSHELL, STEPHEN W. (1906). *Chinese Art*. 2 vols. Victoria and Albert Museum, London, 1914, first published as Victoria and Albert Museum handbooks, 1st ed. 1904, 2nd ed. 1909.

CAJORI, FLORIAN (1893). *A History of Mathematics*. Macmillan, New York; rev. ed. 1919.

CAPRA, F. (1974). 'Modern Physics and Eastern Philosophy', *Human Dimensions*, **3** (2), 3.

CAPRA, F. (1975). *The Tao of Physics*, Wildwood House, London.

CARDWELL, DONALD (1994). *The Fontana History of Technology*. Fontana Press, London.

CARTER, T. F. (1925). *The Invention of Printing in China and its Spread Westward*. Columbia University Press, New York, 1925, 1931; rev. L. Carrington Goodrich. Ronald Press, New York, 1955.

CASTIGLIONI, A. (1947). *A History of Medicine*, tr. and ed. E. B. Krumbhaar. Knopf, New York; 2nd ed. 1954.

CHAMFRAULT, A. & UNG KANG-SAM (1954). *Traité de médecine chinoise: d'après les textes chinois anciens et modernes*, 5 vols., with illustrations by M. Rouhier. Coquemard, Angouleme.

CHAN, WING-TSIT. See CHEN JUNG-CHIEH.

CHANCELLOR, M. tr. of LAVIER, J. (1965).

CHANG CHHENG-SHAO (1945). 'The Present Status of Studies on Chinese Anti-malarial Drugs', *CSMJ*, **63A**, 126.

Chang Chhang-Shao, Fu Feng-Yüng, Huang, K. C. & Wang, C. Y. (1948). 'Pharmacology of chhang-shan (Dichroa febrifuga), a Chinese Antimalarial Herb', *N*, **161**, 400. (With comment by T. S. Work.)

Chang Kwang-Chih (1963). *The Archaeology of Ancient China.* 3rd ed. revised and enlarged. Yale University Press, New Haven, Conn., 1977.

Chase, Stuart (1929). *Men and Machines*, Jonathan Cape, London.

Chemla, Karine (1982). 'Etude du livre reflets des mesures du cercle sur la mer, de Li Ye'. 4 vols. Doctoral thesis, University of Paris XIII.

Chemla, K. (1997–8). 'Fractions and Irrationals between Algorithm and Proof in Ancient China', *Studies in the History of Science and Medicine*, **15** (1–2), 31–54.

Chhen Cheng-Yih (Joseph) ed. (1987). *Science and Technology in Chinese Civilisation.* World Scientific Publishing, Singapore.

Chhen Cheng-Yih (Joseph) ed. (1994). *Two-tone Set-bells of Marquis Yi.* World Scientific Press, Singapore.

Chen Chii-Thien (1938). *Lin Tse-Hsü; Pioneer Promoter of the Adoption of Western Means of Maritime Defence in China.* Dept. of Economics, Yenching University, Vetch, Peiping, 1934. ([Studies in] Modern Industrial Technique in China, no. 1.)

Chhen Jung-Chieh (Chan Wing-Tsit) (1975). 'Chinese and Western Interpretations of Jen (Humanity, Love, Humaneness)', *JCP*, **2** (2), 107.

Chhen Kho-Khuei, Mukerji, E. & Volicer, L. ed. (1965). *The Pharmacology of Oriental Plants.* Pergamon Press, London; Czechoslovak Med. Press, Prague. (Proc. 2nd Int. Pharmacological Meeting, Prague, 1963, vol. 7.)

Chhien Tshun-hsün (1985). See Tsien, Hsuen-Hsuin.

Chhien Wen-Yüan (Qian Wen Yuan) (1985). *The Great Inertia: Scientific Stagnation in Traditional China.* Croom Helm, London.

Childe, V. Gordon (1951). *Social Evolution.* Watts, London.

Ciba Foundation Symposium (1968). *Law and Ethics of Transplantation*, ed. Gordon Wolstenholme and Maeve O'Connor, J & A Churchill, London. Originally published in 1966 under the title: *Ethics in Medical Progress with Special Reference to Transplantation.*

Ciba Foundation Symposium (1972). *Civilisation and Science: in Conflict or Collaboration?* Symposium held 28th–30th June 1971, Associated Scientific Publishers (Elsevier, Excerpta Medica, North-Holland), Amsterdam and London.

Ciba Foundation Symposium (1973). *Law and Ethics of A.I.D. and Embryo Transfer*, Symposium on Legal and Other Aspects of Artificial Insemination by Donor – A.I.D. – and Embryo Transfer held at the Ciba Foundation, London, 1st Dec. 1972, Associated Scientific Publishers (Elsevier, Excerpta Medica, North-Holland), Amsterdam and London.

Cicero, '*De Finibus*'. See Rackham (1914).

Cicero, '*De Legibus*'. See Rackham (1928).

Cicero, *M. Tulli Ciceronis Scripta Quae Mansuerant Omnia Fasc. 46: De Divinahone; De fato; Timaeus*, ed. W. Ax, Teubner, Leipzig, 1938.

Clark, D. H. & Stephenson, F. R. (1977). *The Historical Supernovae*, Pergamon Press, Oxford.

Cobb, Richard Charles (1970). *The Police and the People: French Popular Protest 1789–1820.* Clarendon Press, Oxford.

Cohen, H. Floris (1994). *The Scientific Revolution: A Historiographical Inquiry.* University of Chicago Press. Chicago.

Collingwood, R. G. (1924). *Speculum Mentis, or The Map of Knowledge*, Clarendon Press, Oxford.

Collins, Glen (1994). 'Bronze Bells Yield Secrets of Ancient China's Music', *New York Times.* 18 January.

Cooke Johnson, Linda (1993). *Cities of Jiangnan in Late Imperial China.* State University of New York, New York.

Cooper, Arthur tr. (1973). *Li Po and Tu Fu: Poems Selected and Translated with an Introduction and Notes.* (Chinese calligraphy by Shui Chien-tung.) Penguin, Harmondsworth.

Copernicus, Nicholas (1526). *Monete Cudende Ratio* (*Monetae Cudendae Ratio.*) Private paper written at the invitation of Sigismond I, King of Poland. First published by Felix Bentkowski in 1816 in pamietnik Warszawski (Warsaw memorial). Republished with *Tractatus de Origine, Natura, Jure et Mutationibus Monetarum* of Nicolas Oresme by M. L. Wolowski, Guillaumin, Paris, 1864.

Copernicus, Nicholas (1543). *De revolutionibus orbium coelestium.* Petreium, Nurnberg.

Corrin, S. tr. (1962). *The Forge and the Crucible.* Harper, New York.

Cotgrove, Stephen (1973). 'Anti-Science', *NS*, **59** (854), 12 July.

Cotgrove, Stephen (1974). 'Objections to Science', *N*, **250**, 764.

Cotterell, Brian & Kamminga, Johan (1990). *Mechanics of the Pre-Industrial Technology.* Cambridge University Press.

Couling, Samuel (1917). *The Encyclopaedia Sinica.* Oxford University Press.

Couvreur, F. S., tr. (1914). '*Tch'ouen Ts'iou' (Chhun Chhiu) et 'Tso Tchouan' (Tso Chuan); Texte chinois avec traduction française.* 3 vols. Mission Press, Hochienfu; repr. Belles Lettres, Paris, 1951.

Cox, E. H. M. (1945). *Plant-Hunting in China: A History of Botanical Exploration in China and the Tibetan Marches.* Collins, London. (Photolitho edition, Scientific Book Guild, London.)

Cranmer-Byng, J. L. ed. (1958). 'Lord Macartney's Embassy to Peking in 1793 – From Official Chinese Documents', *JOSHK*, **4**.

CRANMER-BYNG, J. L. ed. (1962). *An Embassy to China; being the Journal kept by Lord Macartney during his Embassy to the Emperor Chhien-Lung, +1793 and +1794*. Longmans, London.

CRANMER-BYNG, J. L. (1972). 'The Chinese Attitude towards the Natural World', *Ontario Naturalist*, **12** (4), 28.

CREEL, H. G. (1937a). *Studies in Early Chinese Culture*. (1st series.) Waverly, Baltimore.

CREEL, H. G. (1937b). *The Birth of China*, Fr. tr. M. C. Salles, Payot, Paris. (References are to page numbers of the French ed.)

CRICK, F. (1966). *Of Molecules and Men*, University of Washington Press, Seattle.

CROMBIE, A. C. (1952). *Augustine to Galileo: The History of Science A.D. 400 – 1650*. Falcon, London.

CROMBIE, A. C. (1961). 'Quantification in Mediaeval Physics', in *Quantification; a History of the Meaning of Measurement in the Natural and Social Sciences*, ed. H. Woolf, Bobbs-Merrill, Indianapolis & New York. Published separately in *Isis* (1961), **52**, 143.

CROMBIE, A. C. (1963). 'The Relevance of the Middle Ages to the Scientific Movement', in *Perspectives in Mediaeval History*, ed. K. F. Drew & F. S. Lear, Rice University (Houston, Texas), University of Chicago Press, 35.

CROMBIE, A. C. (1994). *Styles of Scientific Thinking in the European Tradition*. 3 vols. Duckworth, London.

CROMBIE, A. (1994). *Styles of Scientific Thinking in the European Tradition: The History of Argument and Explanation Especially in the Mathematical and Biomedical Sciences and Arts*. 3 vols. Duckworth, London.

CRUMP, THOMAS (1990). *The Anthropology of Numbers*. Cambridge University Press.

CULLEN, C. (1977). Cosmological Discussions in China from Early Times up to the T'ang Dynasty'. Ph.D. dissertation, University of London.

CUROTT, D. R. (1966). 'Earth Deceleration from Ancient Solar Eclipses', *ASTJ*, **71**, 264.

DANIELS, V. G. (1972). 'Research Report on the Termination of Macromolecular Uptake by the Neonatal Intestine after Lactation in the Rat' (Unpub.).

DARWIN, CHARLES (1859). *The Origin of Species by Means of Natural Selection*. 1st ed. Murray, London; 6th ed. repr. 1921.

DARWIN, ERASMUS (1789–91). *The Botanic Garden*. Part I containing 'The Economy of Vegetation', 1791; Part II containing 'The Loves of the Plants', London, 1789.

DATTA, BIBHUTIBHUSAN & SINGH, AVADHESH NARAYAN (1935–1938). *History of Hindu Mathematics: A Source Book*. 1935 (vol. 1), 1938 (vol. 2). Motilal Banarsi Das, Lahore.

DAWSON, RAYMOND ed. (1964). *The Legacy of China*. Clarendon Press, Oxford.

DE BARY, WILLIAM THEODORE (1993). 'Chen Te-hsiu and Statecraft', in HYMES & SHIROKAUER (1993), 349–79.

D'ELIA, P. (1946). 'Echi delle Scoperte Galileiane in Cina vivente ancora Galileo (1612 1640)', *AAL/RSM* (8e ser.), **1**, 125. Republished in enlarged form as 'Galileo in Cina. Relazioni attraverso il Collegio Romano tra Galileo e i gesuiti scienzati missionari in Cina (1610–1640)', *Analecta Gregoriana*, **37** (Series Facultatis Missiologicae A, no. 1), Rome, 1947. Eng. tr. '*Galileo in China*', with emendations and additions by R. Suter & M. Sciascia, Harvard University Press, Cambridge, Mass., 1960.

Dictionary of Christian Spirituality (1983). Ed. Wakefield, Gordon S. SCM Press, London.

Dictionary of Scientific Biography. See GILLISPIE.

Dizionario etimologico della lingua italiana, Cortelazzo & Zolli, Bologna, 1983. ref. to use in +1330 by IACOPO D'ACQUI.

DORN, HAROLD (1991). *The Geography of Science*. John Hopkins University Press, Baltimore and London.

DUBS, H. H. (1947). 'The Beginnings of Alchemy', *Isis*, **38**, 62.

DUBS, H. H. (1961). 'The Origin of Alchemy', *AX*, **9**, 23.

DUDGEON, J. (1895). 'Kung-Fu, or Medical Gymnastics', *JPOS*, **3**, 341.

DUGGAR, B. M. & SINGLETON, V. L. (1953). 'The Biochemistry of Antibiotics', *ARB* **22**, 459.

DUNN, S. P. (1982). *The Fall and Rise of the Asiatic Mode of Production*. RKP, London.

EAMON, WILLIAM (1994). *Science and the Secrets of Nature: Books of Secrets in Mediaeval and Early Modern Culture*. Princeton University Press.

EBERHARD, W. (1933). *Beiträge zur Kosmologischen Spekulation Chinas in der Han-Zeit*. Baessler Archiv. *16*, 1–100. Baessler Instituts, Berlin.

EDWARDS, ROBERT G. & SHARPE, DAVID J. (1971). 'Social Values and Research in Human Embryology', *N*, **231**, 87.

EDWARDS, ROBERT G. (1974). 'Fertilisation of Human Eggs in vitro; Morals, Ethics and the Law', *QRB*, **49**, 3.

EISENSTEIN, ELIZABETH L. (1979). *The Printing Press as an Agent of Social Change*. 2 vols. Cambridge University Press.

EISENSTEIN, E. (1980). *The Printing Press as an Agent of Change*. Cambridge University Press.

ELIADE, M. (1956). *Forgerons et Alchimistes*. M. Flammarion, Paris. English. tr. by S. Corrin, The Forge and the Crucible, 2nd ed., University of Chicago Press, 1978.

ELLUL, JACQUES (1964). *The Technological Society*, tr. John Wilkinson, Alfred A. Knopf, New York; Jonathan Cape, London, 1965.

ELVIN, M. (1972). 'The High-Level Equilibrium Trap: The Causes of the Decline of Invention in the Traditional Chinese Textile Industries', in W. Willmott, ed., *Economic Organization in Chinese Society*, Stanford University Press, Stanford, CA, 137–72; repr. in ELVIN (1996), 20–63.

ELVIN, M. (1973). *The Pattern of the Chinese Past.* Stanford University Press, Stanford, CA.

ELVIN, M. (1975). 'Skills and Resources in Late Traditional China'. in D. Perkins, ed., *China's Modern Economy in Historical Perspective*, Stanford University Press, Stanford, CA, 85–113; repr. in ELVIN (1996), 64–100.

ELVIN, M. (1984). 'Why China Failed to Create an Endogenous Industrial Capitalism: A Critique of Max Weber's Explanation'. *Theory and Society*, **13** (3), 379–91.

ELVIN, M. (1988). 'China as a Counterfactual'. In J. Baechler, J. Hall & M. Mann, eds., *Europe and the Rise of Capitalism.* Blackwell, Oxford, 101–12.

ELVIN, M. (1990). 'The Logic of Logic: A Comment on Mr Makeham's Note', *Papers on Far Eastern History*, **42** (September), 131–4.

ELVIN, M. (1993–4). 'The Man Who Saw Dragons: Science and Styles of Thinking in Xie Zhaozhe's *Fivefold Miscellany*', *Journal of the Oriental Society of Australia*, **25** and **26**, 1–41.

ELVIN, M. (1996). *Another History: Essays on China from a European Perspective.* Wild Peony Press, Sydney.

ELVIN, M. (1997). *Changing Stories in the Chinese World.* Stanford University Press, Stanford, CA.

ELVIN, M. (1999). 'Blood and Statistics: Reconstructing the Population Dynamics of Late Imperial China from the Biographies of Virtuous Women in Local Gazetteers', in H. Zurndorfer, ed., *Chinese Women in the Imperial Past: New Perspectives*, Brill, Leiden, 135–222.

ELVIN, M. (2000). 'Personal Luck: Why Premodern China – Probably – Did Not Develop Probabilistic Thinking'. On deposit in the Library of the Needham Research Institute, 8 Sylvester Road, Cambridge.

ELVIN, M. (2001). Review of K. Pomeranz, *The Great Divergence: China, Europe, and the Making of the Modern World Economy* Princeton University Press, 2000 in *The China Quarterly*, **167**, 754–8.

ELVIN, M. (forthcoming). *The Retreat of the Elephants: The Environmental History of China.* Yale University Press, New Haven.

ELVIN, M. & Liu, T.-J., eds. (1998). *Sediments of Time: Environment and Society in Chinese History*, Cambridge University Press, New York.

ELVIN, M., PETERSON, W., LIBBRECHT, U., & CULLEN, C. (1980). 'Symposium, The Work of Joseph Needham'. *Past and Present*, **87**, May.

ELVIN, M. & Su, N. (1998). 'Action at Distance: The Influence of the Yellow River on Hangzhou Bay since A.D. 1000'. In M. Elvin & T-J. Liu, eds., *Sediments of Time: Environment and Society in Chinese History*, Cambridge University Press, New York, 344–407.

EMPSON, WILLIAM (1961). *Seven Types of Ambiguity.* Penguin Books, Harmondsworth. First published Chatto and Windus, 1930.

ENGLEHARDT, W. A. (1974). 'Hierarchies and Integration in Biological Systems', *Bull. Amer. Acad. Arts & Sciences*, **27** (4), 11.

FAIRCLOUGH, H. RUSHTON tr. (1930). Virgil, with an English Translation. 'Georgics'. (Loeb Classics edition). Heinemann, London, 1960.

VON FALKENHAUSEN, LOTHAR (1993). *Suspended Music – Chime Bells in the Culture of Bronze-Age China.* University of California Press, Berkeley.

FANG HSIEN-CHIH, CHOU YING-CHHING, SHANG THIEN-YÜ, & KU YÜN-WU (1963–4). 'The Integration of Modern and Traditional Chinese Medicine in the Treatment of Fractures', *CSMJ*, **82**, **83**, 411, 419, 425, 493.

FANG, J. (1994). *The Needham Question: Between Mathematics and Sociology I.* Paideia and P.M., Virginia.

FANG, J. (1994a). *The Birth of Exact Sciences: Between Mathematics and Sociology II.* Paideia and P.M., Virginia.

FEDELE, FORTUNATO (1602). *De Relationibus Medicorum.* Libri Quattuor, Palermo; repr. Leipzig, 1674.

FELD, BERNARD T. (1974). 'Doves of the World, Unite!', *NS*, **64** (929), 26 Dec., 910.

FENG, YU-LAN (1922). 'Why China Has No Science – An Interpretation of the History and Consequences of Chinese Philosophy', *IJE*, **32** (3), 237.

FENG YU-LAN (1953). *A History of Chinese Philosophy*, Vol. II: *The Period of Classical Learning (from the Second Century B.C. to the Twentieth Century A.D.)*, tr. Derk Bodde. Princeton University Press.

FEUCHTWANG, STEPHEN (1977). 'School Temple and City God', in SKINNER, WILLIAM (1977), 581ff.

FINNISTON, SIR MONTY, WILLIAMS, TRAVOR & BISSELL, CHRISTOPHER eds. (1992). *Oxford Illustrated Encyclopedia of Inventions and Technology.* Oxford University Press.

FITZHERBERT, JOHN (1523). *The Boke of Husbandrye Verye Profytable and Necessarye for Al Maner of Persons Newlye Corrected and Amended by the Auctor Fitzherbard.* Richard Kele, Lumbard St. London.

FORKE, A. tr. (1907). *'Lun Heng', Philosophical Essays of Wang Chhung.* Vol. 1, 1907 Kelly & Walsh, Shanghai; Luzac, London; Harrasowitz, Leipzig. Vol. 2, 1911 with the addition of Reimer, Berlin. Mitteilungen des Seminars für Orientalisime Sprachen, Jahrgang X und XIV. Photolitho repr., Paragon, New York, 1962.

FRIDAY, ADRIAN & INGRAM, DAVID S., eds. (1985). *The Cambridge Encyclopaedia of Life Sciences.* Cambridge University Press.

FRYE, NORTHROP (1974). 'The Decline of the West by Oswald Spengler', *DAE* (winter), issued as *PAAAS*, **103** (1), 1.

FRYER, JOHN (n.d). 'Scientific Terminology: Present Discrepancies and Means of Securing Uniformity', 1891 in *Miscellaneous Pamphlets* (n.d.), vol. 1, no. 1, p. 11.

FU FENG-YUNG & CHANG CHHANG-SHAO (1948). 'Chemotherapeutic Studies on chhang shan (Dichroa febrifuga): III. Potent antimalarial alkaloids from chhang shan, *STIC*', **1** (3), 56.

GABELENTZ, GEORG VON DER (1881). *Chinesische Grammatik*, T. O. Weigel, Leipzig. New ed. Deutscher Verlag für Wissenschaften, Berlin, 1953.

GABELENTZ, GEORG VON DER (1891). *Die Sprachwissenschaft.* T. O. Weigel, Nachfolger, Leipzig.

GANELIUS, T., ed. (1986). *Progress in Science and Its Social Conditions.* Pergamon for the Nobel Foundation, Oxford. (Nobel Symposium 58.)

GANIERE, NICOLE (1974). *The Process of Industrialisation of China, Primary Elements of an Analytical Bibliography* (OECD Development Centre Working Document CD/TI (74)o), Development Centre of the Organisation for Economic Development, Paris.

GARDNER, F. F. & MILNE, D. K. (1965). 'The Supernova of A.D. 1006', *ASTJ*, **70**, 754.

GARRISON, F. H. (1929). *An Introduction to the History of Medicine; with Medical Chronology, Suggestions for Study, and Bibliographic Data.* 4th ed. Philadelphia and Saunders, London.

GAUBIL, A. (1783). *Histoire de l'Astronomie Chinoise.* In *Lettres Edifiantes et Curieuses, écrites des Missions Etrangères.* Nouvelle Edition – Mémoires des Indes et de la Chine, vol. 26, pp. 65–295. Mérigot, Paris.

GELDNER, FERDINAND (1965). *Das Büchlein von der Fialen Gerechtigkeit (von Matthäus Roriczer).* Faksimile der Originalausgabe, Regensburg, 1486, mit einem Nachwort und Textübertragung. Pressler, Wiesbaden.

'Genetics and the Quality of Life', *Study Encounter*, 1974, **10** (1). (World Council of Churches, 1.)

GERARD, JOHN (1597). *The Herball or General Historie of Plantes Gathered by John Gerard, Master in Chivurgie*, London. *The Herball or Generall Historie of Plantes*, 2nd ed. revised and enlarged by Thomas Johnson, 1633. *The Herbal or General History of Plants.* Dover Publications, New York, 1975.

GERNET, JACQUES (1980). 'Christian and Chinese Visions of the World in the Seventeenth Century', *CSCI*, **4**, 1–17.

GERR, STANLEY (1942). 'Language and Science: The Rational, Functional Language of Science and Technology', *Philosophy of Science.*

GERR, STANLEY (1944). *Scientific and Technical Japanese: A Study of its Efficiency as a Means of Communication, with an Analysis of the Japanese Technical Vocabulary and Practical Suggestions for Translators.* Private publication, New York.

GILES, H. A. (1924). 'The Hsi Yuan Lu or Instructions to Coroners' (tr. from the Chinese), *PRSM*, **17**, 59.

GILLISPIE, CHARLES COULSTON (1970–80). Ed. in Chief, *Dictionary of Scientific Biography.* 16 vols. Charles Scribner's Sons, New York.

GIMPEL, JEAN (1976). *The Medieval Machine: The Industrial Revolution of the Middle Ages.* Holt, Rinehart and Winston, New York.

GOLAS, P. (1999). *Science and Civilisation in China*, Vol. 5, pt. 13, *Mining.* Cambridge University Press.

GOLDSTEIN, B. R. (1965). 'Evidence for a Supernova of A.D. 1006', *ASTJ*, **70**, 105.

GOLDSTEIN, B. R. & HO PING-YÜ (1965). 'The 1006 Supernova in Far Eastern Sources', *ASTJ*, **70**, 748.

GOLDSTONE, J. (1991). *Revolution and Rebellion in the Early Modern World.* University of California Press, Berkeley, CA.

GOODMAN, DAVID & RUSSELL, COLIN A. eds. (1991). *The Rise of Scientific Europe, 1500–1800.* Hodder and Stoughton Ltd., London.

GOODMAN, D., & RUSSELL, C., eds. (1991). *The Rise of Scientific Europe: 1500–1800.* Open University, Milton Keynes.

GOULET, DENIS (1975). 'Le Monde du Sous-Developpement; une Crise de Valeurs', Comptes Rendus de la Reunion de l'Association Canadienne des Etudes Asiatiques, Montreal.

GRAHAM, A. C. (1960). *The Book of Lieh Tzu: A New Translation.* Murray, London. (In the Wisdom of the East Series.)

GRAHAM, A. C. (1965). *Poems of the Late Thang, Translated with an Introduction.* Penguin, London.

GRAHAM, A. C. (1971). 'China, Europe and the Origins of Modern Science', *AM*, **16**, 178; also published as 'China, Europe and the Origin of Modern Science: Needham's "The Grand Titration"' in NAKAYAMA & SIVIN (1973), 45.

GRAHAM, A. C. (1978). *Later Mohist Logic, Ethics and Science.* Chinese University Press, Hong Kong, and the School of Oriental and African Studies, London.

GRAHAM, A. C. (1981). *Chuang Tzu, The Inner Chapters.* Allen and Unwin, London.

GRANET, MARCEL (1934). *La Pensée Chinoise.* Albin Michel, Paris. (L'Evolution de l'Humanité: Synthèse collective no. 25 bis.)

GRANT, E. (1996). *The Foundations of Modern Science in the Middle Ages: Their Religious, Institutional, and Intellectual Contexts.* Cambridge University Press, New York.

GRIMM, T. (1977). 'Academies and Urban Systems in Kwangtung'. In G. W. Skinner, ed., *The City in Late Imperial China.* Stanford University Press, Stanford, CA, 475–98.

GUTMANN, MATTHEW C. (1992). 'Cross Cultural Conceits: Science in China and the West', *Science as Culture*, **3**, (2/15), 208–39.

HAGERTY, M. J. (1948). 'Tai Khai-Chih's *Chu Phu*: A Fifteenth-Century Monograph on Bamboos Written in Rhyme with a Commentary', *HJAS*, **11**, 372.

HALL, A. R. (1963). 'Merton Revisited, or, Science and Society in the Seventeenth Century', *HOSC*, **2**, 1.

HALLPIKE, C. R. (1986). *The Principles of Social Evolution.* Clarendon Press, Oxford.

HARBSMEIER, CHRISTOPH (1981). *Aspects of Classical Chinese Syntax.* Curzon Press, London. (Scandinavian Institute of Asian Studies Monograph Series, No. 45.)

HARBSMEIER, CHRISTOPH (1998). *Science and Civilisation in China*, Vol. 7, pt. 1, *Language and Logic in Traditional China.* Cambridge University Press.

HARLEZ, C. DE (1889). *Le Yih-King [I Ching], Texte primitif, rétabli, traduit et commenté.* Hayez, Brussels.

HARMSWORTH ENCYCLOPAEDIA, Amalgamated Press and Nelson, London, n.d. [*c.* 1900–1905].

HARRINGTON, JAMES (1656). *The Commonwealth of Oceana.* Streater (for Chapman), London.

HARTWELL, ROBERT (1962). 'A Revolution in the Chinese Iron and Coal Industries during the Northern Sung (+960 to +1126)', *JAS*, **21**, 153.

HARTWELL, ROBERT (1966). 'Markets, Technology and Enterprise in the Development of the Eleventh Century Chinese Iron and Steel Industry', *JEH*, **26**, 29.

HARTWELL, ROBERT (1967). 'A Cycle of Economic Change in Imperial China: Coal and Iron in Northeast China, +750 to +1350', *JESHO*, **10** (1), 102.

HAUDRICOURT, A. G. (1962). 'Domestication des animaux, culture des plantes, et traitement d'autrui', *LH*, **40**.

HAUDRICOURT, ANDRÉ (1973). 'Botanical Nomenclature and its Translation', in *Changing Perspectives in the History of Science, Essays in Honour of Joseph Needham*, ed. Teich, Mikulas & Young, Robert. Heinemann, London.

HEAD, H. *et al.* (1920). *Studies in Neurology*, 2 vols. Oxford Medical Publications, London.

HEATH, SIR THOMAS (1921). *A History of Greek Mathematics.* 2 vols. Oxford.

HERON OF ALEXANDRIA (+1st cent.). Artillery Manual in *Mechanici*, Venice, 1572 and in Spiritalium Liber, Urbino, 1575.

HERRMANN, A. (1935). *Historical and Commercial Atlas of China.* Harvard-Yenching Institute, Cambridge, Mass.

HESSEN, B. (1932). 'The Social and Economic Roots of Newton's Principia', in *Science at the Cross-Roads.* Papers read to the 2nd International Congress of the History of Science and Technology, London, 1931. Kniga, London.

HO PING-YÜ & NEEDHAM, J. (1959a). 'Ancient Chinese Observations of Solar Haloes and Parhelia', *W*, **14**, 124.

HO PING-YÜ & NEEDHAM, J. (1959b). 'The Laboratory Equipment of the Early Mediaeval Chinese Alchemists', *AX*, **7**, 57.

HO PING-YÜ & NEEDHAM, J. (1959c). 'Theories of Categories in Early Mediaeval Chinese Alchemy' (with translation of the Tshan Thung Chhi Wu Hsiang Lei Pi Yao, *c.* +6th to +8th cent.), *JWCI*, **22**, 173.

HOBBES, THOMAS (1651). *The Leviathan of the Matter, Forme and Power of a Commonwealth, Ecclesiastical and Civil, 1651*. Ed. M. Oakeshott, Blackwell, Oxford, n.d. (but after 1934). Also in The English Works of T. H., ed. Sir W. Molesworth (vol. 3), Bohn, London, 1939.

HOBSON, R. L. (1915). *Chinese Pottery and Porcelain – An Account of the Potter's Art in China from Primitive Times to the Present Day.* Dover Publications Inc., New York 1976 (unabridged reproduction of 1915 ed.).

HOLMYARD, E. J. (1957). *Alchemy.* Penguin Books, Harmondsworth.

HOMMEL, RUDOLPH P. (1937). *China at Work: An Illustrated Record of the Primitive Industries of China's Masses, Whose Life is Toil, and thus an Account of Chinese Civilisation.* Bucks County Historical Society, Doylestown, Pa., John Day, New York.

HOOPER, WILLIAM DANS & ASH, H. B. tr. (1934). *Marcus Terentius Varro on Agriculture*, with an Eng. tr. by W. D. Hooper, rev. H. B. Ash, Loeb Classical Library, Heinemann, London, repr. 1967.

HOOVER, H. C. & HOOVER, L. H. tr. (1950). *Georgius Agricola – De Re Metallica.* Dover, New York.

HORT, ARTHUR tr. (1916). *Theophrastus' 'Enquiry into Plants' and Minor Works on Odours and Weather Signs.* Loeb classics, 2 vols. First published 1916; Heinemann, London, 1948.

HUANG, JEN-YÜ, RAY (1981). *1587: A Year of No Significance.* Yale University Press, New Haven.

HUANG, JEN-YÜ, RAY (1988). *China: A Macro-History.* M. E. Sharpe, New York and London.

HUARD, P. & HUANG KUANG-MING (M. WONG) (1956). 'La notion de cercle et la science chinoise', *A/AIHS*, **9** (35), 111.

HUARD, P. & HUANG KUANG-MING (M. WONG) (1959). *La Médecine chinoise au cours des siècles.* Dacosta, Paris.

HUBOTTER, F. (1929). *Die chinesische Medizin zu Beginn des XX Jahrhunderts, und ihr historischer Entwicklungsgang*, Schindler, Leipzig (China Bibliothek d. *AM*, 1).

HUFF, T. (1993). *The Rise of Early Modern Science: Islam, China, and the West.* Cambridge University Press.

HUME, E. H. (1940). *The Chinese Way in Medicine.* John Hopkins University Press, Baltimore.

HUMMEL, ARTHUR W. ed. (1943). *Eminent Chinese of the Ch'ing Period (1644–1912).* 2 vols. U. S. Government Printing Office, Washington.

HUNTINGTON, ELLSWORTH (1907). *The Pulse of Asia: A Journey in Central Asia Illustrating the Geographical Basis of History.* Archibald Constable.

HUNTINGTON, ELLSWORTH (1924). *Civilisation and Climate.* 3rd ed. Yale University Press, New Haven.

HUNTINGTON, ELLSWORTH (1945). *Mainsprings of Civilisation.* Wiley, New York; repr. Mentor, New York, 1959.

HUXLEY, ALDOUS (1932). *Brave New World.* Chatto and Windus, London. Reviewed by Joseph Needham in *Scrutiny*, Cambridge 1932, Vol. 1, no. 1, pp. 76–9.

HYMES, ROBERT P. & SCHIROKAUER, CONRAD (1993). *Ordering the World: Approaches to State and Society in Sung Dynasty China.* University of California Press, Berkeley.

ILIFFE, R. (1998). Review of A. C. Crombie, *Styles of Scientific Thinking in the European Tradition: The History of Argument and Explanation Especially in the Mathematical and Biomedical Sciences and Arts*, 3 vols. Duckworth, London, in *History of Science*, **36**, 329–57.

INSTITUTE OF THE HISTORY OF NATURAL SCIENCES, CHINESE ACADEMY OF SCIENCES (1983). *Ancient China's Technology and Science.* Foreign Languages Press, Beijing.

JACOBS, N. (1958). *The Origin of Modern Capitalism and Eastern Asia.* Hong Kong University Press, Hong Kong.

JESSEN, KARL F. W. (1864). 'Botanik der Gegenwart und Vorzeit', in *Culturhistorischer Entwicklung: ein Beitrag zur Geschichte der abendlandischen Volker*, Brockhaus, Leipzig; repr. by Chronica Botanica, Waltham, Mass., 1948.

JOHNS, A. (1998). *The Nature of the Book: Print and Knowledge in the Making.* University of Chicago Press.

JOHNS, ADRIAN (2001). 'The Birth of Scientific Reading', *N*, **407** (6818), 287.

JONES, ALUN & BODMER, W. F. eds. (1974). *Our Future Inheritance: Choice or Chance?* Oxford University Press, London.

JOSEPH, GEORGE GHEVERGHESE (1991). *The Crest of the Peacock: Non-European Roots of Mathematics.* Tauris & Co., London.

JOYCE, JAMES (1939). *The Essential James Joyce, with an Introduction and Notes*, ed. LEVIN, HARRY 'Anna Livia Plurabelle'. Jonathan Cape, London, 508–24.

JULIAN OF NORWICH (*c.* +1373). *Revelations of Divine Love*, tr. Clifton Wolters, Penguin, Harmondsworth, 1966.

KANG TENG (1993). *Development Versus Stagnation – Technological Continuity and Agricultural Progress in Pre-modern China.* Greenwood Press, Connecticut and London.

KARLGREN, B. (1940). *Grammata Serica Script and Phonetics in Chinese and Sino-Japanese. BMFEA*, **12**, 1. (Photographically reproduced as separate volume, 1941.) Rev. ed. *Grammata Serica Recensa*, Stockholm, 1957.

KEELE, K. D. (1963). *The Evolution of Clinical Methods in Medicine.* Pitman, London. (Fitzpatrick Lectures, Royal College of Physicians, 1960–1.)

KEIGHTLEY, D. N. (1978). 'The Religious Commitment: Shang Theology and the Genesis of Chinese Political Culture', *HOR*, **17** (3–4), 211–24.

KING-HELE, DESMOND (1968). *Essential Writings of Erasmus Darwin.* MacGibbon and Kee, London.

KOESTLER, ARTHUR (1960). *The Watershed: A Biography of Johannes Kepler.* Heinemann, London.

KRADER, LAWRENCE ed. and tr. (1974). *The Ethnological Notebooks of Karl Marx.* 2nd ed., Van Gorcum, Assen. (Studies of Morgan, Phear, Maine, Lubbock.)

KRADER, LAWRENCE (1975). *The Asiatic Mode of Production, Sources, Development and Critique in the Writings of Karl Marx.* Van Gorcum, Assen.

KUHN, D. (1988). *SCC, Vol.* 5, pt. 9, *Textile Technology: Spinning and Reeling.* Cambridge University Press.

KUO YU-SHOU (1963). *La Lune sur le fleuve perle.* Bonne, Paris.

KWOK, D. W. Y. (1965). *Scientism in Chinese Thought, 1900 to 1950.* Yale University Press, New Haven.

LAKATOS, I. & FEYERABEND, P. (1999). *For and Against Method.* University of Chicago Press.

LAM LAY-YONG (1987). 'Linkages: Exploring the Similarities Between the Chinese Rod Numeral System and our Numeral System', *Archive for History of Exact Sciences*, **37** (4), 365–92.

LAMARCK, JEAN-BAPTISTE DE MONET (1779). *Flore françoise.* 3 vols. Published by Government, Paris. 2nd ed., 1795; 3rd ed., 1805 in collaboration with A. P. de Candolle, 4 vols.

LAMB, H. H. (1972). *Climate: Present, Past and Future.* Vol. 1*: Fundamentals and Climate Now.* Methuen, London; Barnes & Noble, New York; repr. 1978.

LANGELLIER, J.-P. (2000). 'Guy d'Arezzo, le père de la musique', *Le Monde*, 21 July, pp. 10–11.

LATTIMORE, OWEN (1940). *Inner Asian Frontiers of China.* Oxford University Press, London and New York. (Amer. Geogr. Soc. Research Monograph Series, no. 21.)

LAVERGNE, M. & LAVERGNE, C. (1947). *Précis d'Acupuncture Pratique.* Baillière, Paris.

LAVIER, J. (1964). *Les Bases traditionelles de l'acuponcture chinoise, les definitions essentielles de la bio-energetique chinoise dans la terminologie des acuponcteurs.* Maloine, Paris.

LAVIER, J. (1965). *Points of Chinese Acupuncture*, tr., indexed and adapted by P. M. Chancellor. Health Science Press, Rustington, Sussex.

LAWRENCE, G. H. M. (1951). *Taxonomy of Vascular Plants.* Macmillan, New York.

LAWSON-WOOD, D. & LAWSON-WOOD, J. (1964). *Acupuncture Handbook.* Health Science Press, Rustington, Sussex.

LEACH, G. (1970). *The Biocrats.* Jonathan Cape, London.

LEE, ORLAN (1964). 'Traditionelle chinesische "Rechtsebrauche" und der Bergriff "Orientalischer Despotismus" ', *ZVRW*, **1964**, 66–157.

LEGGE, JAMES tr. (1879). *The Texts of Confucianism, Translated*: Pt. I. *The 'Shu Khing', the Religious Portions of the 'Shih Ching', the 'Hsiao Ching'.* Oxford. (*SBE*, no. 3; reprinted in various eds. Commonwealth Press, Shanghai.)

LEGGE, JAMES tr. (1885). *The Texts of Confucianism*: Pt. III. *The 'Li Chi'.* 2 vols. Oxford; repr. 1926. (*SBE*, nos. 27 and 28.)

LE GOFFE, J. (1960). 'Au moyen Age, temps de l'église et temps du marchand', *Annales*, May–June.

LEICESTER, H. M. (1965). *The Historical Background of Chemistry.* Wiley, New York.

LEISS, W. (1972). *The Domination of Nature,* Braziller, New York.

LEVENSON, JOSEPH (1976). *European Expansion and the Counter-Example of Asia, 1300–1600.* Prentice Hall, Englewood Cliffs.

LEWIS, JOHN (1974). *Beyond Chance and Necessity: A Critical Enquiry into Prof. Jacques Monod's 'Chance and Necessity'.* Garnstone, London.

LI CHHIAO-PHING (1948). 'The Chemical Arts of Old China', *JCE*, Easton, Pa.

LI GUOHAO, ZHANG MENGWEN & CAO TIANQIN eds. (1982). *Explorations in the History of Science and Technology in China: A Special Number of the 'Collections of Essays on Chinese Literature and History'.* Shanghai Chinese Classics Publishing House, Shanghai.

LI HSÜEH-CHHIN (Li Xueqin) (1985). *Eastern Chow and Chhin Civilisations*, tr. K. C. Chang. Yale University Press, New Haven and London.

LI WEN-LIN (1982). *The Chinese Indigenous Tradition of Mathematics Prior to the Introduction of Modern Mathematics in the 19th Century.* n.p.

LIBBRECHT, U. J. (1973). *Chinese Mathematics in the Thirteenth Century: The Shu-shu chiu-chang of Ch'in Chiu-shao.* MIT East Asian Science Series, Cambridge, MA.

LIBBRECHT, ULRICH (1973). *Chinese Mathematics in the Thirteenth Century: The Shu-Shu Chiu-Chang of Chhin Chiu-Shao.* MIT Press, Cambridge, Mass.

LINDBERG, DAVID C. (1990). 'Conceptions of the Scientific Revolution from Bacon to Butterfield: A Preliminary Sketch', in LINDBERG & WESTMAN (1990), 1–26.

LINDBERG, D. C. (1992). *The Beginnings of Western Science: The European Scientific Tradition in Philosophical, Religious, and Institutional Context, 600 B.C. to A.D. 1450.* University of Chicago Press.

LINDBERG, DAVID C. & WESTMAN, ROBERT S. eds. (1990). *Reappraisals of the Scientific Revolution.* Cambridge University Press.

LIU TZU-CHIEN (1957). 'An Early Sung Reformer: Fan Chung-Yen', in *Chinese Thought and Institutions*, ed. J. K. Fairbank. University Chicago Press, 105.

LLOYD, GEOFFREY E. R. (1991). *Methods and Problems in Greek Science.* Cambridge University Press.

LLOYD, G. (1996). *Adversaries and Authorities: Investigations into Ancient Greek and Chinese Science.* Cambridge University Press.

LOCKE, JOHN (1689). *Essay Concerning Human Understanding*, ed. Roger Woodhouse. Penguin, London, 1947.

LOEWE, MICHAEL (1967). *Records of Han Administration.* 2 vols. Cambridge University Press.

LOVINS, AMORY B. (1975). *Nuclear Power: Technical Bases for Ethical Concern*, 2nd ed., Friends of the Earth Ltd. for Earth Resources Ltd., London.

LOW, M. F. (1998). *Beyond Joseph Needham: Science, Technology, and Medicine in East and Southeast Asia.* Editor, *Osiris* **13**.

LOWRY, T. M. (1936). *Historical Introduction to Chemistry.* Macmillan, London.

LOWTHER, WILLIAM (1994). '4,000 Years Old: The White Tribe of China', *The Mail*, 20 March.

LU, GWEI-DJEN & NEEDHAM J. (1963). 'China and the Origin of (Qualifying) Examinations in Medicine', *PRSM*, **56**, 63.

LU, GWEI-DJEN & NEEDHAM, J. (1964). 'Mediaeval Preparations of Urinary Steroid Hormones', *Med. Hist.*, **8**, 101. Prelim. pub. *N*, **200**, 1963, 1047.

LU, GWEI-DJEN & NEEDHAM, J. (1980). *Celestial Lancets: A History and Rationale of Acupuncture and Moxa.* Cambridge University Press.

LU, GWEI-DJEN & NEEDHAM, J. (1988). 'A History of Forensic Medicine in China', *MH*, **32**, 357–400.

LUNDBAEK, KNUD (1986). *Dialogue Between a Fisherman and a Woodcutter.* C. Bell Verlag, Hamburg.

DE MORANT, G. SOULIÉ (1934). *Précis de la variée acuponcture chinoise: doctrine, diagnostique, therapeutique.* Mercure de France, Paris.

DE MORANT, G. SOULIÉ (1939). *L'Acuponcture chinoise*, 4 vols. I. *L'Energie (Points, méridiens, circulation)*, II. *Le Maniement de l'energie*, III. *Les Points et leurs symptomes*, IV. *Les Maladies et leurs traitements.* Mercure de France, Paris.

MAHDIHASSAN, S. (1946a). 'The Chinese Origin of the Word Chemistry', *CS*, **15**, 136.

MAHDIHASSAN, S. (1946b). 'Another Probable Origin of the Word Chemistry from the Chinese', *CS*, **15**, 234.

MAHDIHASSAN, S. (1951). 'The Chinese Origin of Three Cognate Words: Chemistry, Elixir, and Genii', *JUB*, **20**, 107.

MAHDIHASSAN, S. (1953). 'The Chinese Origin of Alchemy', *UNASIA*, **5** (4), 241.

MAHDIHASSAN, S. (1957). 'Alchemy and its Connection with Astrology, Pharmacy, Magic and Metallurgy', *JAN*, **46**, 81.

MAHDIHASSAN, S. (1959a). 'Alchemy in its Proper Setting, with jinn, sufi and suffa as Loan-words from the Chinese', *IQB*, **7** (3), 1.

MAHDIHASSAN, S. (1959b). 'On Alchemy, kimiya and iksir', *Pakistan Philos J.*, **3**, 67.

MAHDIHASSAN, S. (1961a). 'Der Chino-Arabische Ursprung des Wortes Chemikalie', *Pharm. Ind.*, Berlin, **23**, 515.

MAHDIHASSAN, S. (1961b). 'Alchemy in the Light of its Names in Arabic, Sanskrit and Greek', *JAN*, **49**, 79.

MAHDIHASSAN, S. (1961c). 'Alchemy: Its Three Important Terms and their Significance', *MJA*, 227.

MAHDIHASSAN, S. (1962). 'Kimiya and Iksir: Notes on the Two Fundamental Concepts of Alchemy', *MBLB*, **5** (3), 38.

MANN, F. (1962a). *Anatomical Charts of Acupuncture Points, Meridians and Extra Meridians.* Barnet Publications, Barnet, Herts, 1982.

MANN, F. (1962b). *Acupuncture: The Ancient Chinese Art of Healing* (with foreword by Aldous Huxley.) Heinemann, London; 2nd ed. 1971; rev. 1974.

MANN, F. (1963). *The Treatment of Disease by Acupuncture.* Heinemann, London.

MAO TSE-TUNG (1954). *Selected Works.* 5 vols. Laurence and Wishart, London.

MARSDEN, B. G. (1965). 'Summary of Information on the Position of the Supposed Supernova of A.D. 1006'. *ASTJ*, **70**, 126.

MARSDEN, E. W. (1969). *Greek and Roman Artillery: Historical Development.* Oxford University Press.

MARSDEN, E. W. (1971). *Greek and Roman Artillery: Technical Treatises.* Oxford University Press.

MARTINET, LOUIS. (1826). *Manuel de Pathologie*, Paris. Eng. tr. *Manual of Pathology, containing the Symptoms, Diagnosis and Morbid Characters of Diseases together with an Exposition of the Different Methods of Examination, applicable to Affections of the Head, Chest, and Abdomen.* Translated with notes and additions by J. Quain. John Anderson, London, 1826; 2nd ed., ed. R. Norton, London, 1830.

MARTZLOFF, J.-C. (1997). *A History of Chinese Mathematics.* Original ed. 1987; rev. English ed. Berlin and Heidelberg: Springer, 1997. Translated from the French by S. S. Wilson.

MARX, KARL (1973). *Grundrisse. Foundations of a Critique of Political Economy* (Rough Draft). Tr. M. Nichlaus, Penguin/NLB, Harmondsworth.

MAVERICK, L. A. (1946). *China: A Model for Europe* (photo-litho typescript). Anderson, San Antonio, Tex. 6. Vol. 1: *China's Economy and Government Admired by Seventeenth and Eighteenth Century Europeans.* Vol. 2: *Despotism in China, a translation of François Quesnay's Le Despotisme de la Chine* (Paris, 1767), issued bound together in one.

MCMULLIN, ERNAN (1990). 'Conceptions of Science in the Scientific Revolution', in LINDBERG & WESTMAN (1990), 27–92.

MCNEILL, WILLIAM H. (1976). *Plagues and Peoples: A History of Bubonic Plague in the British Isles.* Basil Blackwell, Oxford.

MEDAWAR, SIR PETER BRIAN (1967). *The Art of the Soluble.* Methuen, London.

MERTON, R. K. (1938). 'Science, Technology and Society in Seventeenth Century England', *OSIS*, **4**, 360.

MIKAMI, YOSHIO (1913). *The Development of Mathematics in China and Japan.* Teutner, Leipzig.

MINKOWSKI, R. (1965). 'The Suspected Supernova of A.D. 1006', *ASTJ*, **70**, 755.

MITCHAM, CARL & MACKEY, ROBERT (1973). 'Bibliography of the Philosophy of Technology', *TCULT*, **14** (2/2), 1–205.

MONOD, JACQUES (1971). *Chance and Necessity: An Essay on the Natural Philosophy of Modern Biology,* trans. Austryn Wainhouse. Alfred A. Knopf, New York; Collins, London, 1972.

MORRIS, DESMOND (1967). *The Naked Ape: A Zoologist's Study of the Human Animal.* Jonathan Cape, London.

MORSE, HOSEA BALLOU (1932). *The Gilds of China: With an Account of the Gild Merchant or Co-Hong of Canton.* Kelly & Walsh (2nd ed.). Shanghai.

MORSE, W. R. (1934). *Chinese Medicine.* Hoeber, New York. (Clio Medica Series.)

MOSIG, A. & SCHRAMM, G. (1955). *Der Arzneipflanzen und Drogen-Schatz Chinas; und die Bedeutung des 'Pen Tshao Kang Mu' als Standardwerk der chinesischen Materia Medica.* Volk und Gesundheit, Berlin. (Beihefte der Pharmazie, no. 4.)

MOSS, L. (1964). *Acupuncture and You: A New Approach to Treatment Based on the Ancient Method of Healing.* Elek, London.

MURRAY, JAMES A. H., BRADLEY, HENRY, CRAIGIE, W. A. & ONIONS, C. T. (1933). *The Oxford English Dictionary.* Clarendon Press, Oxford, 1978 (1st ed. 1933).

MURTHY, S. BHANU (1992). *A Modern Introduction to Ancient Indian Mathematics.* Wiley Eastern Ltd., New Delhi.

MUSEUM, THE BRITISH (1972). *Treasures of Tutankhamun* (1972), Thames and Hudson, London.

NAKAYAMA, T. (1934). *Acuponcture et médecine chinoise vérifiées au Japon.* (tr. from the Japanese by T. Sakurazawa and G. Soulie de Morant.) Editions Hippocrate (le François), Paris.

NAKAYAMA, SHIGERU & SIVIN, NATHAN eds. (1973). *Chinese Science: Explorations of an Ancient Tradition.* MIT Press, Cambridge, Mass.

NEEDHAM, JOSEPH (1932). 'Biology and Mr. Huxley' (Review of Aldous Huxley's Brave New World), *Scrutiny*, **1**, 76; repr. in Watt, Donald (1975), ed., *Aldous Huxley, the Critical Heritage*, Routledge, London, 202.

NEEDHAM, JOSEPH (1943). *Time, the Refreshing River.* Allen & Unwin, London.

NEEDHAM, JOSEPH (1945). *Chinese Science*, Pilot, London.

NEEDHAM, JOSEPH (1946). 'The Chinese Contribution to Science and Technology', in *Reflections on our Age* (Lectures delivered at the Opening Session of UNESCO at the Sorbonne, Paris, 1946), ed. D. Hardman & S. Spender, 211. Wingate, London. (tr. from the French 'Conférences de l'UNESCO', Fontaine, Paris, 1947, 203.)

NEEDHAM, JOSEPH (1954). *Science and Civilisation in China*, Cambridge University Press, Cambridge, 1954–.

Volume 1. Introductory Orientations, 1954.
Volume 2. History of Scientific Thought, 1956.
Volume 3. Mathematics and the Sciences of the Heavens and the Earth, 1959.
Volume 4. Physics and Physical Technology
Pt. 1 Physics, 1962.
Pt. 2 Mechanical Engineering, 1965.
Pt. 3 Civil Engineering, 1971.
Volume 5. Chemistry and Chemical Technology
Pt. 2 Spagyrical Discovery and Innovation, 1974.
Pt. 3 Historical Survey, 1976.
Pt. 4 Apparatus, Theories and Gifts, 1980.
Pt. 5 Physiological Alchemy, 1983.
Pt. 1 Paper and Printing, 1985.
Pt. 7 The Gunpowder Epic, 1986.
Volume 6. Biology and Biological Technology
Pt. 1 Botany, 1986.
Volume 5. Chemistry and Chemical Technology
Pt. 9 Textile Technology: Spinning and Reeling, 1988.
Volume 6. Biology and Biological Technology
Pt. 2 Agriculture, 1988.
Volume 5. Chemistry and Chemical Technology
Pt. 6 Missiles and Sieges, 1994.
Volume 7. The Social and Economic Background
Pt. 1 Language and Logic, 1998

NEEDHAM, JOSEPH (1955). 'The Peking Observatory in A.D. 1280 and the Development of the Equatorial Mounting', in *Vistas of Astronomy* (Stratton Presentation Volume), ed. A. Beer, Vol. 1, 67. Pergamon, London.

NEEDHAM, JOSEPH (1956). 'Mathematics and Science in China and the West', *SS*, **20**, 320.

NEEDHAM, JOSEPH (1958a). *Chinese Astronomy and the Jesuit Mission, an Encounter of Cultures*, China Society, London. (China Society Occasional Papers No. 10.)

NEEDHAM, JOSEPH (1958b). *The Development of Iron and Steel Technology in China.* (Dickinson Lecture, 1956). Newcomen Society, London, 1958, repr. Heffers, Cambridge, 1964. Precis in *TNS*, 1960, **30**, 141. Rev. L. C. Goodrich, *Isis*, 1960, **51**, 108. French tr. (unrevised, with some illustrations omitted and others added by the editors.) *RHSID*, 1961, **2**, 187, 235: 1962, **3**, 1, 62.

NEEDHAM, JOSEPH (1959). Review of WITTFOGEL (1957), in *SS*, **23**, 58–65.

NEEDHAM, JOSEPH (1960a). 'The Chinese Contribution to the Development of the Mariner's Compass.' Abstr. in *Resumo das Comunicacoes do Congresso Internacional de Historia dos Descobrimentos*, 273. Lisbon. *Actas*, Lisbon, 1961, **2**, 311. Also *SCI*, 1961 (6e ser), **55** (July).

NEEDHAM, JOSEPH (1960b). 'The Past in China's Present', *CR/MSU*, **4**, 145 and 281; repr. with some omissions, *PV*, 1963, **4**, 115; Fr. tr.: 'Du passé culturel, social et philosophique chinois dans ses rapports avec la chine contemporaine', by G. M. Merkle-Hunziker. *COMP*, 1960, **21–2**, 261; 1962, **23–4**, 113; repr. in *CFC*, 1960, **8**, 26; 1962, **15–16**, 1.

NEEDHAM, JOSEPH (1961). *Classical Chinese Contributions to Mechanical Engineering* (Earl Grey Lecture, University of Durham), Newcastle.

NEEDHAM, JOSEPH (1963). 'The Pre-Natal History of the Steam Engine' (Newcomen Centenary Lecture), *TNS*, **35**, 3.

NEEDHAM, JOSEPH (1964a). 'Science and China's Influence on the West', in *The Legacy of China*, ed. R. N. Dawson. Oxford University Press.

NEEDHAM, JOSEPH (1964b). 'Science and Society in East and West', in J. D. Bernal Presentation Volume, *The Science of Science*, ed. M. Goldsmith & A. McKay. Souvenir, London. Also in *SS*, 1964, **28**, 385 and *CEN*, 1964, **10**, 174.

NEEDHAM, JOSEPH (1968). 'The Development of Botanical Taxonomy in Chinese Culture', in *Actes du XIIme Congrès International d'Histoire des Sciences*, Paris. Vol. 8, 127.

NEEDHAM, JOSEPH (1969a). *The Grand Titration: Science and Society in China and the West.* Allen and Unwin, London.

NEEDHAM, JOSEPH (1969b). *Within the Four Seas: A Dialogue of East and West.* Allen and Unwin, London.

NEEDHAM, JOSEPH (1970). *Clerks and Craftsmen in China and the West.* Cambridge University Press.

NEEDHAM, JOSEPH (1971). 'Do the Rivers Pay Court to the Sea? The Unity of Science in East and West', *TTT*, **5** (2), 68–77.

NEEDHAM, JOSEPH (1974a). 'Dr. Needham's Address', *University of Hong Kong Gazette*, **21** (5/1), 69.

NEEDHAM, JOSEPH (1974b). 'The Nature of Chinese Society: A Technical Interpretation' (with Huang Jen-Yü [Ray Huang]), *University of Hong Kong Gazette*, **21** (5/2), 75. *EW*, n.s. **24** (3–4), 381. *JOSHK*, **12** (1–2), 1.

NEEDHAM, JOSEPH (1975). 'Dilemmas of Modern Science and Medicine – A Chinese Cure?' *IMPAC*, **25** (1), 45.

NEEDHAM, JOSEPH (1986). 'Science, Technology, Progress and the Break-through: China as a Case-study in Human History', in T. GANELIUS (1986).

NEEDHAM, JOSEPH, BEER, A., HO PING-YÜ, LU GWEI-DJEN, PULLEYBLANK, E. G., & THOMPSON, G. I. (1964). 'An 8th-Century Meridian Line: I-Hsing's Chain of Gnomons and the Pre-History of the Metric System', *Vistas in Astronomy*, **4**, 3.

NEEDHAM, JOSEPH, & NEEDHAM, D. eds. (1948). *Science Outpost: Papers of the Sino-British Science Co-Operation Office* (British Council Scientific Office in China), 1942–46, Pilot, London.

NEEDHAM, JOSEPH & LU GWEI-DJEN (1962). 'Hygiene and Preventive Medicine in Ancient China', *J. Hist. Med.*, **17**, 429. (Abridged in *HEJ*, 1959, **17**, 170.)

NEEDHAM, JOSEPH & LU GWEI-DJEN (1966a). *The Optick Artists of Chiangsu.* Roy. Mic. Soc., London (Oxford Symposium Volume). Abstract in *PRMS* **1** (2), 59.

NEEDHAM, JOSEPH & LU GWEI-DJEN (1966b). 'A Korean Astronomical Screen of the Mid-eighteenth Century from the Royal Palace of the Yi Dynasty (Choson Kingdom, +1392 to +1910)', *PHY*, **8**, 137.

NEEDHAM, JOSEPH & LU GWEI-DJEN (1975). 'Problems of Translation and Modernization of Ancient Chinese Technical Terms: Manfred Porkert's Interpretations of Terms in Ancient and Medieval Chinese Natural and Medical Philosophy', *Annals of Science*, **32** (5), 491.

NEEDHAM, JOSEPH & ROBINSON, KENNETH (1991). 'Literary Chinese as a Scientific Language: A Selection from the Concluding Volume', *Comparative Criticism* **13**, ed. E. S. Shaffer (Cambridge University Press), 3–30.

NEEDHAM, JOSEPH, WANG LING & DEREK J. DE S. PRICE (1960). *Heavenly Clockwork: The Great Astronomical Clocks of Medieval China.* Cambridge. (Antiquarian Horological Society Monographs, no. 1.) Prelim. pub. *AHOR*, 1956, **1**, 153.

NEEDLEMAN, JACOB (1974). *The Sword of the Gnosis*, Penguin, Harmondsworth.

NERI, ANTONIO (1612). *L'Arte Vetraria distinta in libri sette . . . Gionti*, Florence. 2nd ed. Rabbviati, Florence, 1661, Batti. Venice, 1663. Latin tr. *De Arte Vitruvia Libri Septem, et in eosdem Christoph Merretti . . . Observationes et Notae.* Amsterdam, 1668. German tr. F. Geissler, Frankfurt and Leipzig, 1678. English tr. C. Merriett, London, 1662. Cf. FERGUSON (1906), Vol. 2, 134 ff.

NETTLESHIP, HENRY & SANDYS, J. E. (1891). Tr. and ed., with revision of SEYFFERT, August Oskar: *A Dictionary of Classical Antiquities, Mythology, Religion, Literature and Art.* Sonnenschein, London, 1904; Meridian Books, New York, 1957.

NEUBAUER, K. & VOGEL, J. (1881). *Anleitung zur qualitativen und quantitativen Analyse des Harns.* 8th ed. Wiesbaden, 1881. Eng. tr. W. O. Markham, *Guide to the Qualitative Analysis of the Urine.* The Sydenham Society. London, 1863; 1st ed. 1854.

NEWNHAM, RICHARD (1971). *About Chinese: With the Help of Tan Lin-tung.* Penguin Books, Harmondsworth.

NEWTON, ISAAC (1687). *Philosophiae Naturalis Principia Mathematica.* London.

NEWTON, ISAAC (1704). *Opticks: or a Treatise of the Reflexions, Refractions, Inflexions and Colours of Light. Also two Treatises of the Species and Magniture of Curvilinear Figures.* London.

NEWTON, I. (1999). *The Principia: Mathematical Principles of Natural Philosophy*, tr. I. Bernard Cohen and Anne Whitman. University of California Press, CA.

NIKIFOROV, V. N. (1970). *Sovetskie istoriki o problemakh Kitaya.* Nauka, Moscow.

NORTHROP, F. S. C. (1946). *The Meeting of East and West: An Inquiry Concerning Human Understanding.* Macmillan, New York.

ODAGIRI, MIZUHO (1955–62). *Science Reports of the Society for Research in Physics-Chemistry (formerly Theoretical Chemistry)*, numerous articles from **1** (no. 1) to **8** (no. 1, serial number 10).

ODAGIRI, MIZUHO (1974). 'The Exhaustion of Possibilities of Theoretical Science in History and its Reason', *Proceedings of the XIVth International Congress of the History of Science, Tokyo and Kyoto, 19–27 August 1974*, Science Council of Japan, Tokyo, **3**, 445.

OGDEN, K. ed. (1929–32). *Psyche* X (1) (2), XI (1), XII (2) (4). See *Psyche* for dates.

OLBY, R. (1974). *The Path to the Double Helix.* Macmillan, London.

OLDHAM, C. H. G. (1973). 'Science and Technology Policies', in *China's Developmental Experience*, ed. Michel Oksenberg, for the Academy of Political Science, Columbia University. Praeger, New York, 80.

ONIONS, C. T. ed. (1936). *The Shorter Oxford English Dictionary on Historical Principles.* 2 vols. Clarendon Press, Oxford. Rev. and ed. 1972.

Oxford Companion to English Literature (1932). Ed. Harvey, Sir Paul. Oxford University Press (3rd ed. 1946).

Oxford Dictionary of the Christian Church (1974). Ed. Cross, F. L. & Livingstone, E. A. London 2nd ed.

Oxford Dictionary of Quotations (1941). Oxford University Press (2nd ed. 1953).

Oxford English Dictionary. See MURRAY, JAMES.

PAGEL, W. (1935). 'Religious Motives in the Medical Biology of the Seventeenth Century', *BIHM*, **3**, 97.

PAGEL, W. (1944). 'The Religious and Philosophical Aspects of van Helmont's *Science and Medicine*', *BIHM*, Suppl. 2.
PAGEL, W. (1960). 'Paracelsus and the Neo-Platonic and Gnostic Tradition', *AX*, **8**, 125.
PAGEL, W. (1961). 'The Prime Matter of Paracelsus', *AX*, **9**, 117.
PAGEL, W. (1962a). *Das medizinische Weltbild des Paracelsus; seine Zusammenhang mit Neuplatonismus und Gnosis.* Steiner, Wiesbaden. (Kosmosophie; Forschungen und Texte zur Geschichte des Weltbildes, der Naturphilosophie, der Mystik und des Spiritualismus vom Spatmittelalter bis zur Romantik, no. 1.)
PAGEL, W. (1962b). 'The "Wild Spirit" (Gas) of John-Baptist van Helmont (+1579 to +1644), and Paracelsus', *AX*, **10**, 2.
PAGEL, W. (1968). 'Paracelsus: Traditionalism and Mediaeval Sources', in *Medicine, Science and Culture, O. Temkin Presentation Volume*, ed. L. G. Stevenson & R. P. Multhauf. Johns Hopkins University Press, Baltimore, 51.
PALLIS, MARCO (1974). 'The Catholic Church in Crisis: Thinking around the Vatican Oracle by the Reverend Brocard Sewell', in *The Sword of Gnosis: Metaphysics, Cosmology, Tradition, Symbolism*, ed. Jacob Needleman. Penguin, Harmondsworth, 57.
PALOS, S. (1963). *Chinesische Heilkunst; Ruckhesinnung auf eine grosse Tradition.* (Tr. from the Hungarian by W. Kronfuss.) Delp, Munich.
PAPPWORTH, M. H. (1967). *Human Guinea-pigs: Experimentation on Man*, Routledge & Kegan Paul, London; Pelican, Harmondsworth, 1969.
PARLETT, D. (1999). *The Oxford History of Board Games.* Oxford University Press.
PARTINGTON, J. R. (1957). *A Short History of Chemistry*, 3rd ed. Macmillan, London.
PARTINGTON, J. R. (1961). *A History of Chemistry.* Vol. l, *Earliest Period to 1500*; Vol. 2, *1500 to 1700*; Vol. 3, *1700 to 1800*; Vol. 4, *1800 to the Present Time.* Macmillan, London.
PAVLOV, IVAN PETROVITCH (1941). *Lectures on Conditioned Reflexes: Twenty-five Years of Objective Study of the Higher Nervous Activity (Behaviour) of Animals*, tr. W. H. Gantt and G. Volborth; Vol. 1, Martin Lawrence, London (1928); Vol. 2: *Conditioned Reflexes and Psychiatry*, Lawrence & Wishart, London 1941; repr. 1963.
PETROSKI, HENRY (1993). *The Evolution of Useful Things.* Pavilion Books Ltd., London.
PFISTER LE PERE LOUIS, S. J. (1932). *Notices biographiques et bibliographiques sur les Jésuites de l'Ancienne Maison de la Chine.* Catholic Mission, Shanghai.
PIRSIG, R. (1974). *Zen and the Art of Motorcycle Maintenance: An Enquiry into Values.* Morrow, New York; Bodley Head, London.
PLATTER, FELIX (1603). *Praxis Medica (Praxeus seu de cognoscendis, praedicendis curandisque affectibus homini incommodantibus tractatus)*, 2 vols. Basel, 1602–3.
PLINY, THE ELDER (CAIUS PLINIUS SECUNDUS) (1950). *Natural History*, books 7, 9, 13, 15, 35 and 65, tr. H. Rackham, 10 vols. Loeb Classical Library, Heinemann, London.
POMERANZ, K. (2000). *The Great Divergence: China, Europe, and the Making of the Modern World Economy.* Princeton University Press.
POPPER, KARL (1945). *The Open Society and its Enemies.* 2 vols. Routledge and Kegan Paul, London.
POPPER, KARL (1972). *Objective Knowledge: An Evolutionary Approach.* Oxford.
POYNTER, F. F. N. L. (with the collaboration of BARBER-LOMAX, J. & CRELLIN, J. J.) (1966). *Chinese Medicine: An Exhibition Illustrating the Traditional System of Medicine of the Chinese People* (catalogue with introduction). Wellcome Historical Medical Museum and Library, London.
PRICE, D. J. DE S. (1961). *Science since Babylon.* Yale University Press, New Haven.
PSYCHE (1929–32). Ed. C. K. Ogden. Vols. X, XI and XII. Orthological Institute, London. 2nd publication from damaged originals, Routledge/Thoemmes Press, London, 1995.
PULLEYBLANK, E. G. (1958a). Review of *Wittfogel's Oriental Despotism: A Comparative Study of Total Power, BLSOAS*, **21**, 657.
PULLEYBLANK, E. G. (1958b). 'The Origins and Nature of Chattel Slavery in China', *JESHO*, **1**, 185.

QUESNAY, FRANCOIS (1767). *Le Despotisme de la Chine.* Paris. Eng. tr. in Maverick (1946, Vol. 2).
QUINE, W. VAN O. (1962). *Mathematical Logic.* Rev. ed. 1951; repr., Harper, New York, 1962.
QUINE, W. VAN O. (1963). *Set Theory and its Logic.* Harvard University Press, Cambridge MA.

RACKHAM, H. tr. (1914). *Cicero, De finibus*, Loeb Classical Library, Heinemann, London.
RACKHAM, H. tr. (1928). *Cicero, De republica and De legibus*, Loeb Classical Library, London.
RACKHAM, H. tr. (1952). *Pliny 'Natural History', IX, with an English Translation.* Heinemann, London, 1938, rev. ed. Loeb Classical Library, Heinemann, London, 1952.
RAINE, KATHLEEN (1968). *Blake and Tradition*, 2 vols. Princeton University Press.
RALL, J. (1962). 'Uber die Warmekrankheiten', *ORE*, **9**, 139.
RANKIN, MARY BACKUS (1986). *Elite Activism and Political Transformation in China: Zhejiang province, 1865–1911*. Stanford University Press, Stanford, CA.
RAPHALS, LISA (1992). *Knowing Words: Wisdom and Cunning in the Classical Traditions of China and Greece.* Cornell University Press, Ithaca.

RAWSKI, EVELYN S. (1979). *Education and Popular Literacy in Ch'ing China.* Michigan Studies on China, University of Michigan, Ann Arbor.

Revue d'histoire des sciences, **42**, 4 (1989), and **43**, 1 (1990). 'Problèmes d'histoire des sciences en Chine', (I) 'Méthodes, contacts et transmissions' and (II) 'Approches spéci fiques'.

RICKETT, H. W. (1944). 'The Classification of Inflorescences', *Bot. Review*, **10**, 187–231.

RICKETT, W. A. (1965). *Kuan-Tzu: A Repository of Early Chinese Thought: A Translation and Study of Twelve Chapters*; with a foreword by Derk Bodde. Hong Kong University Press.

ROBERTS, CATHERINE (1974). *The Scientific Conscience: Reflections on the Modern Biologist and Humanism.* Centaur, Fontwell, Sussex.

ROBINSON, J. (1942). *An Essay on Marxian Economics.* Macmillan, London.

ROBINSON, KENNETH (1980). *A Critical Study of Chu Tsai-yü's Contribution to the Theory of Equal Temperament in Chinese Music.* Sinologica Coloniensia Band 9. Additional notes by Erich F. W. Altwein. Franz Steiner, Wiesbaden.

ROBINSON, KENNETH (1993). *The Way and the Wilderness.* Pentland Press, Durham.

ROBINSON, KENNETH & NEEDHAM, JOSEPH (1991). 'Literary Chinese as a Language for Science', *Comparative Criticism*, **13**, ed. E. S. Shaffer (Cambridge University Press), 3–30.

RONAN, COLIN (1991). 'The Origins of the Reflecting Telescope', *JBASA*, **101**, 6.

RORICZER, MATTHAÜS (1486). *Das Büchlein von der Fialen Gerechtigkeit*, Faksimile der Originalausgabe, Regensburg, 1486, mit einem Nachwort und Textübertragung herausgegeben von Ferdinand Geldner. Pressler, Wiesbaden, 1965.

ROSENBERG, NATHAN & BIRDZELL, L. E. JR (1986). *How The West Grew Rich: The Economic Transformation of the Industrial World.* Basic Books, New York.

ROSZAK, THEODORE (1969). *The Making of a Counter-Culture: Reflections on the Technocratic Society and its Youthful Opposition.* Doubleday, New York; Faber & Faber, London, 1970, repr. 1971.

ROSZAK, THEODORE (1972). *Where the Wasteland Ends: Politics and Transcendence in Post-Industrial Society.* Doubleday, New York; Faber & Faber, London, 1973.

ROSZAK, THEODORE (1974). 'The Monster and the Titan: Science, Knowledge and Gnosis', *DAE* (summer), issued as *PAAAS*, **103** (3), 17.

ROWE, WILLIAM T. (1984). *Hankow: Commerce and Society in a Chinese City, 1796–1889.* Stanford University Press, Stanford, CA.

ROWE, WILLIAM T. (1985). 'Approaches to Modern Chinese Society History', in ZUNZ (1985), 236–96.

ROWE, WILLIAM T. (1989). *Hankow: Conflict and Community in a Chinese City, 1796–1895.* Stanford University Press, Stanford, CA.

ROWE, WILLIAM T. (1993). 'City and Region in the Lower Yangzi', in COOKE JOHNSON (1993), 1–16.

RUHLAND, MARTIN (1612). *Lexicon Alchemiae sive Dicitonarium Alchemisticum cum obscuriorum verborum et rerum Hermeticarum tum Theophrast-Paracelsicarum Phrasium, planam explicationem continens.* Frankfurt 1621. (2nd ed. Frankfurt, 1661.)

RUSSELL, BERTRAND (1922). *The Problem of China.* Allen & Unwin, London.

SAGAN, CARL (1980). *Cosmos.* Random House, New York.

SAID HUSAIN NASR (1968a). *Science and Civilisation in Islam.* Harvard University Press, Cambridge, Mass.

SAID HUSAIN NASR (1968b). *The Encounter of Man and Nature: The Spiritual Crisis of Modern Man.* Allen & Unwin, London.

SAMPSON, GEOFREY (1985). *Writing Systems.* Hutchinson, London.

SARTON, GEORGE (1927–47). *Introduction to the History of Science*, Vol. 1, 1927; Vol. 2, 1931 (2 parts); Vol. 3, 1947 (2 parts). Williams and Wilkins, Baltimore. (Carnegie Institution Publ. no. 376.)

SAWER, MARIAN (1977). *Marxism and the Asiatic Mode of Production.* Nijhoff, The Hague.

SCHLEGEL, G. (1875). *Uranographie Chinoise ou Preuves Directes que l'Astronomie Primitive est originaire de la Chine et qu'elle a été empruntée par les Anciens Peoples Occidentaux a la Sphère chinoise; ouvrage accompagné d'un atlas céleste chinois et grec.* 2 vols. with star-maps in separate folder. Brill, Leiden.

SCHNEIDER, W. (1959). 'Uber den Ursprung des Wortes "Chemie" ', *Pharm. Ind., Berl.*, **21**, 79.

SCHUMPETER, JOSEPH A. (1939). *Business Cycles: A Theoretical, Historical, and Statistical Analysis of the Capitalist Process.* 2 vols. McGraw-Hill, New York and London.

SCHURMANN, H. F. tr. (1956). *Economic Structure of the Yuan Dynasty: A Translation of chs. 93 and 94 of the 'Yuan Shih'.* Harvard University Press, Cambridge, Mass. (Harvard-Yenching Institute Studies, no.16.) Rev. by J. Prusek, *ARO*, 1959, **27**, 479; H. Franke, *RBS*, 1959, **2**, 84.

SCOTT, J. C. (1952). *Health and Agriculture in China.* Faber & Faber, London.

Service Historique De La Marine BB4 1555, 'Rapport Du SoudingéNieur De Division Thibaudin', Juin 1868. (French Navy Archive report), note 42)

SEYFFERT, AUGUST OSKAR (1891). *A Dictionary of Classical Antiquities, Mythology, Religion, Literature and Art*, tr. and rev. NETTLESHIP, HENRY & SANDYS, J. E.

SHAKESPEARE, WILLIAM. *Julius Caesar.*

SHAKESPEARE, WILLIAM. Sonnet LXXIII.

SHAPIN, S. (1996). *The Scientific Revolution.* University of Chicago Press.

SHIBA. Y. (1972). *Commerce and Society in Sung China.*, tr. W. Elvin. Center for Chinese Studies, Michigan University, Ann Arbor.

SHIH, SHENG-HAN (1958). *A Preliminary Survey of the Book 'Chhi Min Yao Shu', an Agricultural Encyclopaedia of the* +6th *Century.* Science Press, Peking.

SHIH, SHENG-HAN (1959). *On the 'Fan Sheng-Chih Shu', an Agricultural Book written by Fan Sheng-Chih in –1* st *Century China.* Science Press, Peking.

Shorter Oxford English Dictionary (1936). See ONIONS (1936).

SHREWSBURY, J. F. D. (1950). '*The Plague of Athens' Bulletin of the History of Medicine*, **24**, 1–25 but also see *The Cambridge World History of Human Disease*, p. 934 for article by Anne G. Carmichael.

SIMON, JOAN (1962). 'Stages of Social Development', *MXTD*, **6** (6), 183. A contribution to a discussion under this title on the comparative sociology of different civilisations which also includes articles by R. Jardine, J. Lindsay, D. Craig, E. G. Pulleyblank, R. Page Arnot, R. Browning, B. R. Mann, E. Hobsbawm, M. Shapiro, M. Dobb and others, extending from **5** (7) to **6** (9).

SIVIN, NATHAN (1965). 'Preliminary Studies in Chinese Alchemy; the "Tan Ching Yao Chueh" attributed to Sun Ssu-Mo (+581? to after +672)'. Inaug. Diss., Harvard University.

SIVIN, NATHAN (1970), review of A. A. BENNETT (1967), in *Isis*, **61** (207), 280–2 and in *Technology Review*, March 1970, 17.

SIVIN, NATHAN (1982). 'Why the Scientific Revolution Did Not Take Place in China – Or Didn't It?', *CSCI*, **5**, 45ff.

SKINNER, B. F. (1971). *Beyond Freedom and Dignity*, Knopf, New York; Jonathan Cape, London, 1972; Penguin, Harmondsworth, 1973.

SKINNER, WILLIAM G. (1977). *The City in Late Imperial China.* Stanford University Press, Stanford, CA.

VAN SLYKE, L. P. (1988). *Yangtze: Nature, History and River.* Addison-Wesley, New York.

SMITH, C. S. & GNUDI, M. T. (1959). *The Pirotechnia of Vannocchio Biringuccio Translated from the Italian with an Introduction and Notes.* Basic Books, New York.

SOFRI, GIANNI (1973). *Il modo di produzione asiatico: Storia di una controversia marxista.* Einaudi, Turin, 1st ed. 1966; 2nd ed. 1973.

SPENCE, JONATHAN (1968). 'Chang Po-hsing and the K'ang-hsi Emperor', *Ch'ing-shih wen-t'i*, **1** (8), 5.

SPENGLER, OSWALD (1926). *The Decline of the West*, tr. C. F. Atkinson. Allen & Unwin, London; repr. 1928.

SPOONER, R. (1942). 'Chinese Chemical Terms', *SM* (July).

SPRAT, THOMAS (1722). *The History of the Royal Society of London, for the Improving of Natural Knowledge.* 3rd ed. Knapton *et al.* London.

STEARN, WILLIAM T. (1966). *Botanical Latin: History, Grammar, Syntax, Terminology and Vocabulary.* Nelson, London.

STEBBING, SUSAN (1942). *A Modern Introduction to Logic.* Methuen, London, 1930; 3rd ed. 1942.

STEWARD, A. N. (1958). *Manual of Vascular Plants of the Lower Yangtse Valley, China.* Oregon State College, Corvallis, Oregon.

STRONG, ANNA LOUISE (1964). *Letters from China.* Coward McCaun, New York.

SWANN, N. L. tr. (1950). *Food and Money in Ancient China: The Earliest Economic History of China to A.D. 25 – '[Chhien] Han Shu' ch. 24, with Related Texts'*, '[Chhien] Han Shu' ch. 91, and 'Shih Chi' ch. 129. Princeton University Press, Princeton.

SYDENHAM, THOMAS (1676). *Observationes medicae circa morborum acutorum historiam et curationem.* London.

SYDENHAM, THOMAS (1683). *Tractatus de Podagra et Hydrope.* London.

SYDENHAM, THOMAS (1685). *Opera Universa.* London.

Szechuan Institute for the Cure and Prevention of Disease from Parasites (Mienchu Prefecture Revolutionary Committee Leadership Group Office for the Cure and Prevention of Disease from Schistosomes), 'Chao-chhi-chhih chhu-li fen-pien tui chi-sheng-chhung tan sha-mieh hsiao-kuo de tiao-chha he shih-yen kuan-chha' (The Digestion of Night-soil for the Destruction of Parasite Ova), Chung-hua I-hsüeh tsa-chih (CSMJ), 1974, **54** (2), 107, Eng. abstr., p. 31.

TAYLOR, F. S. (1951). *The Alchemists.* Heinemann, London.

THEOPHRASTUS, *Enquiry into Plants.* See HORT (1916).

THILO, THOMAS (1980). 'Die Schrift vom Pflug (Leisijing) und das Verhältnis ihres Verfassers Lu Guimeng zur Landwirtschaft', *ALTOF*, **1980**, 7.

THOMAS, IVOR tr. (1957). *Selections Illustrating the History of Greek Mathematics.* I. *From Thales to Euclid*; II. *From Aristarchus to Pappus of Alexandria.* Loeb Classics Series. Heinemann, London.

THORPE, SIR E. (1921). *History of Chemistry*, 2 vols. in 1. Watts, London.

TOGO, TOSHIHIRO (2002). 'Bibliography of the Late Prof. Yabuuti Kiyoshi', *EASTM*, **19**, 67–105.

TOKEI, F. (1959). 'Die Formen der chinesischen patriarchalischen Sklaverei in der Chou-Zeit', in *Opuscula Ethnologica Memoriae Ludovici Biro Sacra.* Budapest, 291.

TONKIN, I. M. & WORK, T. S. (1945). 'A New Anti-malarial Drug', *N*, **156**, 630.

TOYNBEE, ARNOLD (1958). 'Wittfogel's "Oriental Despotism"' in BAILEY & LLOBERA (1981), 164ff.

TSHAO THIEN-CHHIN, HO PING-YÜ & NEEDHAM, J. (1959). 'An Early Mediaeval Chinese Alchemical Text on Aqueous Solutions (the San-shih-liu Shui Fa, early +6th century)', *AX*, **7**, 122.

TSIEN, TSUIN-HSUIN, *Science and Civilisation in China*, Vol. 5, Pt. 1: *Paper and Printing*. Cambridge University Press.

UNESCO (1966). *History of Mankind*. 6 vols. 13 parts. Allen and Unwin, London.

VARRO, MARCUS TERENTIUS. See HOOPER & ASH (1934).

VERMEER, E. (1987). 'P'an Chi-hsün's Solutions for the Yellow River Problems of the Late 16th Century'. *T'oung Pao*, **73**, 33–67.

VERNANT, J. P. (1963). 'Commentary on the Contribution of Dr. B. L. Van der Waerden', *Scientific Change: Historical Studies in the Intellectual, Social and Technical Conditions for Scientific Discovery and Technical Invention, from Antiquity to the Present*, ed. A. C. Crombie. Heinemann, London.

VERNANT, J. P. (1964). 'Les origines de la pensée grecque', Eng. tr. 1981. Paris

VIRGIL, POLYDORE, or VERGILIUS, POLYDORUS (1512). *De Inventoribus Verum Libri*. Tras. M. Schürer, Strassburg, 1512. Noviter Impressus per Ioannem Tacuinum de Tridino, Venice, 1519.

DE WAAL, VICTOR (1974). 'A Renewed Sacramental Theology', *SOB*, **6** (10), 697.

WADDINGTON, C. H. (1975a). 'The New Atlantis Revisited', Bernal Lecture to the Royal Society, *PRSB*, **190**, 301.

WADDINGTON, C. H. (1975b). 'Genetic Engineering', Trueman Wood Lecture, *JRSA*, **123** (5225), 262.

WAGNER, DONALD (1978). 'Liu Hui and Tsu Keng-chih on the Volume of a Sphere', *CSCI*, **3**, 58–79.

WAGNER, DONALD (1979). 'An Early Chinese Derivation of the Volume of a Pyramid: Liu Hui, Third Century A.D.', *Historia Mathematica* **6**, 164–88.

WALEY, ARTHUR (1934). *The Way and Its Power: A Study of the 'Tao Te Ching' [translated] and its Place in Chinese Thought*, Allen & Unwin, London.

WALEY, ARTHUR (1958). *The Opium War Through Chinese Eyes*. Allen and Unwin, London.

WALLERSTEIN, IMMANUEL (1992). 'The West, Capitalism, and the Modern World-System.' *Review*. Ferrand Braudel Center, **15** (4), 561–619.

WANG CHI-MIN & WU LIEN-TE (1932). *History of Chinese Medicine*, Nat. Quarantine Service, Shanghai, 1932. (2nd ed. 1936.)

WANG LING (1947). 'On the Invention and Use of Gunpowder and Firearms in China', *Isis*, **37**, 160.

WATANABE, MASAO (1972). 'The Conception of Nature in Japanese Culture', paper presented at the American Association for the Advancement of Science, Washington, D.C. Later published in *SC*, **1839** (25 Jan. 1974), 279; and *Ekistics*, **38** (226) (Sept. 1974), 188.

WATT, DONALD (1975), ed. Aldous Huxley, *The Critical Heritage*. Routledge, London.

WEBSTER, JOHN (1654). *Academiarum examen, or the Examination of Academies*, Giles Calvert, London.

WEBSTER'S DICTIONARY (1976). Ed. in Chief, Philip Babcock Gove, Merrion Coy., Springfield, Mass., 1976. 1st ed. 1909.

WEINBERG, STEVEN (1974). 'Reflections of a Working Scientist', *DAE* (summer), issued as *PAAAS*, **103** (3), 33.

WEISHEIPL, J. A. (ed.) (1980). *Albertus Magnus and the Sciences: Commemorative Essays 1980*. Pontifical Institute of Medieval Studies, Toronto.

WEISSKOPF, VICTOR FREDERICK (1972). *Physics in the Twentieth Century: Selected Essays*, MIT, Cambridge, Mass.

WEISSKOPF, VICTOR FREDERICK (1975). 'The Frontiers and Limits of Science', *Bull. Amer. Acad. Arts & Sciences*, **28** (6), 15.

WELLS, H. G. (1896). *The Island of Dr. Moreau*. Heinemann, London.

WELTFISH, GENE (1975). 'Work: An Anthropological View', Module 9–065 issued by the Empire State College (University of New York), Saratoga Springs, N.Y.

WHITE, LYNN (1962). *Medieval Technology and Social Change*. Clarendon Press, Oxford.

WHITE, LYNN (1967). 'The Historical Roots of our Oecological Crisis', *SC*, **155**, 1203.

WHITE, LYNN (1968). *Machina ex Deo: Essays in the Dynamism of Western Culture*, MIT, Cambridge, Mass.

WILHELM, HELLMUT (1944). *Gesellschaft und Staat in China: acht vortrage*. Catholic University Press, Peking.

WILKINS, JOHN (1648). *Mathematicall Magick, or, the wonders that may be performed by Mechanical Geometry. In two books*. London.

WILKINSON J. GARDNER (1854). *A Popular Account of the Ancient Egyptians, revised and abridged from his larger work* [of 1836]. John Murray, London.

WITTFOGEL, K. A. (1931). *Wirtschaft und Gesellschaft Chinas; Versuch der wissenschaftlichen Analyse einer grossen asiatischen Agrargesellschaft* – Erster Teil, *Produktivkrafte, Produktions- und Zirkulations- prozess*. Hirschfeld, Leipzig. (Schriften d. Instit. f. Sozialforschung a.d. Univ. Frankfurt a.M. III,1.)

WITTFOGEL, K. A. (1957). *Oriental Despotism: A Comparative Study of Total Power*. Yale University Press, New Haven, and Oxford University Press, London. Reviewed by J. Needham in *SS*, 1959, **23**, 58ff.

WITTFOGEL, K. A. (1958). 'Reply to Arnold Toynbee', in BAILEY & LLOBERA (1981), 168–72.

WITTFOGEL. K. A. (1978). *China und die osteurasische Kavallerie-Revolution*. Harrassowitz, Wiesbaden.

WRIGHT, MARY C. (1957). *The Last Stand of Chinese Conservatism: The T'ung-chih Restoration, 1862–1874.* Stanford University Press, Stanford, CA.

WU HUI-PHING (1962). *Chinese Acupuncture.* (French tr. from the Chinese, with added comments, by J. Lavier. Eng. tr. by P. M. Chancellor.) Health Science Press, Rustington, Sussex.

WU HUI-PHING (1959). *Formulaire d'Acuponcture; la Science des Aiguilles et des Cauterisations Chinoises.* (Tr. from Chinese (Thaipei ed.) and abridged by J. Lavier.) Maloine, Paris.

WYATT, DON (1996). *The Recluse of Loyang: Shao Yung and the Moral Evolution of Early Sung Thought.* University of Hawaii Press, Honolulu.

WYLIE, A. (1874 and 1875) 'History of the Hsiung-Nu' (tr. of the chapter on the Huns in the Chhien Han Shu, ch. 94), *JRAI*, **3**, 401; **5**, 41.

WYLIE, A. (1880). 'The History of the South-Western Barbarians and Chao Sëen' (Chao-Hsien, Korea) (tr. of ch. 95 of the Chhien Han Shu), *JRAI*, 1 **9**, 53.

WYLIE, A. tr. (1881 and 1882). 'Notes on the Western Regions, translated from the Ts'een Han Shoo [Chhien Han Shu], Bk. 96'. *JRAI*, **10**, 20; **11**, 83. (Chs. 96A and B, as also the biography of Chang Chhien in ch. 61, pp. 1–6, and the biography of Chhen Thang in ch. 70.)

WYNDHAM, JOHN (1951). *The Day of the Triffids.* Michael Joseph, London, 1951; Penguin, Harmondsworth, 1954.

YU, ANTHONY (1977). *Journey to the West.* University of Chicago Press.

YULE, HENRY & CORDIER, HENRI (1871). *The Books of Ser Marco Polo the Venetian concerning the Kingdoms and Marvels of the East.* Murray, London; rev. ed. Philo Press, Amsterdam, 1975. 2 vols.

ZACCHIA, PAOLO (1634). *Quaestiones medico-legales, liber vi titulis tribus completus.* Rome.

ZILSEL, E. (1940). 'Copernicus and Mechanics'. *JHI*, **1**, 113.

ZILSEL, E. (1941). 'The Origin of William Gilbert's Scientific Method', *JHI*, **2**, 1.

ZILSEL, E. (1942a). 'The Genesis of the Concept of Physical Law', *PHR*, **51**, 245. Comment by M. Taube, *PHR*, 1943, **52**, 304.

ZILSEL, E. (1942b). 'The Sociological Roots of Science', *AJS*, **47**, 544.

ZILSEL, E. (1945). 'The Genesis of the Concept of Scientific Progress', *JHI*, **6**, 325.

ZUNZ, OLIVIER ed. (1985). *Reliving the Past.* University of North Carolina Press, Chapel Hill, N.C.,

GENERAL INDEX

Note: Alphabetical arrangement is word by word. For Chinese words, chh-, hs- and ss- follow normal sequence; ü is treated as u; ê as e. Bold page numbers refer to tables and lists; italics refer to illustrations.